HOLT
PRE-ALGEBRA

Authors

Eugene D. Nichols
Robert O. Lawton Distinguished
Professor of Mathematics Education
Mathematics Department
Florida State University
Tallahassee, Florida

Mervine L. Edwards
Chairman of Mathemathics Department
Shore Regional High School
West Long Branch, New Jersey

Sylvia A. Hoffman
Resource Consultant in Mathematics
Illinois State Board of Education
State of Illinois

Albert Mamary
Superintendent of Schools for Instruction
Johnson City Central School District
Johnson City, New York

HOLT
PRE-ALGEBRA

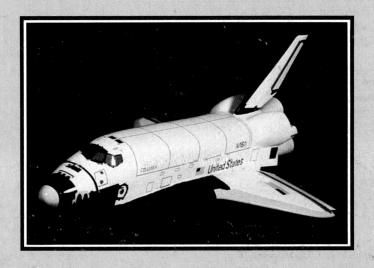

Eugene D. Nichols • Mervine L. Edwards
Sylvia A. Hoffman • Albert Mamary

HOLT, RINEHART AND WINSTON, PUBLISHERS
New York • Toronto • Mexico City • London • Sydney • Tokyo

Photo Credits

p. x Sam C. Pierson, Jr./Photo Researchers; **p. 13** Mike Mazzaschi/Stock, Boston; **p. 17** Van Bucher/Photo Researchers; **p. 20** Freda Leinwand; **p. 25** Patrick Donehue/Photo Researchers; **p. 35** BASF Systems; **p. 43** Courtesy, Anna Kopczynski/HRW Photo by Richard Haynes; **p. 52** Hans Wendler/Image Bank; **p. 54** Stacy Pick/Stock, Boston; **p. 56** Ronny Jacques/Photo Researchers; **p. 60** Bruce Roberts/Photo Researchers; **p. 68** Grace Moore/Imagery; **p. 74** Cliff Feulner/Image Bank; **p. 78** Hans Wendler/Image Bank; **p. 98** Richard Hutchings/Photo Researchers; **p. 110** Marc Solomon/Image Bank; **p. 125** Barbara Burnes/Photo Researchers; **p. 131** Tim Bieber/Image Bank; **p. 139** Peter Gridley/Alpha; **p. 145** Ann Hagen Griffiths/DPI; **p. 153** George Dodge/DPI; **p. 154** Tom Tracy/Image Bank; **p. 155** Dave Schaefer/Picture Cube; **p. 164** Joe Azzara/Image Bank; **p. 168** HRW Photo by Ken Lax; **p. 175** Bill Stanton/International Stock Photo; **p. 184** Mimi Forsyth/Monkmeyer; **p. 189** Andy Caufield/Image Bank; **p. 190** HRW Photo by Russell Dian; **p. 198** Freda Leinwand; **p. 202** HRW Photo by William Hubbell; **p. 208** Richard Hutchings/Photo Researchers; **p. 215** Jim Anderson/Peter Arnold; **p. 222** Focus on Sports; **p. 229** Freda Leinwand/Monkmeyer; **p. 232** Tom Pantages; **p. 234** Gabe Palmer/Image Bank; **p. 239** Tony Linck; **p. 244** Michal Heron; **p. 247** HRW Photo by Paul Light; **p. 254** Michael Melford/Peter Arnold; **p. 259** Peter and Georgina Bowater/Image Bank; **p. 263** Wil Blanche/DPI; **p. 273** Andy Levin/Black Star; **p. 275** Robert Phillips/Image Bank; **p. 280** Larry Dale Gordon/Image Bank; **p. 287** Pete Turner/Image Bank; **p. 294** Suzanne Szasz/Photo Researchers; **p. 298** Michal Heron/Monkmeyer; **p. 301** Lionel Brown/Image Bank; **p. 305** HRW Photo by William Hubbell; **p. 311** Grant Heilman; **p. 319** James H. Karales/Peter Arnold; **p. 325** Rhoda Sidney/Monkmeyer; **p. 334** Gary Ladd/Photo Researchers; **p. 342** Ann Hagen Griffiths/Omni-Photo Communications; **p. 350** Harvey Lloyd/Image Bank; **p. 367** Alvis Upitis/Image Bank; **p. 372** Scott Thode/International Stock Photo; **p. 381** Tom Pantages; **p. 392** Dan McCoy/Black Star; **p. 399** HRW Photo by Paul Light; **p. 400** HRW Photo by Paul Light; **p. 401** HRW Photo by Mac Shaffer; **p. 406** 3M Company, Traffic Control Materials Division; **p. 409** Sepp Seitz/Woodfin Camp; **p. 410** William Rivelle/Image Bank; **p. 419** Jim Harrison/Stock, Boston.

Cover Photo by: Erik Simonsen/Image Bank

Art Credits: Vantage Art, Inc., Karin Batten, Lisa Herring, Kathy Zieman

Copyright © 1986 by Holt, Rinehart and Winston, Publishers
All rights reserved
Printed in the United States of America

ISBN: 0-03-001858-7

567890 039 987654321

Contents

1 Equations – Adding and Subtracting

Introduction to Variables	1	Formulas	18
Properties of Addition	3	Problem Solving—Careers	
Open Sentences	5	Hotel Management	20
Solving Equations	8	Non-Routine Problems	21
Algebraic Phrases	11	Computer Activities	22
Solving Word Problems	13	Chapter Review	23
Problem Solving—Applications		Chapter Test	24
Jobs for Teenagers	16		

2 Equations – Multiplying and Dividing

Multiplication and Division	26	Mixed Types of Equations	44
Properties of Multiplication	28	Problem Solving: Mixed Types	45
Order of Operations	30	Evaluating Geometric Formulas	46
Order of Operations: Variables	33	Mathematics Aptitude Test	48
Problem Solving—Careers		Computer Activities	49
Computer Technicians	35	Chapter Review	50
Solving Equations	36	Chapter Test	51
Solving Word Problems	39		
Problem Solving—Applications			
Jobs for Teenagers	42		

3 Expressions and Formulas

The Distributive Property	53	Word Problems: Two Operations	65
Combining Like Terms	56	Problem Solving—Careers	
Simplifying and Evaluating		Dietitians	68
Expressions	58	Factors and Exponents	69
Problem Solving—Applications		Exponents in Formulas	72
Jobs for Teenagers	60	Non-Routine Problems	74
Solving Equations: Using Two		Computer Activities	75
Properties	61	Chapter Review	76
English Phrases to Algebra	63	Chapter Test	77

Multiplying and Dividing Fractions

Introduction to Fractions	79
Multiplying by a Whole Number	82
Multiplying by a Fraction	85
Prime Factorization	88
Greatest Common Factor (GCF)	90
Simplest Form	92
Simplifying Products	95
Problem Solving—Applications Jobs for Teenagers	98
Dividing Fractions	99
Evaluating Expressions	101
Equations	102
Evaluating Formulas	104
Computer Activities	106
Chapter Review	107
Chapter Test	108
Cumulative Review	109

Adding and Subtracting Fractions

Fractions with the Same Denominator	111
The Least Common Multiple (LCM)	114
Fractions with Unlike Denominators	116
Addition and Subtraction	119
Mixed Numbers	122
Problem Solving—Applications Jobs for Teenagers	125
Renaming in Subtraction	126
Evaluating Expressions	129
Problem Solving—Careers Bookkeepers	131
Equations	132
Non-Routine Problems	135
Computer Activities	136
Chapter Review	137
Chapter Test	138

Organizing Data

Reading and Making Tables	140
Pictographs	143
Problem Solving—Applications Jobs for Teenagers	145
Bar Graphs	147
Line Graphs	150
Problem Solving—Careers Appliance-Repair Specialists	153
Using Graphs	154
Applying Equations	157
Mathematics Aptitude Test	160
Computer Activities	161
Chapter Review	162
Chapter Test	163

7 Decimals

Introduction to Decimals	165	Decimal Equations	182
Changing Fractions to Decimals	168	Problem Solving—Careers	
Adding and Subtracting Decimals	171	Machinists	184
Multiplying Decimals	173	Non-Routine Problems	185
Problem Solving—Applications		Computer Activities	186
Jobs for Teenagers	175	Chapter Review	187
Dividing Decimals	177	Chapter Test	188
Scientific Notation and Large Numbers	180		

8 Measurement

Centimeters and Millimeters	190	Problem Solving—Careers	
Units of Length	192	Stonemasons	208
Changing Units of Length	195	Temperature: Celsius and Fahrenheit	209
Problem Solving—Applications		Computer Activities	211
Jobs for Teenagers	198	Chapter Review	212
Area	200	Chapter Test	213
Capacity	202	Cumulative Review	214
Weight	205		

9 Percents

Introduction to Percent	216	Problem Solving—Careers	
Percent of a Number	219	Sales Representatives	234
Finding Percents	222	Non-Routine Problems	235
Finding the Number	224	Computer Activities	236
Commission	226	Chapter Review	237
Discounts	229	Chapter Test	238
Problem Solving—Applications			
Jobs for Teenagers	232		

10 Statistics

Range, Mean, Median, Mode	240	Counting Principle	252
Mean of Grouped Data	243	Problem Solving—Careers	
Analyzing Data	246	Manufacturing Inspectors	254
Problem Solving—Applications		Mathematics Aptitude Test	255
Jobs for Teenagers	247	Computer Activities	256
Simple Events	248	Chapter Review	257
Independent and Dependent Events	250	Chapter Test	258

11 Ratio and Proportion

Ratios	260	Scale Drawing	274
Proportions	263	Similar Triangles	277
Applying Proportions	266	Trigonometric Ratios	280
Inverse Variation	269	Non-Routine Problems	283
Circle Graphs	270	Computer Activities	284
Problem Solving—Careers		Chapter Review	285
Electric Sign Service	273	Chapter Test	286

12 Adding and Subtracting Integers

Integers on a Number Line	288	Subtracting by Adding	301
Adding Integers	291	Using Integers	303
Problem Solving—Applications		Equations	305
Jobs for Teenagers	294	Computer Activities	307
Subtracting Integers	295	Chapter Review	308
Problem Solving—Careers		Chapter Test	309
Farm Equipment Mechanics	298	Cumulative Review	310
Opposites	299		

13 Multiplying and Dividing Integers

Multiplying Integers	312	Problem Solving—Careers	
Multiplying Negative Integers	314	Industrial Machinists	325
Properties of Multiplication	316	Simplifying Expressions	326
Problem Solving—Applications		Solving Equations	328
Jobs for Teenagers	319	Non-Routine Problems	330
Dividing Integers	320	Computer Activities	331
From Integers to Rationals	323	Chapter Review	332
		Chapter Test	333

14 Equations and Inequalities

Solving Equations	335	Exponents	351
Equations with Parentheses	337	Properties of Exponents	353
Number Problems	339	Polynomials	356
Problem Solving—Applications		Simplifying Polynomials	358
Jobs for Teenagers	342	Common Factors	360
Properties for Inequalities	344	Mathematics Aptitude Test	363
Solving Inequalities	347	Computer Activities	364
Problem Solving—Careers		Chapter Review	365
Railroad Brakers	350	Chapter Test	366

15 Some Ideas From Geometry

Points, Lines, and Planes	368	Problem Solving—Careers	
Angles	370	Tile Setters	381
Problem Solving—Applications		Square Roots	382
Jobs for Teenagers	372	The Pythagorean Theorem	385
Parallel Lines and Angles	373	Non-Routine Problems	388
Perpendicular Lines	375	Computer Activities	389
Triangles	377	Chapter Review	390
Polygons	379	Chapter Test	391

16 Coordinate Geometry

Graphing on a Number Line	393	Problem Solving—Careers	
More on Graphing Inequalities	396	Hospital Technicians	409
Problem Solving—Applications		Slope and Y-Intercept	410
Jobs for Teenagers	399	Solving Systems of Equations	413
Graphing Points	401	Computer Activities	415
Horizontal and Vertical Lines	404	Chapter Review	416
Graphing Equations and		Chapter Test	417
Inequalities	406	Cumulative Review	418

COMPUTER SECTION	420
EXTRA PRACTICE	436
ANSWERS TO PRACTICE EXERCISES	455
ANSWERS TO ODD-NUMBERED EXTRA PRACTICE	466
TABLES	471
GLOSSARY	474
INDEX	479

Equations – Adding and Subtracting

1

Up to 62,000 people can be seated in the Houston Astrodome, the first domed stadium.

Introduction to Variables

◇ OBJECTIVES ◇

To identify the variables, constants, and terms in an expression

To evaluate algebraic expressions for given values of variables

◇ RECALL ◇

Compute $18 - 5 + 6$.
$18 - 5 + 6$
$13 + 6$
19

Think of a number. Add 2 to the number.

Number + 2
↓ ↓
x + 2
↑
Variable

Use x to represent the number.

A *variable* is a letter that may be replaced by different values.

Example 1

To evaluate means to find the value.

Substitute 8 for y.

Evaluate $y + 7$ if $y = 8$.
$y + 7$
↓
$8 + 7$
15

So, the value of $y + 7$ is 15 if $y = 8$.

practice ▷ Evaluate if $x = 2$, $m = 3$, $y = 5$, and $a = 2$.

1. $x + 5$ 2. $7 + m$ 3. $6 - y$ 4. $a + 9$

Example 2

$\begin{array}{r}90\\-23\end{array}$ $\begin{array}{r}{}^{8}{\not{9}}{}^{10}{\not{0}}\\-2\;3\\\hline 7\end{array}$ $\begin{array}{r}{}^{8}{\not{9}}{}^{10}{\not{0}}\\-2\;3\\\hline 6\;7\end{array}$

Evaluate $c - d$ if $c = 90$ and $d = 23$.
$c - d$
$90 - 23$
67

So, the value of $c - d$ is 67 if $c = 90$ and $d = 23$.

practice ▷ Evaluate if $x = 40$, $y = 26$, $a = 56$, $b = 39$, and $d = 13$.

5. $x - y$ 6. $a + b$ 7. $a - b$ 8. $b + d$

Example 3

Evaluate $a + b - c + d$ if $a = 17$, $b = 48$, $c = 19$, and $d = 8$.

$$\begin{array}{r} \overset{1}{17} \\ +48 \\ \hline 65 \end{array} \qquad \begin{array}{r} 65 \\ -19 \\ \hline 46 \end{array} \qquad \begin{array}{r} \overset{5}{\cancel{6}}\overset{15}{\cancel{5}} \\ -19 \\ \hline 46 \end{array}$$

$$\begin{aligned} a + b &- c + d \\ \underbrace{17 + 48}_{} &- 19 + 8 \\ \underbrace{65 - 19}_{} &+ 8 \\ 46 &+ 8 \\ &54 \end{aligned}$$

So, the value of $a + b - c + d$ is 54.

practice ▷ Evaluate if $x = 13$, $y = 59$, $z = 32$, $w = 6$, $a = 48$, $b = 44$, $c = 17$, and $d = 5$.

9. $x + y - z + w$ 10. $a + b - c + d$ 11. $y - a + z - d$

The expression $x + y - 6$ has three *terms*: x, y, and 6.
Terms are added or subtracted.
The *variables* are x and y; 6 is a *constant*.

Example 4

Name the terms, variables, and constants of $x - 7 + y$.
The terms are x, 7, and y.
The variables are x and y. The constant is 7.

practice ▷ Name the terms, variables, and constants of each expression.

12. $11 - p$ 13. $t + 4 - u$ 14. $x + 8 - y$ 15. $x - 9 + y - 4$

◇ EXERCISES ◇

Evaluate the following expressions if $r = 7$, $s = 9$, $t = 4$, and $w = 5$.

1. $r + 5$
2. $9 + t$
3. $11 - w$
4. $s + 6$
5. $17 - t$
6. $r + t$
7. $s - w$
8. $w + r$
9. $s + 22$
10. $19 - w$

Evaluate the following expressions if $a = 70$, $b = 42$, $c = 73$, and $d = 49$.

11. $b + 19$
12. $c - 19$
13. $a + b$
14. $c - d$
15. $a + b - 13$
16. $a - b + d$
17. $d - b + a$
18. $39 + a - d$
19. $a + c - b + d$
20. $c - d + a - b$
21. $d - b + c - a$

Name the terms, variables, and constants of each expression.

22. $a - 4$
23. $p - 12 + t$
24. $5 + a + 7 - n$

Properties of Addition

◇ OBJECTIVES ◇

To recognize the commutative, associative, and additive identity properties

To use the commutative and associative properties to make computation easier

◇ RECALL ◇

8 + 3 = 11 and 3 + 8 = 11

So, 8 + 3 = 3 + 8.

When adding two numbers, you can change the order.

The recall suggests this.

commutative property of addition
For all numbers a and b, $a + b = b + a$.

Example 1

Add inside parentheses first.

Show that (8 + 2) + 5 = 8 + (2 + 5).

| (8 + 2) + 5 | = | 8 + (2 + 5) |
| 10 + 5 | | 8 + 7 |

Answers are the same.   15 = 15

So, (8 + 2) + 5 = 8 + (2 + 5)

When adding, you can change the way numbers are grouped.

---

associative property of addition
For all numbers a, b, and c,
$(a + b) + c = a + (b + c)$.

Example 2

Add.
6 + 0   0 + 4   5 + 0   0 + 5
  6   4   5   5

The result of addition is called the sum.   So, the sum of any number and 0 is that number.

---

property of additive identity
For any number a, $a + 0 = a$ and $0 + a = a$.

Example 3

Which property is illustrated?

8 + 9 = 9 + 8   commutative property of addition

(6 + 4) + 2 = 6 + (4 + 2)   associative property of addition

12 + 0 = 12   property of additive identity

practice ▷   Which property of addition is illustrated?

1. $(7 + 6) + 9 = 7 + (6 + 9)$ 2. $6 + 8 = 8 + 6$ 3. $7 + 0 = 7$

Example 4

Rewrite the sum $297 + 18 + 3$ to make the computation easier. Then compute.

Group 297 and 3 together since their sum is easier to find.

$297 + 18 + 3 = 297 + 3 + 18$ commutative property
$= (297 + 3) + 18$ associative property
$= 300 + 18$
$= 318$

If 18 and 3 are grouped together, the answer is the same, but the computation is more difficult.

$297 + 18 + 3 = 297 + (18 + 3)$
$= 297 + 21$
$= 318$

---

practice ▷   Rewrite to make the computation easier. Then compute.

4. $49 + 28 + 1$ 5. $93 + 62 + 7$ 6. $2 + 19 + 198$

◇ EXERCISES ◇

Which property of addition is illustrated?

1. $16 + 9 = 9 + 16$
2. $(14 + 7) + 26 = 14 + (7 + 26)$
3. $38 + 0 = 38$
4. $19 + (5 + 14) = (19 + 5) + 14$
5. $0 + 49 = 49$
6. $72 + 28 = 28 + 72$

Rewrite to make the computation easier. Then compute.

7. $99 + 66 + 1$
8. $3 + 79 + 97$
9. $47 + 50 + 3$
10. $8 + 37 + 2$
11. $199 + 39 + 1$
12. $68 + 40 + 2$
13. $25 + 49 + 75$
14. $72 + 20 + 8$
15. $87 + 36 + 3$
16. $195 + 49 + 5$
17. $149 + 29 + 1$
18. $139 + 47 + 1$
19. $7 + 62 + 93$
20. $398 + 76 + 2$
21. $55 + 89 + 45$
22. $175 + 89 + 25$
23. $17 + 38 + 3 + 2$
24. $18 + 49 + 2 + 1$
25. $78 + 19 + 2 + 1$
26. $1 + 98 + 79 + 2 + 399$

27. Does the associative property hold for subtraction?
    Hint: Is $(13 - 5) - 4 = 13 - (5 - 4)$ true?

Open Sentences

 **OBJECTIVES**  **RECALL**

To recognize open sentences
To recognize the solution(s) of an open sentence

Evaluate $x + 5$ if $x = 3$.
$x + 5$
$3 + 5$ Substitute 3 for x.
8

A sentence like "She was prime minister of India." is neither true nor false. It is called an *open sentence*.

Example 1

In the open sentence "She was prime minister of India." replace *She* with the name of a person to make a true sentence. Then replace *She* to make a false sentence.

true   *Indira Ghandi* was prime minister of India.
false *Margaret Thatcher* was prime minister of India.

It is true or false depending upon the value of x.

A sentence like $x + 4 = 10$ is an open sentence. It contains a variable x.

Example 2

Identify the sentence as open or not open. For each sentence that is not open, indicate whether it is true or false.

$7 + 2 = 9$ | $x - 4 = 11$ | $8 - 5 = 4$
not open; true | open | not open; false

practice ▷ Identify the sentence as open or not open. For each sentence that is not open, indicate whether the sentence is true or false.

1. $7 - 3 = 2$ 2. $x + 9 = 14$ 3. $5 + 11 = 16$

Example 3

Substitute the indicated values for x in $x + 5 = 12$. Are the resulting sentences true or false?

Substitute 3 for x.   Substitute 7 for x.
$x + 5 = 12$ $x + 5 = 12$

Substitute 3 for x. $3 + 5$ | 12 $7 + 5$ | 12
                        8 | 12 12 |
                        8 $= 12$ false $12 = 12$ true

OPEN SENTENCES 5

A value of a variable that makes an open sentence true is called a *solution* of the open sentence.

Example 4

Which of the values 4, 6, or 8 is a solution of $7 + x = 13$?

Substitute 4 for x.	Substitute 6 for x.	Substitute 8 for x.
$7 + x = 13$	$7 + x = 13$	$7 + x = 13$
$7 + 4 \mid 13$	$7 + 6 \mid 13$	$7 + 8 \mid 13$
11	13	15
$11 = 13$	$13 = 13$	$15 = 13$
false	true	false

So, 6 is a solution of $7 + x = 13$.

practice ▷ Which of the given values of the variables is a solution of the open sentence?

4. $5 + x = 14$; 3, 9, 11

5. $x - 12 = 9$; 14, 20, 21

Sometimes an open sentence contains an inequality symbol, $<$ or $>$.

$4 < 8$ $7 > 1$
4 *is less than* 8.   7 *is greater than* 1.

Example 5

Which of the values 0, 1, or 3 are solutions of $x + 6 < 8$?

Substitute 0 for x.	Substitute 1 for x.	Substitute 3 for x.
$x + 6 < 8$	$x + 6 < 8$	$x + 6 < 8$
$0 + 6 \mid 8$	$1 + 6 \mid 8$	$3 + 6 \mid 8$
6	7	9
$6 < 8$	$7 < 8$	$9 < 8$
true	true	false

So, 0 and 1 are solutions of $x + 6 < 8$.

practice ▷ Which of the given values of the variable is a solution of the open sentence?

6. $x + 9 < 11$; 0, 1, 2

7. $7 > m + 4$; 0, 1, 2, 5

◇ ORAL EXERCISES ◇

Which are open sentences? For each sentence that is not open, indicate whether it is true or false.

1. $5 + 4 = 3$
2. $x - 7 = 10$
3. $7 + 2 = 9$
4. $3 + x > 8$
5. $14 = x + 8$
6. $4 + 3 > 1$
7. $x < 12$
8. $7 = 4 + 3$

◇ EXERCISES ◇

Identify the sentence as open or not open. For each sentence that is not open, indicate whether the sentence is true or false.

1. $7 - 3 = 2$
2. $x + 9 = 14$
3. $8 < 10$
4. $x + 5 > 3$
5. $9 + 1 > 2$
6. $x + 2 < 6$
7. $4 + 5 < 2$
8. $10 = x + 5$

Which of the given values of the variable, if any, is a solution of the open sentence?

9. $x + 8 = 17$; 6, 9, 10
10. $11 = x - 8$; 9, 12, 19
11. $x + 5 = 14$; 6, 8, 9
12. $8 = y - 9$; 4, 11, 17
13. $7 + p = 25$; 5, 12, 18
14. $a + 9 = 34$; 16, 27, 29
15. $y + 5 = 16$; 10, 11, 12
16. $14 = t + 7$; 6, 7, 9
17. $p - 7 = 15$; 18, 20, 22
18. $8 + b = 39$; 30, 31, 32
19. $t + 5 < 14$; 6, 8, 10
20. $y + 9 < 12$; 0, 1, 2, 4
21. $t + 4 > 7$; 2, 4, 5
22. $m + 2 > 15$; 6, 7, 14
23. $a - 5 > 6$; 8, 9, 10
24. $x + 5 < 8$; 4, 5, 6
25. $x + 5 = 7 + 9$; 9, 10, 11
26. $19 - 5 = x + 6$; 5, 8, 9
27. $14 - 9 = 2 + g$; 1, 3, 7
28. $t + 5 > 6 + 4$; 5, 8, 9
29. $7 + g > 3 + 6$; 2, 4, 6
30. $y - 6 < 12 + 5$; 7, 9, 20

★ 31. Are there any whole numbers for which $x + 5 < 5$ is true? Why or why not?

Challenge

Add the numbers in the horizontal row. Add the numbers in the diagonal. Supply the numbers in the empty boxes so that the sum of the numbers in any row, column, or diagonal is the same.

16	10	28	6
30	4		8
		14	
	22		26

OPEN SENTENCES

Solving Equations

◇ OBJECTIVES ◇

To solve equations by using the addition property of equations

To solve equations by using the subtraction property of equations

◇ RECALL ◇

Evaluate $x + 9$ if $x = 4$.
$$4 + 9$$
$$13$$

Mavis weighs 120 lb. She gains or adds 5 lb. She can *undo* the addition of 5 lb by losing or subtracting 5 lb.

Think: $120 + 5 - 5$ takes her back to her original weight of 120 lb.

Addition and subtraction are *inverse* operations. Inverse operations undo each other.

Example 1

Subtracting 6 from 7, then adding 6 to the result gets back to 7.

Simplify each expression.

$7 - 6 + 6$ $x + 5 - 5$
↓   ↓
7   x

practice ▷ Simplify each expression.

1. $9 + 2 - 2$
2. $8 - 4 + 4$
3. $x + 13 - 13$

The sentence $9 + 6 = 15$ is an *equation*. A sentence containing $=$ is an equation.

Example 2

The equation $9 + 6 = 15$ is true. Show that each of the following operations on the equation results in a true equation.

Add 4 to each side.	Subtract 2 from each side.
$9 + 6 + 4 = 15 + 4$	$9 + 6 - 2 = 15 - 2$
$19 = 19$	$13 = 13$
true	true

So, adding 4 to each side of $9 + 6 = 15$ or subtracting 2 from each side results in a true equation.

You can add or subtract the same number from each side of an equation.

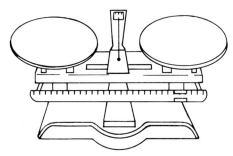

addition/subtraction properties of equations
For all numbers a, b, and c,

$$\text{if } a = b$$
$$\text{then } a + c = b + c$$
$$\text{and } a - c = b - c.$$

In the equation $x - 5 = 14$, x is not alone on the left side of the $=$ symbol. 5 is subtracted from x. To solve, *undo* the subtraction by adding 5 to each side, as shown below.

Example 3

Solve $x - 5 = 14$. Check the solution.

5 is subtracted from x.
Undo the subtraction by adding 5 to each side.

$$x - 5 = 14$$
$$x - 5 + 5 = 14 + 5$$
$$x = 19$$

To check, substitute 19 for x.

Check.
$$x - 5 = 14$$
$$19 - 5 \mid 14$$
$$14$$
$$14 = 14 \quad \text{true}$$

So, 19 is the solution of $x - 5 = 14$.

---

practice ▷ Solve and check.

4. $x - 3 = 5$ 5. $t - 18 = 3$ 6. $g - 9 = 1$

Example 4

Solve $a + 6 = 27$. Check the solution.

Undo the addition of 6 to a. Subtract 6 from each side.

$$a + 6 = 27$$
$$a + 6 - 6 = 27 - 6$$
$$a = 21$$

To check, substitute 21 for a.

Check.
$$a + 6 = 27$$
$$21 + 6 \mid 27$$
$$27$$
$$27 = 27 \quad \text{true}$$

So, 21 is the solution of $a + 6 = 27$.

---

practice ▷ Solve and check.

7. $m + 7 = 13$ 8. $x + 5 = 25$ 9. $b + 3 = 18$

SOLVING EQUATIONS

Example 5

Solve $29 = y + 8$. Check the solution.

Undo the addition of 8.
Subtract 8 from each side.

$29 = y + 8$
$29 - 8 = y + 8 - 8$
$21 = y$
or $y = 21$

Check.
$29 = y + 8$
$29 \mid 21 + 8$
$ 29$
$29 = 29$ true

So, 21 is the solution of $29 = y + 8$.

Example 6

Solve $28 = g - 6$. Check the solution.

Undo the subtraction of 6.
Add 6 to each side.

$28 = g - 6$
$28 + 6 = g - 6 + 6$
$34 = g$
or $g = 34$

Check.
$28 = g - 6$
$28 \mid 34 - 6$
$ 28$
$28 = 28$ true

So, 34 is the solution of $28 = g - 6$.

practice ▷ Solve and check.

10. $x - 8 = 11$
11. $y + 4 = 14$
12. $27 = y + 8$

◇ ORAL EXERCISES ◇

To solve each equation, what must be done to each side?

1. $x + 7 = 19$
2. $a - 14 = 8$
3. $m + 9 = 13$
4. $y - 15 = 14$
5. $9 = x + 4$
6. $14 = t + 13$
7. $25 = b - 8$
8. $29 = h + 14$

◇ EXERCISES ◇

Simplify each expression.

1. $8 + 12 - 12$
2. $17 - 4 + 4$
3. $27 - 19 + 19$
4. $46 + 13 - 13$

Solve each equation. Check the solution.

5. $x + 7 = 19$
6. $y - 14 = 7$
7. $z - 18 = 2$
8. $w + 11 = 19$
9. $9 = a + 7$
10. $11 = b - 17$
11. $c - 12 = 14$
12. $y + 3 = 17$
13. $x + 14 = 16$
14. $17 = b + 13$
15. $19 = p - 16$
16. $t - 7 = 21$
17. $x + 6 = 38$
18. $32 = f - 9$
19. $g + 8 = 23$
20. $43 = u - 7$
21. $x - 27 = 18$
22. $y + 13 = 45$
23. $33 = a + 13$
24. $41 = y + 19$
★ 25. $7 + x + 5 = 49 + 17$
★ 26. $18 + 16 = 12 - 4 + g + 1$

Algebraic Phrases

◇ OBJECTIVE ◇
To write English phrases in mathematical form

◇ RECALL ◇

English Phrase	Mathematical Symbol
increase	+
decrease	−

Example 1

Write in mathematical form.

$\underbrace{7 \text{ increased by } 6}$ $\quad$ $\underbrace{x \text{ increased by } 9}$
$\;7\quad+\quad 6\quad\quad\quad\; x\quad+\quad 9$

$\underbrace{5 \text{ decreased by } 2}$ $\quad$ $\underbrace{y \text{ decreased by } 4}$
$\;5\quad-\quad 2\quad\quad\quad\; y\quad-\quad 4$

practice ▷ Write in mathematical form.

1. 6 increased by 7
2. 5 decreased by 3
3. x increased by 11

Example 2

Write in mathematical form.
the sum of x and 12
$x + 12$

Example 3

8 less than 14 does not mean $8 - 14$.
It means $14 - 8$.

Write in mathematical form.

8 less than 14  7 less than x
$14 \leftarrow - \rightarrow 8 \quad\quad x \leftarrow - \rightarrow 7$

9 more than 3 means 3 made greater by 9.

9 more than 3   2 more than y
$3 \leftarrow + \rightarrow 9 \quad\quad y \leftarrow + \rightarrow 2$

practice ▷ Write in mathematical form.

4. the sum of x and 4
5. the sum of 6 and y
6. 5 less than b
7. 8 more than t

Example 4

Let x represent the number.

Write in mathematical form.

a number decreased by 2   6 less than a number
$\quad\downarrow$
$x\quad$ decreased by 2   6 less than $\;x$
$x\quad\quad -\quad\quad 2 \quad\quad x \leftarrow - \rightarrow 6$

ALGEBRAIC PHRASES

practice ▷ **Write in mathematical form.**

8. a number decreased by 13
9. 5 less than a number

Summary

English Phrase	Mathematical Form
x increased by 4	$x + 4$
the sum of y and 3	$y + 3$
1 more than w	$w + 1$
x decreased by 4	$x - 4$
4 less than x	$x - 4$

Reading in Math

Match each English phrase with its mathematical form.

1. x increased by 10
2. 10 less than x
3. the sum of x and 15
4. 10 more than x
5. x decreased by 15

a. $x - 15$
b. $x + 10$
c. $x + 15$
d. $x - 10$
e. $15 - x$
f. $10 - x$

◇ EXERCISES ◇

Write in mathematical form.

1. 5 increased by 4
2. 7 increased by 2
3. 15 increased by 7
4. 8 increased by 7
5. x decreased by 6
6. 16 increased by g
7. the sum of x and 2
8. 4 less than t
9. 5 more than w
10. 8 decreased by m
11. the sum of y and 12
12. 3 more than b
13. 15 less than w
14. u increased by 14
15. x decreased by 20
16. the sum of 6 and 12
17. 19 less than w
18. y increased by 13
19. a number increased by 4
20. 7 more than a number
21. 29 less than a number
22. 17 increased by a number
23. the sum of a number and 14
24. a number decreased by 23
25. 15 less than a number
26. 18 more than a number
27. 45 increased by a number
28. 42 more than a number
★ 29. 7 increased by 3 more than x
★ 30. 2 increased by 6 less than a number
★ 31. 15 added to 3 more than a number
★ 32. 14 increased by the sum of a number and 4

Solving Word Problems

◇ OBJECTIVE ◇

To solve word problems by using a four-step problem-solving method

◇ RECALL ◇

Four steps for solving a word problem are *Read, Plan, Solve,* and *Interpret.*

Example 1

If the cost of a shirt is decreased by $4, the sale price is $13. Find the cost of the shirt.

Read

Identify the given.   Given: The sale price is $13.
                                       The cost is decreased by $4.

Identify the unknown.   Find: the cost.

Plan

Choose a variable for the unknown.   Let c = the cost.

The cost decreased by $4 is $13.

Write an equation. $c - 4 = 13$

Solve

Solve the equation.

$$c - 4 = 13$$
$$c - 4 + 4 = 13 + 4 \quad \text{Undo the subtraction.}$$
$$c = 17 \quad \text{Add 4 to each side.}$$

Interpret

Check your solution.   Check: cost decreased by 4 is 13.

$$17 - 4 \;\Big|\; 13 \quad \text{Substitute 17 for } c \text{ in the equation.}$$
$$13$$
$$13 = 13 \quad \text{true}$$

Reread the word problem. Is the answer reasonable?

So, the cost of the shirt is $17.

1. Hank's weight decreased by 4 lb is 148 lb. Find his weight.

2. Tanya's salary increased by $40 is $115. Find her salary.

Example 2

Read

10 less than Leroy's age is 14. How old is Leroy?

Given: 10 less than Leroy's age is 14.
Find: Leroy's age.

Plan

Let a = Leroy's age.

10 less than Leroy's age is 14.
10 less than a is 14.

Write an equation. $a - 10 = 14$

Solve

Solve the equation. $a - 10 + 10 = 14 + 10$
$a = 24$

Interpret

Check your solution. Is the answer reasonable?

Check: 10 less than Leroy's age is 14.
10 less than 24 | 14
24 − 10
14
14 = 14
true

So, Leroy's age is 24.

practice

3. 5 more than Susan's age is 20. How old is Susan?

4. 6 less than Bernard's age is 8. How old is Bernard?

Example 3

The sum of Fay's salary and a $15 commission is $175. What is Fay's salary?

Given: commission is $15; total pay is $175.
Find: the salary.

Let s = the salary.

Sum of salary and $15 commission is $175.
$s + 15 = 175$
$s + 15 - 15 = 175 - 15$
$s = 160$

Check the answer. Is the answer reasonable?

◇ EXERCISES ◇

Solve these problems.

1. A number increased by 14 is 39. What is the number?
2. A number decreased by 17 is 43. What is the number?
3. 30 less than Maria's age is 25. How old is Maria?
4. The cost of a shirt increased by $5 is $19. Find the cost.
5. 40 more than Herbie's bowling score is 310. What is his score?
6. 10 less than Matt's age is 24. How old is Matt?
7. The sum of a number and 14 is 38. What is the number?
8. 32 is the same as 18 less than some number. Find the number.
9. The sum of a $3 sales tax and a restaurant bill is $60. What is the bill?
10. The sum of Rob's salary and a $25 commission is $175. What is Rob's salary?
11. The $360 selling price of a stereo is $50 more than the cost. What is the cost of the stereo?
12. The $65 selling price of a camera is the cost increased by a profit of $15. What is the cost?
13. 12 lb less than Joe's weight is 139 lb. How much does he weigh?
14. 5°F less than the temperature is 45°F. What is the temperature?
★ 15. If 6 is increased by 5 more than a number, the result is 38. What is the number?
★ 16. The sum of 7 more than a number, and 19, is the same as 16 increased by 38. What is the number?

Calculator

When using a calculator to compute, mistakes are made by hitting a wrong key. A good practice is to estimate the answer before using the calculator. Then see if your calculator result is close to the estimate.

For example: Estimate 79 + 31 + 199 as

$$\begin{array}{ccc} \downarrow & \downarrow & \downarrow \\ 80 & + 30 & + 200 \\ \downarrow & & \downarrow \\ 110 & + 200 & \text{or} \quad 310. \end{array}$$

Estimate each sum. Use a calculator to check your estimate.

1. 399 + 19 + 188
2. 3,999 + 2,107 + 33,022
3. 39 + 11 + 58 + 89 + 798

SOLVING WORD PROBLEMS

Problem Solving – Applications
Jobs for Teenagers

In solving word problems, key words or phrases can help you understand what to do in each problem-solving step. For example, the following key words and phrases tell you when to add or subtract.

Phrases that tell you to *add*:
  Find the *total*
  How many *combined*
  How many *in all*
  How many *altogether*

Phrases that tell you to *subtract*:
  Find the *change*
  How much *greater than* ⎫
  How much *less than*   ⎬ Phrases that compare
  Find the *difference*   ⎭

Example 1

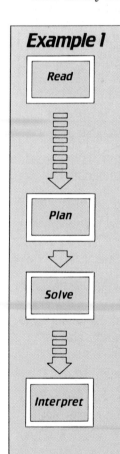

A customer bought a $1.95 hamburger, juice for $0.95, and salad for $1.05. The check was paid with a $20 bill. How much change should the customer receive?

Given: The cost of three items; check was paid with a $20 bill.
Find: The change returned to the customer.

Look for key words to decide what operations to use.
Find the total: *Add* the cost of each item.
Find the change: *Subtract* the total from $20.00.

First: Find the total of the purchases.

```
             $ 1.95
               0.95
     Add     + 1.05
             $ 3.95   total
```

Second: Find the change from $20.00.

```
             $20.00
  Subtract   − 3.95
             $16.05   change
```

Was the total bill correct?
Check the addition.
Was the change correct?
Check the subtraction by adding.

```
     $16.05   change
   + 3.95     total bill
     $20.00   paid for with $20 bill
```

So, the change returned should be $16.05.

16   CHAPTER ONE

Solve these problems.

1. Earl works part time at the luncheonette. He earned $15.35 on Monday, $14.35 on Friday, and $21.55 on Saturday. How much did he earn altogether?

2. Last week, the cooks prepared 257 breakfasts, 197 lunches, and 244 dinners. How many meals did they prepare in all?

3. Mary's tips totaled $14.74 last weekend. This weekend her tips totaled $23.49. How much more did she receive in tips this weekend?

4. How much change from a twenty-dollar bill must the cashier give to a customer who ordered a meal for $4.95 and a cup of coffee for $0.75?

5. One restaurant can seat 235 people. Another restaurant can seat 179 people inside and an additional 72 people in its sidewalk cafe. What is the difference between the number of people the two restaurants can seat?

6. Wanda is working part time at the luncheonette. She saved $375 in the fall, $410 in the spring, and $975.59 in the summer. How much did she save in all?

7. Abe treated his two friends to lunches, which cost $4.75 and $4.79. His lunch cost $4.35. How much was the total bill?

8. The Monday luncheon special costs $4.95. This same lunch on other days costs $6.50. How much less does the lunch cost on Monday?

9. Juice costs $0.50, a hamburger $1.65, and salad $1.35. How much greater than the $3.50 daily special is the total cost of these three items?

10. The local bakery delivered 78 loaves of bread to the luncheonette during one week, 102 during the next, and 97 during the third. The bakery billed the luncheonette owner for 267 loaves. Was this correct?

PROBLEM SOLVING

Formulas

OBJECTIVES
To evaluate formulas
To write formulas describing mathematical relationships

RECALL
Evaluate $x + 7$ if $x = 13$.
$13 + 7$
20

Example 1

A formula for the perimeter of a triangle is $P = a + b + c$. Find the perimeter of triangle STW if $a = 8$ in., $b = 3$ in., and $c = 9$ in.

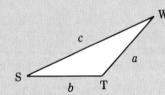

Perimeter P of a triangle is the distance around the triangle; a, b, and c represent the lengths of the sides of the triangle.

$P = a + b + c$
$P = 8 + 3 + 9$
$P = 20$

So, the perimeter of triangle STW is 20 in.

Example 2

A formula for the perimeter of a rectangle is $P = l + w + l + w$. Find the perimeter of a rectangle if the length (l) is 14 cm and the width (w) is 5 cm.

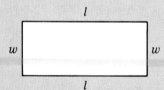

To find the perimeter, replace l by 14 and w by 5 in the formula and add.

$P = l + w + l + w$
$P = 14 + 5 + 14 + 5$
$P = 38$

So, the perimeter of the rectangle is 38 cm.

---

practice ▷ Find P.

1. $P = a + b + c$

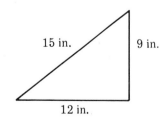

2. $P = l + w + l + w$

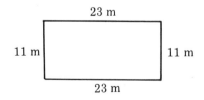

18   CHAPTER ONE

Example 3

An important formula in business is selling price = cost plus profit.

$$S = C + P$$

Find S if C is $14.00 and P is $6.00.

$S = C + P$
$S = 14 + 6$
$S = 20$

So, the selling price, S, is $20.00.

practice ▷ 3. Find S in $S = C + P$ if C is $47 and P is $13.

4. Find C in $C = S - P$ if S is $73 and P is $19.

Example 4

Write a formula.

Customer charge is selling price plus tax.

$$C = S + T$$

practice ▷ **Write a formula.**

5. The winning record is the total games minus the losses.

6. The class income is the dues added to the profit from the dances.

◇ **EXERCISES** ◇

Find P.

1. $P = a + b + c$
2. $P = l + w + l + w$
3. $P = a + b + c + d$

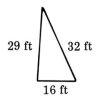
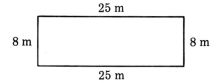
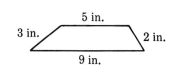

4. Find S in $S = C + P$ if C is $19 and P is $5.
5. Find C in $C = S - P$ if S is $43 and P is $7.
6. Find T in $T = A + B - C$ if A is 8, B is 7, and C is 3.

Write a formula.

7. Money left in a checking account is the balance minus the withdrawal.

Problem Solving – Careers
Hotel Management

1. The manager of the Lowatt Regency Hotel advertises the following weekend special in the newspaper: "For the month of May stay in a deluxe room for two nights, Saturday and Sunday, for only $120 total." If the regular room rate is $85.00 per night, how much less will the hotel make on each weekend special?

2. The manager of the restaurant of a hotel must check on the performance of a new waitress. The waitress wrote a check for $34.65, including tax. The customer gave her $40. She gave him back $5.45. Is the arithmetic correct?

3. Ms. Joyce Hammer is a hotel chain manager. She attends the Hotel Administrators Convention for two days. She is allowed an expense account for the meeting. Joyce submits the following expense voucher: room $135, food $44, and transportation $23. Find the total of her expenses.

4. Frank Childs is a hotel clerk at the Happyway Inn. The manager at the Harriet Motel has called to find out if rooms are available for an over-booking of guests. Frank knows that 57 of the 85 rooms at Happyway are occupied. How many rooms does he have available?

5. The administrator of a large hotel wants to compare this summer's total guest registration with that of last summer. This summer there were 755 guests in June; 1,055 in July; and 1,145 in August. Last summer there was a total of 3,415 guests. How many fewer guests were there this summer than last?

Non-Routine Problems

There is no simple method for solving these problems. You may need to try different methods until you find one that works.

1. The teller's drawer has some $10 bills and some $50 bills, 20 bills altogether. These 20 bills are worth $520. How many $10 bills and how many $50 bills are there?

2. You decided to save money the painless way: 1¢ the first week, 2¢ the second week, 4¢ the third week, and so on, for 16 weeks. Your friend says that you will save more this way than if you save $60 every week for 16 weeks. Is your friend right? First guess the answer, then compute to see if your friend was right.

3. Your friend was buying something. The clerk told your friend that one of these would cost 10¢, fifty-four would cost 20¢, seven hundred would cost 30¢, and five thousand would cost 40¢. What was your friend buying?

4. Henry and Peter started working for two different companies at the same salary. Last year Henry had a raise of one-tenth of his salary. Peter's company had hard financial times, and Peter had to have his salary reduced by one-tenth. This year Henry's company had to reduce his salary by one-tenth. Peter was lucky, and his salary was raised by one-tenth. Who is making more this year?

5. You bought one math book, one science book, and one social studies book. You paid $31 for the three books. The science book costs $1.50 more than the math book, and the social studies book cost $1.00 more than the science book. How much did each of the three books cost?

NON-ROUTINE PROBLEMS

A computer program is a set of written instructions for a computer. If you have access to a computer that uses the language of **BASIC**, you may use the computer to type in the program in the Example below.

Example Write a program to find S for S = 219 − 43 + 124
Then type the program into a computer and **RUN** it. (At the end of each line, press the **RETURN** key to get to the next line.)

Write a formula telling the computer how to find S. This uses the LET command.	10 LET S = 219 − 43 + 124
The PRINT command tells the computer to print S.	20 PRINT S
END tells the computer that the program is finished.	30 END
The computer will do nothing until you tell it to **RUN** the program. Type **RUN**	]RUN
The computer now displays the value of S.	300

Notice that the lines of the program are numbered by 10's. This allows you to go back and insert more lines in between any two lines if necessary. To clear the present program from the computer's memory:

Press **RETURN**

Type **NEW**

Press **RETURN**

Exercises

Find the value of each by first doing your own computation. Then write and **RUN** a program to check that the computer's results are the same as your own.

1. Find A. A = 76 + 49 − 18
2. Find B. B = 82 − 54 + 16
3. Find A. A = 69 + 49
4. Find Y. Y = 114 − 39 + 73
5. Find P. P = 89 − 62 + 43 − 26
6. Find B. B = 95 + 46 − 7

Chapter Review

Name the terms, variables, and constants of each expression. [1]

1. $m + 19$
2. $23 + y - 14$
3. $x + 17 - w + 39$

Evaluate if $a = 19$, $b = 25$, $c = 17$, and $d = 32$. [1]

4. $a + 29$
5. $42 - c$
6. $a + b + d - c$

Which property of addition is illustrated? [3]

7. $43 + 0 = 43$
8. $(8 + 6) + 4 = 8 + (6 + 4)$
9. $142 + 63 = 63 + 142$

Identify whether the sentence is open or not. For each sentence that is not open, indicate whether the sentence is true or false. [5]

10. $x + 8 = 23$
11. $12 - 9 = 1$
12. $25 = 18 + 7$

Which of the given values of the variable is a solution of the open sentence? [5]

13. $x + 9 = 16$; 5, 7, 9
14. $t + 8 < 11$; 0, 1, 2, 3
15. $9 - 5 = x + 3$; 1, 4, 5

Solve each equation. Check the solution. [8]

16. $x - 6 = 17$
17. $y + 8 = 29$
18. $39 = b + 7$
19. $33 = k - 7$

Solve these problems. [13]

20. 7 more than Tina's age is 29. How old is Tina?
21. 49 is the same as a number decreased by 12. What is the number?

Find P. [18]

22. $P = a + b + c$

(Triangle with sides 18 in., 17 in., and 20 in.)

23. Find y in $y = a + b - c$ if $a = 27$, $b = 63$, and $c = 19$.

Write a formula. [18]

24. The amount paid by a customer is the selling price minus the discount.

Chapter Test

Name the terms, variables, and constants of each expression.

1. $t + 14$
2. $x - 19 + y$
3. $c - d + 4 + f$

Evaluate if $a = 15$, $b = 29$, and $c = 19$.

4. $a + 17$
5. $51 - b$
6. $a + b - c$

Which property of addition is illustrated?

7. $(7 + 5) + 28 = 7 + (5 + 28)$
8. $19 + 129 = 129 + 19$
9. $0 + 32 = 32$

Identify whether the sentence is open or not. For each sentence that is not open, indicate whether the sentence is true or false.

10. $8 + 4 = 12$
11. $t - 3 = 11$
12. $7 + 3 < 4$

Which of the given values is a solution of the open sentence?

13. $a + 12 = 32$; 18, 20, 24
14. $m + 5 < 10$; 2, 3, 5, 6

Solve each equation. Check the solution.

15. $x + 7 = 12$
16. $19 = a + 14$
17. $y - 9 = 23$
18. $14 = m - 6$

Solve these problems.

19. $20 less than John's salary is $130. Find his salary.
20. The sum of a number and 19 is 32. What is the number?

Find P.

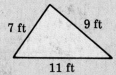

21. $P = a + b + c$

22. Find g in $g = x - y + z$, if $x = 32$, $y = 19$, and $z = 4$.

Write a formula.

23. The score in a game is the earned points minus the penalty points.

Equations – Multiplying and Dividing

2

Diners atop Seattle's Space Needle can view the city from a height of 607 ft.

Multiplication and Division

◇ OBJECTIVE ◇

To evaluate expressions like $6x$, $\frac{x}{3}$, $\frac{18}{y}$, and $\frac{5a}{2}$ for given value(s) of the variable(s)

◇ RECALL ◇

$4 \times 7 = 28$ ↑ multiply

$36 \div 4$ means $4\overline{)36}$ (= 9) ↑ divide

Let n represent any number.
You can represent 5 times any number as
↓
$5 \cdot n$

or, you can omit the centered dot.
$5n$

Example 1   Evaluate.

$4a$ if $a = 3$ | $16b$ if $b = 24$

$4a$ means $4 \cdot a$. $4 \cdot a$ | $16 \cdot b$
↓ | ↓
Substitute 3 for a. $4 \cdot 3$ | $16 \cdot 24$
12 | 384

$$\begin{array}{r} 24 \\ \times 16 \\ \hline 144 \\ 24 \\ \hline 384 \end{array}$$

---

practice ▷ Evaluate if $x = 5$, $b = 12$, $m = 32$, and $p = 23$.

1. $7x$ 2. $9b$ 3. $13m$ 4. $14p$

Example 2   Evaluate.

$\frac{x}{3}$ if $x = 24$ | $\frac{438}{m}$ if $m = 6$

Substitute 24 for x. $\frac{24}{3}$ | $\frac{438}{6}$

$\frac{24}{3}$ means $24 \div 3$. 8 | 73

$$\begin{array}{r} 73 \\ 6\overline{)438} \\ \underline{42} \\ 18 \\ \underline{18} \end{array}$$

practice ▷ Evaluate if $a = 8$, $b = 5$, $c = 6$, and $d = 28$.

5. $\frac{a}{4}$
6. $\frac{20}{b}$
7. $\frac{204}{c}$
8. $\frac{d}{7}$

An expression like $\frac{5a}{2}$ involves both multiplication ($5 \cdot a$) and division ($5 \cdot a$ divided by 2).

Example 3 Evaluate $\frac{5a}{2}$ if $a = 4$.

Substitute 4 for a. $\quad \frac{5 \cdot a}{2} = \frac{5 \cdot 4}{2}$

Multiply. $5 \cdot 4 = 20 \quad\quad\quad = \frac{20}{2}$

Divide. $20 \div 2 = 10 \quad\quad\quad = 10$

practice ▷ Evaluate if $x = 4$, $y = 9$, $a = 8$, and $b = 5$.

9. $\frac{3x}{2}$
10. $\frac{4y}{3}$
11. $\frac{5a}{4}$
12. $\frac{6b}{10}$

◇ EXERCISES ◇

Evaluate if $a = 6$, $b = 12$, $c = 26$, $d = 36$, $x = 5$, and $y = 7$.

1. $9a$
2. $3b$
3. $7y$
4. $5a$
5. $12c$
6. $13d$
7. $23c$
8. $12d$
9. $14x$
10. $32c$
11. $15d$
12. $18b$
13. $\frac{a}{2}$
14. $\frac{35}{x}$
15. $\frac{b}{6}$
16. $\frac{138}{a}$
17. $\frac{252}{a}$
18. $\frac{315}{x}$
19. $\frac{d}{9}$
20. $\frac{405}{x}$
21. $\frac{3b}{4}$
22. $\frac{13a}{2}$
23. $\frac{7b}{4}$
24. $\frac{3d}{2}$
25. $\frac{3a}{6}$
26. $\frac{5b}{4}$
27. $\frac{15b}{2}$
28. $\frac{3c}{2}$
29. $\frac{25a}{3}$
30. $\frac{17b}{3}$
★ 31. $\frac{5d}{3}$
★ 32. $\frac{14ad}{3y}$

MULTIPLICATION AND DIVISION

Properties of Multiplication

◇ OBJECTIVES ◇

To recognize the commutative, associative, and identity properties of multiplication

To use the commutative and associative properties to make computation easier

◇ RECALL ◇

$4 \cdot 5 = 20$ and $5 \cdot 4 = 20$

So, $4 \cdot 5 = 5 \cdot 4$

When multiplying two numbers, you can change the order.

commutative property of multiplication
For all numbers a and b, $a \cdot b = b \cdot a$.

Example 1

Multiply inside parentheses first.
Both answers are the same.

Show that $(4 \cdot 2) \cdot 3 = 4 \cdot (2 \cdot 3)$.

$(4 \cdot 2) \cdot 3$	$4 \cdot (2 \cdot 3)$
$8 \cdot 3$	$4 \cdot 6$
24	24

So, $(4 \cdot 2) \cdot 3 = 4 \cdot (2 \cdot 3)$.
When multiplying, you can change the way numbers are grouped.

associative property of multiplication
For all numbers a, b, and c, $(a \cdot b) \cdot c = a \cdot (b \cdot c)$.

Example 2

Product means multiply.

Multiply.

$6 \cdot 1$	$1 \cdot 6$	$7 \cdot 1$	$1 \cdot 7$
6	6	7	7

So, the product of any number and 1 is that number.

property of multiplicative identity
For any number a, $a \cdot 1 = a$ and $1 \cdot a = a$.

Example 3

Which property is illustrated?

$6 \cdot 8 = 8 \cdot 6$	commutative property of multiplication
$(7 \cdot 5) \cdot 3 = 7 \cdot (5 \cdot 3)$	associative property of multiplication
$1 \cdot a = a$	property of multiplicative identity

practice ▷ Which property of multiplication is illustrated?

　　　1. $(8 \cdot 7) \cdot 9 = 8 \cdot (7 \cdot 9)$　　2. $16 \cdot 43 = 43 \cdot 16$　　3. $39 \cdot 1 = 39$

Example 4

Rewrite the product $25 \cdot 6 \cdot 4$ to make the computation easier. Then compute.

Group 25 and 4 together, since their product is easier to find.

$$25 \cdot 6 \cdot 4 = 25 \cdot 4 \cdot 6 \quad \text{commutative property}$$
$$= (25 \cdot 4) \cdot 6 \quad \text{associative property}$$
$$= 100 \cdot 6$$
$$= 600$$

If 6 and 4 are grouped together, the answer is the same, but the computation is more difficult.

$$25 \cdot 6 \cdot 4 = 25 \cdot (6 \cdot 4)$$
$$= 25 \cdot 24$$
$$= 600$$

---

practice ▷ Rewrite to make the computation easier. Then compute.

　　　4. $25 \cdot 17 \cdot 4$　　5. $2 \cdot 19 \cdot 50$　　6. $5 \cdot 33 \cdot 20$

◇ ORAL EXERCISES ◇

Which two factors would you combine to make the computation easier?

1. $50 \cdot 63 \cdot 2$　　2. $4 \cdot 39 \cdot 25$　　3. $2 \cdot 86 \cdot 50$　　4. $25 \cdot 165 \cdot 4$

◇ EXERCISES ◇

Which property of multiplication is illustrated?

1. $38 \cdot 1 = 38$　　2. $(25 \cdot 4) \cdot 7 = 25 \cdot (4 \cdot 7)$　　3. $1 \cdot 49 = 49$
4. $(35 \cdot 6) \cdot 5 = 35 \cdot (6 \cdot 5)$　　5. $249 \cdot 1 = 249$　　6. $76 \cdot 63 = 63 \cdot 76$

Rewrite to make the computation easier. Then compute.

7. $4 \cdot 23 \cdot 5$　　8. $50 \cdot 14 \cdot 2$　　9. $2 \cdot 47 \cdot 50$　　10. $8 \cdot 3 \cdot 50$
11. $5 \cdot 26 \cdot 2$　　12. $4 \cdot 17 \cdot 25$　　13. $50 \cdot 43 \cdot 2$　　14. $20 \cdot 62 \cdot 5$
15. $25 \cdot 28 \cdot 4$　　16. $150 \cdot 7 \cdot 2$　　17. $4 \cdot 7 \cdot 50$　　18. $125 \cdot 3 \cdot 4$
19. $2 \cdot 165 \cdot 25 \cdot 2$　　20. $500 \cdot 78 \cdot 2 \cdot 2$　　21. $5 \cdot 39 \cdot 2 \cdot 5 \cdot 2$

★ 22. Does the associative property hold for division?
(*Hint:* Does $(12 \div 6) \div 2 = 12 \div (6 \div 2)$?)

PROPERTIES OF MULTIPLICATION

Order of Operations

◇ OBJECTIVES ◇

To compute numerical expressions using the rules for order of operations

To compute numerical expressions containing parentheses

◇ RECALL ◇

$8 \cdot 4$ means $8 \times 4 = 32$.
$12 \div 3$ means $3\overline{)12} = 4$.

Example 1

What is the value of $4 \cdot 6 + 3$?

There are two possibilities.

Multiply first.	Add first.
Then add.	Then multiply.
$\underline{4 \cdot 6} + 3$	$4 \cdot \underline{6 + 3}$
$24 + 3$	$4 \cdot 9$
27	36

Which is correct, 27 or 36?

We make this agreement in order to avoid confusion.

order of operations

When several operations occur,
1. Compute all multiplications and divisions first in order from left to right.
2. Then compute all additions and subtractions in order from left to right.

Thus, in Example 1 above, $\underline{4 \cdot 6} + 3$ Multiply first.
$24 + 3$ Then add.
27

Example 2

Compute.

Multiply first. $9 + \underline{4 \cdot 2}$ $14 - \underline{6 \div 2}$ Divide first.
Then add. $9 + 8$ $14 - 3$ Then subtract.
                 17 11

practice ▷ Compute.

1. $4 \cdot 7 + 2$ 2. $20 \div 4 - 3$ 3. $7 + 8 \cdot 5$ 4. $2 + 12 \div 6$

Sometimes all four basic operations are involved in the same expression.

Example 3

Compute $4 \cdot 3 + 9 - 12 \div 4 + 2 \cdot 5$.

Do all multiplications and divisions first.

Then do all additions and subtractions.

$$\underline{4 \cdot 3} + 9 - \underline{12 \div 4} + \underline{2 \cdot 5}$$
$$12 + 9 - 3 + 10$$
$$21 - 3 + 10$$
$$18 + 10$$
$$28$$

---

practice ▷ **Compute.**

5. $5 \cdot 2 + 6 - 10 \div 2 + 4 \cdot 6$
6. $12 + 6 \cdot 3 - 8 \div 4 - 1 \cdot 3$
7. $8 \cdot 3 - 21 \div 7 + 5 \cdot 7$

If operations occur within parentheses, they are to be done first. Then the rules for order of operations are followed.

Example 4

Compute $5 + (6 + 3) \cdot 2$.

Compute within parentheses first. $5 + \underline{(6 + 3)} \cdot 2$

Multiply. $5 + 9 \cdot 2$

Then add. $5 + 18$
23

When parentheses are used as a grouping symbol, the multiplication symbol may be omitted.

Example 5

Compute $8(5 - 1) + 18 \div 6$.

Compute within parentheses first. $8\underline{(5 - 1)} + 18 \div 6$ $\quad 8(5-1)$ means $8 \cdot (5 - 1)$.

Do multiplications and divisions next. $\underline{8(4)} + \underline{18 \div 6}$

Then add. $32 + 3$
35

ORDER OF OPERATIONS

practice ▷ Compute.

8. $8 + (5 + 2)3$ 9. $7(3 - 1) + 4$ 10. $6(2 + 3) - 16 \div 4$

◇ ORAL EXERCISES ◇

Tell the order of operations you must perform to compute each.

1. $5 + 7 \cdot 3$ 2. $4 \cdot 7 - 1$ 3. $6 + 9 \div 3$ 4. $3 \cdot 6 + 7$ 5. $9 - 8 \div 4$

6. $7 - 15 \div 5$ 7. $8 \cdot 2 + 5$ 8. $45 - 4 \cdot 6$ 9. $10 + 14 \div 7$ 10. $11 + 9 \cdot 3$

◇ EXERCISES ◇

Compute.

1. $5 \cdot 2 + 7$
2. $6 + 5 \cdot 3$
3. $10 - 2 \cdot 3$
4. $16 - 8 \div 4$
5. $7 \cdot 6 - 5$
6. $9 \cdot 4 + 5$
7. $2 \cdot 9 - 8$
8. $8 - 12 \div 6$
9. $9 - 14 \div 2$
10. $8 - 20 \div 5$
11. $16 - 15 \div 3$
12. $13 - 4 \cdot 3$
13. $9 \div 3 + 5$
14. $7 + 4 \cdot 6$
15. $9 \cdot 7 - 3$
16. $4 \cdot 2 + 8 - 15 \div 5 + 8$
17. $8 - 12 \div 2 + 7 \cdot 6$
18. $30 \div 5 - 1 \cdot 3 + 9$
19. $13 - 16 \div 2 + 7 \cdot 4$
20. $6 \cdot 8 - 21 \div 3 + 4 \cdot 9$
21. $3 + 7 \cdot 5 - 25 \div 5$
22. $4 \cdot 12 - 28 \div 14 + 7 \cdot 3$
23. $4 \cdot 9 - 35 \div 7 + 7 \cdot 7$
24. $9 \cdot 5 + 27 \div 3 - 1$
25. $7 + (6 + 4)2$
26. $6(11 - 5) + 3$
27. $7(5 - 2) - 45 \div 9$
28. $(9 - 6)2 + 7$
29. $7(8 - 6) + 3$
30. $11 - 2(18 - 14)$
31. $4(9 - 7) + 49 \div 7$
32. $3(6 + 5) - 18 \div 2$
33. $35 \div 7 + (7 + 2)2$

Compute. (*Hint:* Start within the parentheses, then work within the brackets.)

★ 34. $24 \div [14 - 2(3 + 1)] + 17$ ★ 35. $48 - 6[20 - 3(54 \div 9)] - 4 \cdot 9$

Calculator

Copy and complete.
$1 \cdot 9 + 2 =$ _____
$12 \cdot 9 + 3 =$ _____
$123 \cdot 9 + 4 =$ _____
$1{,}234 \cdot 9 + 5 =$ _____

If the same pattern is continued, what is the next line?
Continue the pattern as far as you can.

Order of Operations: Variables

◇ OBJECTIVE ◇

To use the rules for order of operations to evaluate algebraic expressions for given value(s) of the variable(s)

◇ RECALL ◇

Compute $7 \cdot 4 + 8 - 3 \cdot 2$
$ 28 + 8 - 6 \leftarrow$ Multiply first.
$ 36 - 6 \leftarrow$ Then add and subtract.
$ 30$

Example 1

Evaluate each for the given value of the variable.

$6x + 4$ if $x = 2$
$6 \cdot x + 4$
Substitute 2 for x. $\quad 6 \cdot 2 + 4$
Multiply first. $\quad 12 + 4$
Then add. $\quad 16$

$\frac{x}{3} + 2$ if $x = 15$
$\frac{15}{3} + 2$
$5 + 2$
7

practice ▷ Evaluate if $x = 3$, $y = 2$, and $a = 4$.

1. $7x + 3$
2. $6 + 2y$
3. $\frac{a}{2} + 3$

Example 2

Evaluate $9a - 4 + 3b$ if $a = 6$ and $b = 2$.
$ 9a - 4 + 3b$
$ 9 \cdot a - 4 + 3 \cdot b$
Substitute 6 for a and 2 for b. $\quad 9 \cdot 6 - 4 + 3 \cdot 2$
$ 54 - 4 + 6$
$ 50 + 6$
$ 56$

practice ▷ Evaluate if $x = 5$ and $y = 2$.

4. $9x - 2 + 13y$
5. $7 + 5x + 2y$
6. $5 + 4x + \frac{1}{2}y$

Example 3

Evaluate $4xy - 10$ if $x = 2$ and $y = 5$.
$ 4xy - 10$
$4xy$ means $4 \cdot x \cdot y$. $\quad 4 \cdot x \cdot y - 10$
Substitute 2 for x and 5 for y. $\quad 4 \cdot 2 \cdot 5 - 10$
$4 \cdot 2 \cdot 5 = 8 \cdot 5 = 40.$ $\quad 40 - 10$
$ 30$

ORDER OF OPERATIONS: VARIABLES

practice Evaluate if $a = 3$, $b = 2$, and $c = 5$.

7. $3ab - 5$
8. $6 + 5bc$
9. $2ac + 7$

◇ ORAL EXERCISES ◇

Evaluate each for the indicated value of x.

1. $2x + 1$ if $x = 3$
2. $1 + 3x$ if $x = 3$
3. $2x - 1$ if $x = 4$
4. $\frac{x}{4} + 5$ if $x = 12$
5. $\frac{x}{6} - 1$ if $x = 24$
6. $7 + \frac{x}{2}$ if $x = 6$

◇ EXERCISES ◇

Evaluate if $x = 2$, $y = 4$, $z = 1$, $a = 3$, $b = 9$, and $c = 5$.

1. $3x + 7$
2. $8 + 5x$
3. $\frac{b}{3} + 6$
4. $5y - 1$
5. $4 + 3y$
6. $8y - 1$
7. $7z - 2$
8. $\frac{y}{2} + 5$
9. $3b - 8$
10. $6 + \frac{c}{5}$
11. $3 + 8b$
12. $4b + 12$
13. $2a + 3b$
14. $5x + 3z$
15. $4b + 3c$
16. $2x + 4y$
17. $7a + \frac{x}{2}$
18. $5a + 3c$
19. $2x + 4c$
20. $7x + \frac{b}{3}$
21. $\frac{y}{2} + 6z$
22. $3a - 5 + 7c$
23. $5y + 2z - 4$
24. $3 + 8x + 2y$
25. $7z + 8 + 3y$
26. $3 + 5a + 6b$
27. $9x + 3b - 8$
28. $5bz - 4$
29. $7bc - 2$
30. $11 + 2yz$
31. $5xy - 3$
32. $7 + 9bc$
33. $7xy - 9$
★ 34. $ab(3x + 2y)$
★ 35. $xyz(5ab + 4c)$
★ 36. $3bc(4xyz - 2a)$
★ 37. $7bc(5xyz - 3a)$
★ 38. $2ac(3ab - 2xy)$
★ 39. $(8ab - 3y)2az$

Challenge

Step 1 Choose four digits. Write them in descending order.

Step 2 Reverse the order.

Step 3 Subtract.

Step 4 Reverse the order again.

Step 5 Add. What is the result?

Repeat the five steps above with four different digits. Guess what the answer will be.

Problem Solving – Careers
Computer Technicians

Most microcomputers store information on *floppy* disks. Floppy disks are made from a flexible plastic material called *mylar*. One letter of the alphabet takes one *byte* of memory on a disk. Thus, the address 137 Abbott Avenue uses up 17 bytes of memory. (Allow one byte of memory for each space between words.) The number of bytes available on a disk depends upon how the disk is manufactured. For example, a $5\frac{1}{4}$ inch single-sided, single-density disk will contain 125,000 bytes. A double-sided, double-density disk can hold 4 times that amount.

1. The words *computer program* require how many bytes of disk memory?

Use the following facts to answer Exercises 2–5 below.
A particular kind of disk is made up of 40 *tracks*. Three of these tracks are used for information required to operate the computer. This leaves 37 tracks available for use by the programmer. Each *track* contains 16 *sectors*. Each sector can store 256 bytes of information.

2. A computer technician needs 5 sectors for a program. How many bytes of memory will be used?

3. How many sectors will be required for 768 bytes of memory?

4. Find the total number of bytes of usable memory available on the disk.

5. Use the result of Exercise 4 above to answer the following: José writes programs for games. His new Space Wars game requires 3,265 bytes of memory. How many bytes are left for another program on the same disk?

6. Matt has stored a program that requires 9,285 bytes of memory. How many programs of the same size could he store on a $5\frac{1}{4}$ inch single-sided, single-density disk?

7. The letter K is used to represent approximately 1,000 bytes of memory. A computer having $48K$ memory size can hold approximately 48,000 bytes. Represent the number of bytes that can be stored on a $5\frac{1}{4}$ inch double-sided, double-density disk in terms of K.

PROBLEM SOLVING

Solving Equations

◇ OBJECTIVES ◇

To solve equations by using the division property of equations

To solve equations by using the multiplication property of equations

◇ RECALL ◇

Inverse operations are operations that undo each other.
Addition and subtraction are inverse operations.

Multiplication and division are inverse operations.
Inverse operations undo each other.
Division will undo multiplication.

$$\frac{4 \cdot 8}{4} = \frac{32}{4} = 8 \quad \text{8 remains unchanged.}$$

Similarly, multiplication will undo division.

$$2 \cdot \frac{6}{2} = 2 \cdot 3 = 6$$

Example 1   Simplify each expression.

Multiplying 3 by 7, then dividing the result by 7 leaves 3 unchanged.

$$\frac{7 \cdot 3}{7} \qquad \frac{6x}{6} \qquad 8 \cdot \frac{16}{8} \qquad 5 \cdot \frac{x}{5}$$

$$\frac{7 \cdot 3}{7} \qquad \frac{6 \cdot x}{6} \qquad 8 \cdot \frac{16}{8} \qquad 5 \cdot \frac{x}{5}$$

$$\downarrow \qquad \downarrow \qquad \downarrow \qquad \downarrow$$

$$3 \qquad x \qquad 16 \qquad x$$

practice ▷   Simplify each expression.

1. $\frac{6 \cdot 8}{6}$ 2. $\frac{5k}{5}$ 3. $9 \cdot \frac{25}{9}$ 4. $8 \cdot \frac{x}{8}$

Example 2

The equations $5 \cdot 2 = 10$ and $\frac{12}{3} = 4$ are true.
Show that each of the following operations on each equation results in a true equation.

Dividing or multiplying each side of a true equation by the same number results in a true equation.

$5 \cdot 2 = 10$

Divide each side by 5.

$\frac{5 \cdot 2}{5} = \frac{10}{5}$

$2 = 2$

true

$\frac{12}{3} = 4$

Multiply each side by 3.

$3 \cdot \frac{12}{3} = 3 \cdot 4$

$12 = 12$

true

You can multiply or divide each side of an equation by the same number.

multiplication property of equations:

For all numbers a, b, and c, if $a = b$, then $c \cdot a = c \cdot b$.

division property of equations:

For all numbers a, b, and c ($c \neq 0$), if $a = b$, then $\frac{a}{c} = \frac{b}{c}$.

Example 3   Solve $4x = 20$. Check the solution.

Undo multiplication by 4. $4x = 20$

Divide each side by 4. $\frac{4 \cdot x}{4} = \frac{20}{4}$

$x = 5$

Check: $4x = 20$
            $4 \cdot 5 \;|\; 20$

Substitute 5 for x. 20

               $20 = 20$ true

So, the solution is 5.

---

practice ▷ Solve each equation. Check the solution.

5. $3t = 15$ 6. $9x = 27$ 7. $7p = 14$ 8. $4m = 32$

Example 4   Solve $\frac{a}{5} = 6$. Check the solution.

Undo division by 5. $\frac{a}{5} = 6$

Multiply each side by 5. $5 \cdot \frac{a}{5} = 5 \cdot 6$

$a = 30$

Check: $\frac{a}{5} = 6$

Substitute 30 for a. $\frac{30}{5} \;|\; 6$
               6
               $6 = 6$ true

So, the solution is 30.

---

practice ▷ Solve each equation. Check the solution.

9. $\frac{x}{8} = 7$ 10. $\frac{b}{4} = 4$ 11. $\frac{m}{5} = 9$ 12. $\frac{w}{6} = 3$

SOLVING EQUATIONS

Example 5

Solve each equation.

Undo multiplication by 7.	$28 = 7p$	$5 = \frac{b}{3}$	Undo division by 3.
Divide each side by 7.	$\frac{28}{7} = \frac{7 \cdot p}{7}$	$3 \cdot 5 = 3 \cdot \frac{b}{3}$	Multiply each side by 3.
	$4 = p$	$15 = b$	
Check the solution.	or $p = 4$	or $b = 15$	

practice Solve each equation. Check the solution.

13. $6g = 30$ 14. $\frac{a}{4} = 10$ 15. $45 = 9k$ 16. $7 = \frac{y}{9}$

Summary

To solve an equation that shows multiplication, divide each side by the same number.

To solve an equation that shows division, multiply each side by the same number.

◇ ORAL EXERCISES ◇

To solve each equation, what must be done to each side?

1. $6x = 60$ 2. $\frac{m}{12} = 2$ 3. $7z = 63$ 4. $\frac{a}{4} = 15$
5. $35 = \frac{b}{4}$ 6. $19c = 38$ 7. $48 = 16y$ 8. $17 = \frac{w}{3}$

◇ EXERCISES ◇

Simplify each expression.

1. $\frac{7 \cdot 9}{7}$ 2. $\frac{14t}{14}$ 3. $7 \cdot \frac{12}{7}$ 4. $13 \cdot \frac{m}{13}$

Solve each equation. Check the solution.

5. $11t = 22$ 6. $7m = 56$ 7. $44 = 11b$ 8. $27 = 9m$
9. $24 = 12f$ 10. $5b = 45$ 11. $72 = 9j$ 12. $3p = 33$
13. $\frac{x}{8} = 4$ 14. $\frac{m}{6} = 3$ 15. $8 = \frac{y}{3}$ 16. $5 = \frac{w}{12}$
17. $3 = \frac{s}{12}$ 18. $\frac{r}{13} = 2$ 19. $24 = \frac{t}{2}$ 20. $6 = \frac{u}{13}$
21. $16k = 320$ 22. $13 = \frac{b}{14}$ 23. $245 = 7u$ 24. $23 = \frac{a}{11}$

★ 25. $8 - 3 \cdot 2 + 24 = 13x$ ★ 26. $3(4 + 2) = \frac{x}{5}$

Solving Word Problems

OBJECTIVES
To write English phrases in mathematical form
To solve word problems involving multiplication or division

RECALL

Solve $5n = 35$
$$\frac{5 \cdot n}{5} = \frac{35}{5}$$
$$n = 7$$

Solve $\frac{x}{3} = 8$
$$3 \cdot \frac{x}{3} = 3 \cdot 8$$
$$x = 24$$

Example 1
Write in mathematical form.

8 times n	twice y	x divided by 3
$8 \cdot n$	2 times y	$\frac{x}{3}$
or $8n$	$2y$	

practice
Write in mathematical form.

1. twice 9
2. p divided by 9
3. 13 times w

In Chapter 1, you learned four basic steps for solving word problems. These steps are used below in solving word problems involving multiplication or division.

Example 2
5 times a number is 35. What is the number?

Read
Identify the given. Given: Result of 5 times a number is 35.
Identify the unknown. Find: the number.
Choose a variable for the unknown. Let $n =$ the number.

Plan
5 times a number is 35.

Write an equation. $5n = 35$ Undo the multiplication.

Solve
Solve the equation. $\frac{5 \cdot n}{5} = \frac{35}{5}$ Divide each side by 5.
$n = 7$

Check your solution. Check: 5 times a number is 35
$5 \cdot 7$ | 35
35
35 = 35 true

Interpret
Is the answer reasonable?

So, the number is 7.

SOLVING WORD PROBLEMS   39

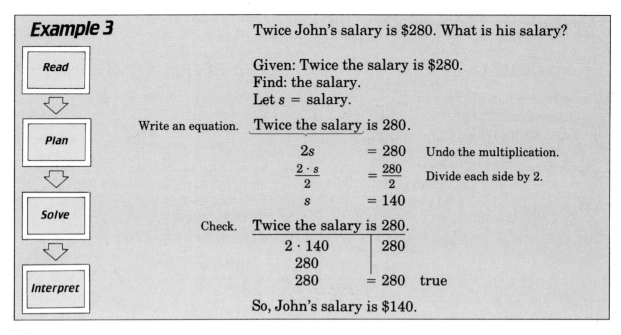

practice ▷ Solve.

4. 7 times a number is 21. Find the number.

5. Twice Mary's age is 40. How old is Mary?

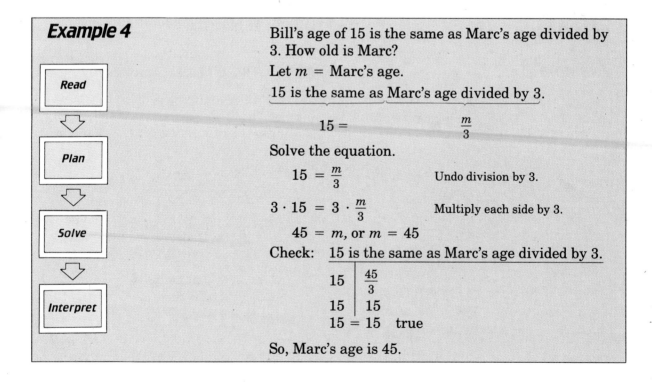

practice

6. 8 is the same as a number divided by 5. Find the number.

7. Bill's age divided by 2 is 14. How old is Bill?

◇ ORAL EXERCISES ◇

Give an equation for each sentence.

1. 6 times n is the same as 18.
2. 40 is twice b.
3. x divided by 4 is the same as 3.
4. 25 is the same as y divided by 6.
5. 27 is the same as 9 times t.
6. r divided by 6 is the same as 12.
7. 9 times w is 45.
8. 32 is the same as u divided by 4.
9. q divided by 6 is 14.
10. 39 is the same as 13 times k.

◇ EXERCISES ◇

Write in mathematical form.

1. 12 times t
2. n divided by 14
3. twice r
4. twice u
5. 24 times g
6. p divided by 8

Solve these problems.

7. 4 times a number is 44. Find the number.
8. Twice a number is 18. Find the number.
9. 5 times a number is 35. Find the number.
10. 42 is twice a number. Find the number.
11. Ruiz's 12 home runs this year is 4 times his record of last season. Find the number of home runs last season.
12. Henry's savings of $120 is the same as 4 times his sister's savings. How much did his sister save?
13. 12 is the same as a number divided by 4. Find the number.
14. A number divided by 7 is 8. Find the number.
15. Mona's age of 13 is the same as Joan's age divided by 3. How old is Joan?
16. Pearl's age of 12 is the same as her grandfather's age divided by 6. How old is her grandfather?
17. Car insurance of $800 is twice what it was a decade ago. Find the cost of insurance 10 yr ago.
18. The cost of a pizza divided among 4 people is $3.00 each person. Find the cost of the pizza.
★ 19. 7 times a number is the same as 24 decreased by 3. Find the number.
★ 20. 8 more than twice 3 is the same as 7 times a number. Find the number.

SOLVING WORD PROBLEMS

Problem Solving – Applications
Jobs for Teenagers

Certain key phrases or situations tell you whether to multiply or divide.

Situations calling for multiplication:	Situations calling for division:
Given the cost of one item, find the cost of *several*.	Given the cost of several items, find the cost of *each* or *one*.
Given the number of items in one carton, find the number in *several* cartons.	Given a total length to be cut into a number of equal parts, find the length of *each* part.

Example 1

Read

The senior class sells juice at football games.
One can costs $0.35.
Find the cost of 275 cans.

Plan

Given: the cost of 1 can is $0.35.

Find: the cost of 275 cans.

Solve

To find the cost of *several*, multiply.

$$\begin{array}{r} \$275 \\ \times 0.35 \\ \hline 1375 \\ 825 \\ \hline \$96.25 \end{array}$$

Interpret

Notice the operation. Multiplication is called for since you have to find the cost of *several* items. Check the multiplication.

So, the cost of 275 cans is $96.25.

42   CHAPTER TWO

Example 2

The senior class has decided to sell greeting cards. A committee of 15 volunteers is to sell 630 boxes. How many boxes must each student sell?

Given: 630 boxes to be sold by 15 students.
Find: the number of boxes to be sold by each student.

To find the number to be sold by *each*, use division.

$$\begin{array}{r} 42 \\ 15\overline{)630} \\ \underline{60} \\ 30 \\ \underline{30} \end{array}$$

Check: 15 students to sell 42 boxes each.

$$\begin{array}{r} 42 \\ \times 15 \\ \hline 210 \\ 42 \\ \hline \end{array}$$

Total to be sold is 630. The answer checks. 630

So, each student must sell 42 boxes.

Solve these problems.

1. Tickets are ordered for a senior activity. A box contains 125 tickets. How many tickets are there in all if 25 boxes are ordered?

2. 800 yearbooks are delivered in 50 cartons. How many yearbooks should be in each carton?

3. Hamburgers come 36 to a box. The fund-raising committee ordered 15 boxes. How many hamburgers were ordered in all?

4. A decorating committee is building shelves for a booth. A 60-in. board is to be cut into 4 equal pieces. How long is each piece?

5. Sandwiches were sold at a game for $1.35 each. The senior class sold 315. How much did the class collect in all for the day?

6. The total cost of the senior class prom dinner is $2,400. There are 120 seniors. How much should each senior be charged?

7. The class sells pretzels for $0.35 each. If 145 are sold at a football game, how much money is collected in all?

8. In a fund-raising raffle, $125 in prize money is to be shared equally by 5 winners. How much money must be given to each winner?

PROBLEM SOLVING

Mixed Types of Equations

◆ OBJECTIVE ◆
To solve equations when different types are mixed

◆ RECALL ◆
You have already learned to solve equations like

$$14 = c + 9 \qquad 3d = 39$$
$$x - 7 = 2 \qquad 6 = \frac{m}{2}$$

Recall that in order to solve equations like those above, you must use one of the following strategies:

1. Undo addition — Subtract the same number from each side.
2. Undo subtraction — Add the same number to each side.
3. Undo multiplication — Divide each side by the same number.
4. Undo division — Multiply each side by the same number.

Example 1
Solve each equation.

Undo subtraction.
$$x - 7 = 2$$
$$x - 7 + 7 = 2 + 7$$
$$x = 9$$

$$14 = x + 9$$
$$14 - 9 = x + 9 - 9$$
$$5 = x \text{ or } x = 5$$
Undo addition.

Example 2
Solve each equation.

Undo multiplication.
Divide each side by 3.
$$3x = 39$$
$$\frac{3 \cdot x}{3} = \frac{39}{3}$$
$$x = 13$$

$$6 = \frac{x}{7}$$
$$7 \cdot 6 = 7 \cdot \frac{x}{7}$$
$$42 = x \text{ or } x = 42$$
Undo division.
Multiply each side by 7.

◆ EXERCISES ◆

Solve.

1. $5x = 25$
2. $x + 9 = 13$
3. $\frac{a}{4} = 10$
4. $11 = y - 6$
5. $14 = p + 3$
6. $x - 14 = 16$
7. $6 = \frac{b}{9}$
8. $17 = k - 3$
9. $48 = 4g$
10. $24 = m + 9$
11. $\frac{t}{7} = 8$
12. $17 = y - 9$
13. $84 = 4w$
14. $w + 12 = 25$
15. $72 = 6n$
16. $x - 13 = 46$
17. $38 = g - 4$
18. $16 = \frac{y}{3}$
★ 19. $2(3 + 7) = 4t$
★ 20. $x - 7 = 6 \cdot 3 + 4$

Problem Solving: Mixed Types

◇ OBJECTIVE ◇

To solve mixed types of word problems

Review the four-step strategy for solving word problems on page 39. Also review the key phrases. This should help you to solve the word problems that follow.

Key Phrases

ENGLISH PHRASE	MATHEMATICAL FORM
x increased by 4	$x + 4$
x decreased by 4	$x - 4$
4 more than x	$x + 4$
4 less than x	$x - 4$
the sum of x and 4	$x + 4$
twice x	$2x$
4 times x	$4x$
x divided by 4	$\frac{x}{4}$

Solve these problems.

1. A number decreased by 7 is 18. Find the number.

2. 7 more than a number is 12. Find the number.

3. 45 is 5 times some number. Find the number.

4. A number divided by 8 is the same as 9. What is the number?

5. The sum of a number and 12 is 18. What is the number?

6. 42 is the same as 7 times a number. What is the number?

7. Hazel's weight decreased by 30 lb is 100 lb. Find her weight.

8. Twice Rod's age is 46. How old is Rod?

9. Henry's score of 12 baskets is 6 baskets less than Harry's. What is Harry's score?

10. The per person cost for a dinner is $8.00. Find the total charge for 7 people at a dinner party.

11. Tina's age of 55 is 11 times her nephew's age. How old is her nephew?

12. The sum of a restaurant bill of $35 and its sales tax is $38. What is the sales tax?

13. Sonya's salary of $215 is $35 more than her husband's salary. Find her husband's salary.

14. If Bill can increase his bowling score by 40 pins his score will be 225. Find his score.

Evaluating Geometric Formulas

◇ OBJECTIVE ◇

To evaluate geometric formulas involving multiplication and division

◇ RECALL ◇

Evaluate xy if $x = 5$ and $y = 6$.
$5 \cdot 6$
30

Example 1

A formula for the area of a rectangle is $A = lw$. Find the area of rectangle $ABCD$ if $l = 7$ in. and $w = 5$ in.

$A = lw$ or $l \cdot w$
↑   ↑   ↑
Area   length   width

$A = l \cdot w$
$A = 7 \cdot 5$
$A = 35$

in.² means *square inches*. So, the area of the rectangle is 35 in.²

Example 2

A formula for the area of a triangle is $A = \frac{bh}{2}$. Find the area of a triangle if $b = 14$ cm and $h = 7$ cm.

$A = \frac{bh}{2}$
↑
Area
   base   height

$A = \frac{b \cdot h}{2}$

$A = \frac{14 \cdot 7}{2}$

$A = \frac{98}{2}$

$A = 49$

So, the area is 49 cm².

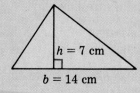

---

practice ▷   **Find the area.**

1. $A = lw$ 2. $A = \frac{bh}{2}$ 3. $A = \frac{bh}{2}$

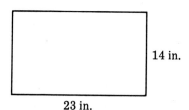
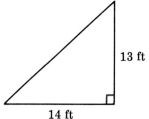
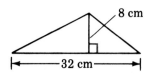

46   CHAPTER TWO

Example 3

A formula for the perimeter of a rectangle is $P = 2l + 2w$. Find the perimeter of a rectangle if $l = 12$ in. and $w = 4$ in.

$$P = 2l + 2w$$

Substitute 12 for l, 4 for w. $\quad P = 2 \cdot l + 2 \cdot w$

Do the multiplications first. $\quad P = 2 \cdot 12 + 2 \cdot 4$
Then add. $\quad P = 24 + 8$
$\quad P = 32$

So, the perimeter of the rectangle is 32 in.

practice ▷ $P = 2l + 2w$. Find P for the given values of l and w.

4. $l = 9$ cm, $w = 3$ cm
5. $l = 13$ ft, $w = 19$ ft

◇ **EXERCISES** ◇

$A = lw$. Find A for the given values of l and w.

1. l is 15 ft, w is 8 ft
2. l is 20 cm, w is 14 cm
3. l is 13 in., w is 17 in.

$A = \frac{bh}{2}$. Find A for the given values of b and h.

4. b is 14 in., h is 20 in.
5. b is 21 m, h is 8 m
6. b is 6 ft, h is 24 ft

$P = 2l + 2w$. Find P for the given values of l and w.

7. l is 18 yd, w is 14 yd
8. l is 24 m, w is 13 m
9. l is 40 in., w is 24 in.

The formula for the volume of a rectangular solid is $V = lwh$ or $V = l \cdot w \cdot h$. Find V for the given values of l, w, and h.

10. l is 4 in., w is 7 in., h is 8 in.
11. l is 9 cm, w is 12 cm, h is 4 cm

Solve these problems.

12. The circumference of a circle is the distance around it. d is the diameter. Evaluate $C = 3.14d$ if d is 8.2
13. The circumference of a circle is $C = 3.14d$ where d is the diameter. Find C if d is 5.6.
★ 14. The area of a rectangle is 54 in.^{2} The length is 9 in. Find the width.
★ 15. The area of a right triangle is 24 ft^2. The height is 8 ft. Find the base.

EVALUATING GEOMETRIC FORMULAS

Mathematics Aptitude Test

This test will prepare you for taking standardized tests. Choose the best answer for each question.

1. How many numbers between 1 and 100 end in 8?
   a. 9
   b. 10
   c. 19
   d. 20

2. Six people meet in a room and each shakes hands with everyone. How many handshakes are there?
   a. 6
   b. 21
   c. 30
   d. 720

3. $6 + a + b = 18$, where a and b are 1-digit whole numbers. What is the lowest value of a?
   a. 0
   b. 1
   c. 2
   d. 3

4. The first number is 12. The second number is less than 8. Which of the following is true?
   a. The first number is greater than the second.
   b. The second number is greater than the first.
   c. The sum of the numbers is 20.
   d. Cannot be determined which number is greatest.

5. Snow is falling at the rate of 15 cm/h. How many centimeters of snow will fall in 40 minutes?
   a. 10
   b. 45
   c. 60
   d. 80

6. Which number is 10 more than 999?
   a. 1,000
   b. 989
   c. 1,009
   d. 1,109

7. Find the missing number in the following sequence.
   7, 13, 25, 49, ___?___
   a. 67
   b. 97
   c. 73
   d. 81

8. The outer diameter of a plastic pipe is 3.05 inches. The inner diameter is 2.45 inches. What is the thickness of the plastic?
   a. 1.20 in.
   b. 0.60 in.
   c. 5.50 in.
   d. 0.30 in.

9. How many whole numbers between 10 and 100 have two identical digits?
   a. 8
   b. 9
   c. 10
   d. 20

COMPUTER ACTIVITIES

You have seen that the **BASIC** symbols for addition and subtraction are the same as in arithmetic: + and −. The **BASIC** symbols for × and ÷ are:

Multiply by: *
Divide by: /

The order of operations rule you have used in this chapter also holds for expressions written in **BASIC**.

Example   A = 12 + 8 / (4 − 2) − 3 * 5. Find A first by computation. Then write a program to find A. Type and **RUN** it.

First find A by computation.

A = 12 + 8 / (4 − 2) − 3 * 5   Compute within parentheses first.
A = 12 + 8 / 2 − 3 * 5   Do all multiplications and divisions next.
A = 12 + 4 − 15   Do all additions and subtractions last.
A = 16 − 15
A = 1

Type in this program.

```
10  LET A = 12 + 8 / (4 - 2) - 3 * 5
20  PRINT A
30  END
```

Now **RUN** the program. No line number is needed. The screen displays the answer.

```
]RUN
1
```

Exercises

Compute.

1. 6 * 4   2. 18 / 9   3. 8 * 3 + 2   4. 13 + 4 / 2   5. 10 + 8 / 4 − 2
6. 5 + 4 * 2 − 12 / 3   7. (10 + 8) / (4 − 2)   8. 10 + 8 / (4 − 2)

Write a program to find the value of each variable. Then type and **RUN** the program. Remember to type **NEW** when you have finished **RUN**ning each program and are ready to begin the next.

9. A = 4 + 6 * 7 − 24 / 6   10. B = 6 * 7 − 8 + 15 / 5   11. K = 45 / 9 + 8 − 2 * 5
12. T = (40 + 8) / (3 + 1)   13. K = 40 + 8 / (3 + 1)   14. W = (40 + 8) / 4 + 1

Chapter Review

Evaluate if $x = 9$, $a = 2$, $k = 6$, and $t = 10$. [26]

1. $7x$
2. $\frac{32}{a}$
3. $\frac{258}{k}$
4. $\frac{3t}{2}$

What property of multiplication is illustrated? [28]

5. $(9 \cdot 3) \cdot 4 = 9 \cdot (3 \cdot 4)$
6. $1 \cdot 7 = 7$
7. $16 \cdot 8 = 8 \cdot 16$
8. $24 = 24 \cdot 1$

Compute. [30]

9. $8 + 4 \cdot 3$
10. $18 - 14 \div 7$
11. $19 - 6 \div 2$
12. $9 + 3(4 + 7)$
13. $32 \div 8 + 4(2 + 5)$
14. $2(7 + 4) - 35 \div 7$

Evaluate if $x = 3$, $y = 2$, $a = 6$, and $b = 4$. [33]

15. $7x + 3$
16. $\frac{a}{3} + 7$
17. $6a - 2 + 5b$
18. $3 + 2xy$
19. $8ab + 14$
★ 20. $(6xy - 2a)4ay$

Solve each equation. Check the solution. [36]

21. $9x = 45$
22. $7 = \frac{a}{10}$
23. $68 = 34b$
24. $\frac{y}{6} = 14$
25. $408 = 8m$
★ 26. $8(2 + 7) = 18p$

Write in mathematical form. [39]

27. 12 times x
28. twice q
29. m divided by 18

Solve these problems. [39]

30. 8 times a number is 56. Find the number.
31. The $40 selling price of a radio is twice the cost. Find the cost.
32. Mr. Ruiz's savings of $4,000 is the same as his father's divided by 4. Find his father's savings.
33. This year's basketball team has 30 wins. This is twice last year's. How many games were won last year?
34. Maria's age of 3 is the same as her cousin's age divided by 6. How old is her cousin?
★ 35. 7 times a number is the same as 20 increased by 8. Find the number.
36. [46] $A = lw$. Find A if l is 22 in. and w is 14 in.
37. [46] $A = \frac{bh}{2}$. Find A if b is 4 m and h is 18 m.
38. [46] $P = 2l + 2w$. Find P if l is 32 ft and w is 14 ft.
39. [46] $P = 2l + 2w$. Find P if l is 28 cm and w is 42 cm.

Chapter Test

Evaluate if $x = 8$, $a = 3$, and $t = 12$.

1. $4x$
2. $\dfrac{168}{a}$
3. $\dfrac{5t}{2}$

What property of multiplication is illustrated?

4. $1 \cdot 9 = 9$
5. $17 \cdot 23 = 23 \cdot 17$
6. $(8 \cdot 4) \cdot 7 = 8 \cdot (4 \cdot 7)$

Compute.

7. $11 + 6 \cdot 3$
8. $44 - 33 \div 11$
9. $6 + 4(7 + 2)$

Evaluate if $a = 8$ and $b = 2$.

10. $7a + 9$
11. $3a - 4 + 2b$
12. $5 + 4ab$

Solve each equation. Check the solution.

13. $7m = 42$
14. $48 = 4t$
15. $\dfrac{a}{5} = 12$

Write in mathematical form.

16. 15 times g
17. p divided by 29
18. twice r

Solve these problems.

19. Three times Maria's age is 45. How old is Maria?

20. Jared's 42 baskets for the season is the same as O'Ruark's divided by 2. Find O'Ruark's baskets for the season.

21. $A = lw$. Find A if l is 13 in. and w is 9 in.

22. $A = \dfrac{bh}{2}$. Find A if b is 16 m and h is 4 m.

23. $P = 2l + 2w$. Find P if l is 43 ft and w is 19 ft.

CHAPTER TEST

Expressions and Formulas

3

New England's covered bridges are supported by triangular trusses.

The Distributive Property

◇ OBJECTIVE ◇
To rewrite expressions using the distributive property

◇ RECALL ◇
$5(4 + 2)$ means $5 \cdot (4 + 2)$.

Example 1

Show that $5(2 + 4)$ is the same as $5 \cdot 2 + 5 \cdot 4$.

$5(2 + 4)$	$5 \cdot 2 + 5 \cdot 4$
$5 \cdot 6$	$10 + 20$
30	30

Compute $5(2 + 4)$ and $5 \cdot 2 + 5 \cdot 4$.
Answers are the same.

So, $5(2 + 4) = 5 \cdot 2 + 5 \cdot 4$.

Example 2

Show that $(5 - 2)7 = 5 \cdot 7 - 2 \cdot 7$.

$(5 - 2)7$	$5 \cdot 7 - 2 \cdot 7$
$3 \cdot 7$	$35 - 14$
21	21

Answers are the same.

So, $(5 - 2)7 = 5 \cdot 7 - 2 \cdot 7$.

---

distributive property
For all numbers a, b, and c,
$a(b + c) = a \cdot b + a \cdot c$ and
$(b + c)a = b \cdot a + c \cdot a$;
$a(b - c) = a \cdot b - a \cdot c$ and
$(b - c)a = b \cdot a - c \cdot a$.

Example 3

Rewrite $6(7 + 1)$ using the distributive property. Then compute both expressions.

Distribute the 6 over 7 and 1. $\quad 6(7 + 1) = 6 \cdot 7 + 6 \cdot 1$

$6(7 + 1)$	$6 \cdot 7 + 6 \cdot 1$
$6 \cdot 8$	$42 + 6$
48	48

Use the rules for order of operations.

So, the results are the same, 48.

Example 4

Rewrite $(9 - 6)8$ using the distributive property.

Distribute 8 over 9 and 6. $\quad (9 - 6)8 = 9 \cdot 8 - 6 \cdot 8$

practice ▷ **Rewrite using the distributive property. Then compute both expressions.**

1. $9(7 + 5)$
2. $(5 - 3)4$
3. $6(7 + 9)$

You can use the distributive property in reverse to rewrite $5 \cdot 6 + 5 \cdot 8$.

$5 \cdot 6 + 5 \cdot 8$

The 5 is distributed as a multiplier over 6 and 8.

So, $5 \cdot 6 + 5 \cdot 8$ is the same as

$5(6 + 8)$.

Example 5

Rewrite using the distributive property.

The 8 is distributed over 4 and 2. $8 \cdot 4 + 8 \cdot 2$
$8(4 + 2)$

The 5 is distributed over 7 and 1. $7 \cdot 5 - 1 \cdot 5$
$(7 - 1)5$

practice ▷ **Rewrite using the distributive property.**

4. $8 \cdot 6 + 8 \cdot 11$
5. $13 \cdot 3 - 10 \cdot 3$
6. $14 \cdot 12 + 14 \cdot 9$

You can use the distributive property to simplify expressions that contain variables.

Example 6

Simplify.

Distribute the 4. $4(3a + 5)$ $7(2b - 3)$ Distribute the 7.
$3a$ means $3 \cdot a$. $4 \cdot 3a + 4 \cdot 5$ $7 \cdot 2b - 7 \cdot 3$
$4 \cdot 3a = 12 \cdot a = 12a$ $12a + 20$ $14b - 21$

practice ▷ **Simplify.**

7. $8(2x + 3)$
8. $7(2b - 5)$
9. $5(4x - 6)$

◇ ORAL EXERCISES ◇

What number must be distributed as a multiplier?

1. $6(3 + 12)$
2. $(8 + 9)4$
3. $5(14 - 3)$

What number has been distributed as a multiplier?

4. $8 \cdot 7 + 8 \cdot 12$
5. $9 \cdot 6 - 7 \cdot 6$
6. $4 \cdot 8 + 4 \cdot 13$

◇ EXERCISES ◇

Rewrite using the distributive property. Then compute both expressions.

1. $4(7 + 1)$
2. $6(8 - 3)$
3. $3(9 + 4)$
4. $(9 + 6)2$
5. $(5 - 4)7$
6. $(9 + 3)8$

Rewrite using the distributive property.

7. $5(4 + 6)$
8. $6(4 + 9)$
9. $3(6 - 2)$
10. $(3 + 8)6$
11. $4 \cdot 6 + 4 \cdot 2$
12. $9 \cdot 6 - 9 \cdot 4$

Simplify.

13. $3(2a + 5)$
14. $4(6m - 7)$
15. $7(4a + 3)$
16. $8(6b - 3)$
17. $4(3x - 5)$
18. $6(2m + 10)$
19. $12(2a - 4)$
20. $11(2m - 3)$
21. $14(2a + 3)$

★ 22. Is division distributive over addition?
  Does $36 \div (6 + 3) = (36 \div 6) + (36 \div 3)$?

★ 23. Is addition distributive over multiplication?
  Does $8 + (6 \cdot 3) = (8 + 6) \cdot (8 + 3)$?

Algebra Maintenance

Evaluate if $x = 4$, $m = 7$, $k = 9$, and $d = 10$.

1. $k - x$
2. $4d + m$
3. $\dfrac{121}{x + m}$

Solve each equation.

4. $7a = 56$
5. $42 = x + 9$
6. $34 = 2x$
7. $\dfrac{m}{5} = 12$
8. $45 = c - 12$
9. $5 = \dfrac{r}{3}$

Solve these problems.

10. The $32 selling price of a set of computer disks is twice the cost. What is the cost?

11. Tom's age of 7 is the same as his sister's age divided by 3. What is his sister's age?

THE DISTRIBUTIVE PROPERTY

Combining Like Terms

◇ OBJECTIVES ◇

To simplify algebraic expressions by combining like terms

To simplify algebraic expressions using the property of multiplicative identity

◇ RECALL ◇

Rewrite using the distributive property.

$5 \cdot 6 + 2 \cdot 6$ ← 6 is distributed over 5 and 2.
$(5 + 2)6$

The expression $3x + 4y$ has two *terms*, $3x$ and $4y$. In the term $3x$, 3 is the *coefficient* of x. A *coefficient* is a multiplier of a variable.

$5a + 5a$ $7m - 5m$ $3x + 4y$
like terms   like terms   *un*like terms

Like terms are terms that are exactly alike or differ only by their *coefficients*.

Example 1

Rewrite $5a + 4a$ using the distributive property. Simplify the result.

distributive property $5a + 4a = 5 \cdot a + 4 \cdot a$ a is distributed over 5 and 4.
$= (5 + 4) \cdot a$
$= 9 \cdot a$
$9a$

You can use a shortcut to combine like terms. Like terms can be combined by adding or subtracting their coefficients. Thus, $5a + 4a = 9a$ and $7x - 5x = 2x$.

Example 2

Simplify if possible.

Rearrange to group like terms.

$3a + 4 + 7a$
$\underline{3a + 7a} + 4$
$10a \;\;\;\;\;\;\; + 4$

$5x + 2y + 3$
The terms $5x$, $2y$, and 3 are *unlike terms*. They cannot be combined.

So, $3a + 4 + 7a = 10a + 4$; $5x + 2y + 3$ cannot be simplified.

Example 3

Simplify $7 + 3x + 5y + 4 - 2y + 6x$.

$$7 + 3x + 5y + 4 - 2y + 6x$$

Rearrange to group like terms. $\quad 3x + 6x + 5y - 2y + 4 + 7$

$$9x \quad + \quad 3y \quad + \quad 11$$

practice ▷ Simplify if possible.

1. $5t + 4t$
2. $8x + 3y + 5$
3. $6 + 2m + 9n + 1 + 11m - 5n$

Recall the property of multiplicative identity. Multiplication of any number by 1 gives that number.

Example 4

Simplify $8y + 3 + y$.

The coefficient of $8y$ is 8. $\quad 8y + 3 + y$

Rewrite y as $1 \cdot y$. $\quad 8y + 3 + 1 \cdot y \quad 1 \cdot y = y$

$$8y + 1y + 3$$

Add the coefficients of the *like* terms. $\quad 9y \quad + 3$

practice ▷ Simplify.

4. $6a + 4 + a$
5. $p + 8 + 7p$
6. $2 + 3a + 7 + a$

◇ ORAL EXERCISES ◇

For each, name the like terms. Name the coefficient of each of these terms.

1. $7x + 5y + 3x$
2. $3a + 4b + 7a + 6b$
3. $12y + 4t + 2y$
4. $9g + 3r + 7g + 13r$
5. $x + 5c + 3x$
6. $2t + 3g + t + g$

◇ EXERCISES ◇

Simplify if possible.

1. $3x + 5x$
2. $8y + 7y$
3. $6m - 4m + 3$
4. $9a + 3b - 5$
5. $7 + 4x + 3y$
6. $8 + 11x + 3 - 4x$
7. $3 + 5x + 4 + 2x$
8. $2a + 3x + 4a + 8x$
9. $6k + 9y - 2k + 4y$
10. $3x + x + 4$
11. $5 + y + 13y$
12. $h + 21 + 6h - 4$
13. $3 + 8g + 1 + g$
14. $y + 14 + 3y + 2$
15. $x + 6y + 7x + y$
★ 16. $1 + 19x + x + 4 \cdot 5$
★ 17. $(2 + 3 + 4)y + 23y$
★ 18. $x + 3(3 + 2) + 5x + 8$

COMBINING LIKE TERMS

Simplifying and Evaluating Expressions

◇ **OBJECTIVE** ◇

To simplify and then evaluate algebraic expressions for given values of the variables

◇ **RECALL** ◇

Evaluate $4x + 6$ if $x = 2$.
$4 \cdot 2 + 6$ ←— Substitute 2 for x.
$8 \;\; + 6$
14

Example 1 Simplify $3a + 4 + 5a$. Then evaluate if $a = 2$.

	$3a + 4 + 5a$
Rearrange to group like terms.	$3a + 5a + 4$
Combine like terms.	$8a \;\;\;\;+ 4$
Substitute 2 for a.	$8 \cdot 2 \;\;+ 4$
Multiply first.	$16 \;\;\;+ 4$
Then add.	20

Example 2 Simplify $7 + x - 6 + 4x$. Then evaluate if $x = 6$.

	$7 + 1x - 6 + 4x$
First rewrite x as $1x$.	
Rearrange to group like terms.	$1x + 4x + 7 - 6$
Combine like terms.	$5x \;\;\;+ 1$
Substitute 6 for x.	$5 \cdot 6 \;\;+ 1$
Use the rules for order of operations.	$30 \;\;\;+ 1$
	31

practice ▷ Simplify. Then evaluate for $a = 2$, $b = 3$, and $c = 4$.

1. $4a + 5 + 3a$ 2. $2b + 5 + 6b$ 3. $6c + 8 + c + 9$

Example 3 Simplify $3a + b + a + b$. Then evaluate if $a = 3$ and $b = 7$.

	$3a + b + a + b$
Coefficient of 1 is understood.	$3a + 1b + 1a + 1b$
	$3a + 1a + 1b + 1b$
	$4a \;\;\;+\;\;\; 2b$
Substitute 3 for a, 7 for b.	$4 \cdot 3 \;\;+\;\; 2 \cdot 7$
	$12 \;\;\;+\;\;\; 14$
	26

CHAPTER THREE

Example 4

Simplify $4 + 3a + a + 6b + 5 + b$. Then evaluate if $a = 6$ and $b = 3$.

$$4 + 3a + a + 6b + 5 + b$$
$$4 + 3a + 1a + 6b + 5 + 1b$$
$$\underbrace{4 + 5} + \underbrace{3a + 1a} + \underbrace{6b + 1b}$$
$$9 \quad + \quad 4a \quad + \quad 7b$$

Substitute 6 for a, 3 for b.
$$9 \quad + \quad 4 \cdot 6 \quad + \quad 7 \cdot 3$$
$$9 \quad + \quad 24 \quad + \quad 21$$
$$54$$

practice ▷ Simplify. Then evaluate if $x = 5$ and $y = 6$.

4. $x + y + x + y$
5. $3x + 5y + 4x + 2y$
6. $x + y + 5x + y$
7. $7 + 2y + y + 3x + 4 + x$
8. $x + 2x + 5 + 9y + 4 - 2y$

◇ EXERCISES ◇

Simplify. Then evaluate for these values of the variables:
$a = 4, b = 3, c = 2, x = 7, y = 9, z = 6$.

1. $2x + 4 + 7x$
2. $8a + 3 - 5a$
3. $6c + 9 + 3c$
4. $8a + 1 + 3a$
5. $7b - 2 + 5b$
6. $8z - 2z + 5$
7. $9 + 2y + 3y$
8. $9c + 4 + 2c$
9. $8 + 4a + 4a$
10. $8 + y - 4 + 3y$
11. $2b + 5 + b + 4$
12. $7 + z + 3 + z$
13. $x + 6 + x + 9$
14. $3 + b + 9 + 2b$
15. $8z + 1 + z + 6$
16. $8 + 11x + 3 - 4x$
17. $8x + 5 + x + 1$
18. $4 + 3c + 7 + c$
19. $b + z + b + 2z$
20. $4x + y + 3x + y$
21. $z + 3a + a + 5z$
22. $c + 9b + c + b$
23. $7y + a + a + y$
24. $a + 3a + 4b + b$
25. $9a + c + c + a$
26. $5a + a + y + 4y$
27. $9z + 3y + y + 4z$
28. $7x + 3y + x + 4y + y$
29. $a + 3b + 4a + b + 5a$
30. $z + 3y + z + 4y + 5z + y$
31. $4 + 3x + 5 - 2x + y + y$
32. $6a + a + 5 + 3b + 2 + 2b$
33. $5a + b + 2 + 3a + 4b + 6$

SIMPLIFYING AND EVALUATING EXPRESSIONS

Problem Solving – Applications
Jobs for Teenagers

1. A musical group started playing at 8:15 P.M. They finished at 1:15 A.M. How many hours did the group play?

2. One evening the Keynotes played for three hours. Each song lasted an average of 4 minutes. How many songs did they play?

3. A band earns an average of $450 a weekend. They wanted to buy new equipment that costs $2,700. How many weekends must they work in order to buy the new equipment?

4. A three-piece band played from 9:00 P.M. to midnight on Friday, from 8:00 P.M. to 2:00 A.M. on Saturday, and from 9:00 P.M. to 11:45 P.M. on Sunday. How many hours did they play altogether on that weekend?

5. A rock group charges a flat fee of $620. They played for 5 hours. How much did they make per hour?

6. The sophomore class hires a band that charges $350. Students are charged $3.00 a ticket. 275 students attend the dance. Find the class profit.

7. A piano player earned $20 per hour for the first three hours of playing and $24 per hour for each hour over 3 hours. How much did she earn for a 4-hour engagement?

8. A drummer earns $19.50 per hour for playing until midnight. She is then paid $22.00 per hour for playing after midnight. She played from 8:00 P.M. to 2:00 A.M. How much did she earn?

Solving Equations: Using Two Properties

◇ OBJECTIVE ◇

To solve equations using more than one property of equations

◇ RECALL ◇

Solve $x + 3 = 8$.   Undo the addition.
$x + 3 - 3 = 8 - 3$ Subtract 3 from each side.
$x = 5$

Solve $3y = 15$.   Undo the multiplication.
$\dfrac{3y}{3} = \dfrac{15}{3}$ Divide each side by 3.
$y = 5$

So far you have solved equations using only one operation. In the equation $2x + 3 = 11$, x is multiplied by 2. Then 3 is added to the result. This is an equation with two operations. Solving this equation requires undoing the addition and then undoing the multiplication.

Example 1

Solve $2x + 3 = 11$. Check the solution.

First undo addition of 3.	$2x + 3 = 11$
Subtract 3 from each side.	$2x + 3 - 3 = 11 - 3$
Undo multiplication of x by 2.	$2x = 8$
Divide each side by 2.	$\dfrac{2x}{2} = \dfrac{8}{2}$
	$x = 4$

Check: $2x + 3 = 11$
$2 \cdot 4 + 3 \mid 11$
$8 + 3$
$11 = 11$ true

So, 4 is the solution of $2x + 3 = 11$.

Example 2

Solve $13 = 5k - 2$.   Check the solution.

Undo subtraction of 2.	$13 = 5k - 2$
Add 2 to each side.	$13 + 2 = 5k - 2 + 2$
Undo multiplication of k by 5.	$15 = 5k$
Divide each side by 5.	$\dfrac{15}{5} = \dfrac{5k}{5}$
	$3 = k$

Check: $13 = 5k - 2$
$13 \mid 5 \cdot 3 - 2$
$15 - 2$
$13 = 13$ true

So, 3 is the solution of $13 = 5k - 2$.

practice ▷ Solve each equation. Check the solution.

1. $3y + 4 = 13$ 2. $2t + 5 = 15$ 3. $13 = 5g - 2$

Example 3

Solve $\frac{x}{2} + 3 = 6$. Check the solution.

Undo addition of 3. Subtract 3 from each side.	$\frac{x}{2} + 3 - 3 = 6 - 3$	Check: $\frac{x}{2} + 3 = 6$
Undo division by 2.	$\frac{x}{2} = 3$	$\frac{6}{2} + 3 \;\big|\; 6$
Multiply each side by 2.	$2 \cdot \frac{x}{2} = 3 \cdot 2$	$3 + 3$
	$x = 6$	$6 = 6$ true

So, 6 is the solution of $\frac{x}{2} + 3 = 6$.

---

practice ▷ Solve each equation. Check the solution.

4. $\frac{x}{3} + 2 = 7$ 5. $\frac{x}{4} - 1 = 11$ 6. $9 = \frac{x}{6} + 4$

◇ **ORAL EXERCISES** ◇

Tell what must be done to solve each equation.

1. $2z + 4 = 14$ 2. $12 = 3d - 6$ 3. $8a + 4 = 28$
4. $6 = 2b - 4$ 5. $\frac{x}{7} + 5 = 8$ 6. $\frac{x}{2} - 4 = 8$

◇ **EXERCISES** ◇

Solve each equation. Check the solution.

1. $2x + 1 = 5$ 2. $3y + 1 = 10$ 3. $14 = 4u + 6$
4. $2m - 2 = 8$ 5. $6k - 4 = 8$ 6. $12 = 5y - 3$
7. $\frac{a}{8} + 2 = 6$ 8. $\frac{b}{6} - 2 = 1$ 9. $9 = \frac{x}{3} + 2$

Solve each equation.

10. $8x - 3 = 29$ 11. $24 = 5y - 1$ 12. $14 = 8c - 2$
13. $3y - 9 = 12$ 14. $15 = 5m - 10$ 15. $12 = 4z + 4$
16. $2t - 8 = 24$ 17. $18 = 7g - 3$ 18. $11u - 8 = 14$
19. $3k + 8 = 23$ 20. $28 = 9y + 1$ 21. $44 = 6p + 2$
22. $\frac{g}{5} + 7 = 17$ 23. $8 = \frac{k}{4} + 2$ 24. $\frac{b}{7} - 2 = 3$
25. $4 = \frac{i}{6} - 3$ 26. $\frac{w}{5} + 6 = 15$ 27. $\frac{d}{12} - 5 = 2$

★ 28. $4x - 1 = 5 \cdot 4 + 3$ ★ 29. $2(5 + 3) = 4x - 12$ ★ 30. $\frac{y}{7} + 2 = 2(6 - 3)$

English Phrases to Algebra

◇ **OBJECTIVE** ◇

To rewrite in mathematical form English phrases involving two operations

◇ **RECALL** ◇

English Phrases	Mathematical Form
n increased by 6	$n + 6$
6 more than n	$n + 6$
n decreased by 6	$n - 6$
6 less than n	$n - 6$
6 times n	$6n$
n divided by 6	$\frac{n}{6}$

You can easily extend the ideas of the recall to English phrases involving two operations.

Example 1   Write in mathematical form.

<u>3 times n</u>, <u>increased by 5</u>.

$\quad 3 \cdot n \qquad\quad + 5$

or $3n + 5$

Example 2   Write in mathematical form.
Twice a number, decreased by 4.

Let n represent the number.   <u>twice n</u>   <u>decreased by 4</u>

$\qquad\qquad 2 \cdot n \qquad\qquad - 4$

or $2n - 4$

practice ▷ Write in mathematical form.

1. 4 times x, increased by 3
2. twice a number, decreased by 9

Example 3   Write in mathematical form.
8 less than a number divided by 4.

Let x represent the number.   8 less than x divided by 4.

8 less than $\frac{x}{4}$.

$\frac{x}{4} - 8$

ENGLISH PHRASES TO ALGEBRA

practice Write each in mathematical form.

3. 7 less than a number divided by 6
4. 6 more than a number divided by 7

◇ ORAL EXERCISES ◇

Express in mathematical form.

1. $5x$ increased by 3
2. 9 less than $2b$
3. $12k$ decreased by 5
4. 12 more than twice y
5. 15 less than $\frac{y}{7}$
6. $\frac{t}{2}$ increased by 14
7. 18 more than $\frac{b}{3}$
8. 1 less than $\frac{y}{6}$

◇ EXERCISES ◇

Write in mathematical form.

1. 5 times y, increased by 4
2. twice x, increased by 7
3. 9 less than 8 times b
4. 4 more than 6 times p
5. twice g, increased by 12
6. 16 times y, decreased by 13
7. 18 less than 5 times k
8. 11 times w, decreased by 1
9. 12 times a number, increased by 6
10. 15 more than 3 times a number
11. 4 times a number, decreased by 2
12. twice a number, increased by 19
13. 2 less than a number divided by 5
14. 17 more than a number divided by 3
15. 6 more than a number divided by 13
16. 14 less than a number divided by 2
17. n increased by 7 times n
18. 4 times x, decreased by twice x
19. 8 times a number, increased by that number
20. 6 times a number, decreased by that number
★ 21. 4 more than twice a number, increased by 8
★ 22. 9 increased by 5 less than 3 times a number

Calculator

Simplify. Then evaluate for these values of the variables: $a = 349$, $b = 75$, $c = 1{,}234$, $d = 992$.

1. $13a + 29b + 16b + 112a$
2. $195a + 315c + 58a + 77c$
3. $249d + 154c + 8c + 77d$
4. $1{,}345d + 77c + 879d + 145c$
5. $456b + 762a + 448b + 1{,}123a$
6. $4{,}675a + 158 + 59a + 789$

Word Problems: Two Operations

◇ OBJECTIVE ◇

To solve word problems involving two operations

◇ RECALL ◇

There are four basic steps for solving all word problems.

READ: Identify what is given.
Identify what is to be found.

PLAN: Analyze the information.
Represent the data.
Write an equation.

SOLVE: Solve the equation.

INTERPRET: Check the solution.
Decide if the answer is reasonable.

The four basic steps above can now be applied to solving word problems involving *two* operations.

Example 1

Read

8 less than twice a number is 4. What is the number?

Given: 8 less than twice a number is 4.
Find: the number.

Represent the number. Let n = the number.
8 less than twice a number is 4.

Plan

Write an equation.

8 less than $2n$ is 4
$2n$ — 8   = 4

Solve

Undo subtraction of 8. $2n - 8 = 4$
Add 8 to each side. $2n - 8 + 8 = 4 + 8$
Undo multiplication. $2n = 12$
Divide each side by 2. $\dfrac{2 \cdot n}{2} = \dfrac{12}{2}$
                           $n = 6$

Interpret

Check: 8 less than twice a number is 4
        8 less than twice 6 | 4
        12 − 8 |
        4 | = 4   true

So, the number is 6.

Example 2

Rosa's age divided by 4, increased by 5, is 10. How old is Rosa?

Represent her age with a variable. Let a = Rosa's age.
age divided by 4, increased by 5 is 10

Write an equation.
$$\frac{a}{4} + 5 = 10$$

Undo addition of 5.
Subtract 5 from each side.
$$\frac{a}{4} + 5 = 10$$
$$\frac{a}{4} + 5 - 5 = 10 - 5$$

Undo division.
$$\frac{a}{4} = 5$$

Multiply each side by 4.
$$4 \cdot \frac{a}{4} = 5 \cdot 4$$
$$a = 20$$

Check: age divided by 4, increased by 5, is 10

$\frac{a}{4} + 5$	10
$\frac{20}{4} + 5$	
10	= 10 true

So, Rosa's age is 20.

practice ▷

1. 6 less than twice a number is 14. Find the number.

2. If Tina's age divided by 3 is decreased by 2, the result is 10. How old is Tina?

Example 3

The $300 selling price of a stereo is the same as the cost increased by 3 times the cost. What is the cost?

Let c = the cost.
300 is the same as cost increased by 3 times the cost.

Write an equation. $300 = c + 3c$

$c = 1c$. $300 = 1c + 3c$

Combine like terms. $300 = 4c$

Undo multiplication by 4. $\frac{300}{4} = \frac{4c}{4}$

Divide each side by 4. $75 = c$

Check on your own. So, the cost of the stereo is $75.

practice ▷ 3. The $90 selling price of a camera is the same as the cost increased by twice the cost. What is the cost?

4. Dan's score increased by 4 times his score is 120. What is his score?

◇ ORAL EXERCISES ◇

Give an equation for each sentence.

1. 7 more than 3 times x is 28.
2. 4 times n, decreased by 8 is 12.
3. 24 is 6 less than 3 times y.
4. t, increased by twice t, is 27.
5. 48 is 8 more than y divided by 10.
6. 15 is 10 less than p divided by 3.

◇ EXERCISES ◇

Solve these problems.

1. 10 less than twice a number is 14. Find the number.
2. 16 is 4 more than twice a number. Find the number.
3. 2 more than Lisa's age divided by 5 is 8. How old is Lisa?
4. A number divided by 3, increased by 4, is 7. Find the number.
5. 26 is 4 less than 5 times a number. Find the number.
6. 50 more than 3 times Jo's bowling score is 500. Find her score.
7. If the cost of a shirt is increased by twice the cost, the result is $27. Find the cost.
8. The $80 selling price of a cassette radio is the cost increased by 3 times the cost. Find the cost.
9. 40 lb more than twice Royal's weight is 280 lb. Find his weight.
10. If 5 times Renee's age is decreased by 20, the result is 30. Find her age.
11. If Lou's savings were increased by 5 times his savings, the result would be $30,000. How much has he saved?
12. 8 times the temperature decreased by the temperature was 14°. Find the temperature.
★ 13. If 3 times Pete's age is increased by 4 more than twice his age, the result is 34. How old is Pete?
★ 14. 5 times a number, increased by 2 less than twice the number, is 40. Find the number.

WORD PROBLEMS: TWO OPERATIONS

Problem Solving – Careers
Dietitians

Breakfast	Calories	Lunch	Calories	Dinner	Calories
orange juice	54	1 hamburger	425	roast beef	250
2 scrambled eggs	212	French fries	225	mashed potatoes with butter	180
1 slice of buttered toast with jam	168	milk	166	peas	35
		apple	66	salad with dressing	80
		vegetable soup	74	2 rolls with butter	163
				apple pie	331
				skim milk	87

1. How many calories are there in one slice of apple pie?
2. Find the total number of calories in the dinner.
3. Find the total number of calories in all 3 meals together.
4. An average teenage boy needs about 3,000 calories each day. How many more calories is this than the total number of calories for the three meals?
5. In order to lose 2 lb of fat in one week, your diet should be about 7,000 calories a week below your body needs. How many fewer calories is this each day?
6. The starting salary of a hospital dietitian is about $13,000 a year. How much is this per week?

Factors and Exponents

◇ OBJECTIVES ◇

To write expressions like
  $2 \cdot 2 \cdot 4 \cdot 4 \cdot 4$ using
  exponents
To evaluate expressions like x^4
  and $6x^2$ if $x = 2$

◇ RECALL ◇

Compute 7×6 and $6 \cdot 4$.
$7 \times 6 = 42$; $6 \cdot 4 = 24$
Multiplication can be shown using "$\times$"
or a raised dot "$\cdot$."

In the product $7 \cdot 3$, 7 and 3 are *factors*. *Factors* are numbers that are *multiplied*. The product $2 \cdot 2 \cdot 2 \cdot 2 \cdot 2$ consists of the factor 2 used as a multiplier 5 times.

The exponent indicates the number of times the base is used as a factor.

A convenient way of writing $2 \cdot 2 \cdot 2 \cdot 2 \cdot 2$ is
base $\to 2^5$. $\leftarrow$ exponent

Example 1

3^4 can be read as the 4th power of 3, or 3 to the 4th power.

Find the value of 3^4.
$3^4 = 3 \cdot 3 \cdot 3 \cdot 3$
$ = 9 \cdot 9$
$ = 81$
So, $3^4 = 81$.

practice ▷ Find the value.

1. 2^4 2. 3^2 3. 5^3 4. 4^3 5. 8^1 6. 1^4 7. 6^3

Example 2

Write $6 \cdot 6 \cdot 6 \cdot 6$ using an exponent.
6 is used 4 times as a factor.
base $\to 6^4$ $\leftarrow$ exponent
So, $6 \cdot 6 \cdot 6 \cdot 6 = 6^4$.

practice ▷ Write using an exponent.

8. $5 \cdot 5 \cdot 5$ 9. $8 \cdot 8 \cdot 8 \cdot 8$ 10. $9 \cdot 9$ 11. $7 \cdot 7 \cdot 7$ 12. $6 \cdot 6 \cdot 6 \cdot 6$

Example 3 Write $4 \cdot 4 \cdot 4 \cdot 5 \cdot 5$ using exponents.

$$4 \cdot 4 \cdot 4 \cdot 5 \cdot 5$$

4 is used 3 times   5 is used 2 times

$$4^3 \cdot 5^2$$

So, $4 \cdot 4 \cdot 4 \cdot 5 \cdot 5 = 4^3 \cdot 5^2$

practice ▷ Write using exponents.

13. $5 \cdot 5 \cdot 5 \cdot 5$ 14. $7 \cdot 7 \cdot 7 \cdot 8 \cdot 8$ 15. $a \cdot a \cdot a \cdot a \cdot a \cdot b \cdot b$

Example 4 Evaluate a^3 if $a = 4$.

a^3
Use a as a factor 3 times. $a \cdot a \cdot a$
Substitute 4 for a. $4 \cdot 4 \cdot 4$
$16 \cdot 4$
64

So, $a^3 = 64$ if $a = 4$.

practice ▷ Evaluate.

16. n^3 if $n = 2$ 17. a^2 if $a = 8$ 18. b^5 if $b = 3$

Example 5 Evaluate $5x^2$ if $x = 3$.

$5x^2$
Substitute 3 for x. $5 \cdot 3^2$
Use 3 as a factor twice. $5 \cdot 3 \cdot 3$
$5 \cdot 9$
45

So, $5x^3 = 45$ if $x = 3$.

practice ▷ Evaluate.

19. $4x^5$ if $x = 2$ 20. $7n^4$ if $n = 1$ 21. $5x^2$ if $x = 9$

◇ ORAL EXERCISES ◇

State as a product of factors.

1. x^4 2. b^3 3. y^5 4. t^8 5. g^1 6. w^2 7. m^6

State using exponents.

8. $5 \cdot 5 \cdot 5 \cdot 3 \cdot 3$
9. $3 \cdot 3 \cdot 3 \cdot 3 \cdot 2 \cdot 2$
10. $a \cdot a \cdot a \cdot a \cdot b \cdot b$
11. $3 \cdot 3 \cdot 3 \cdot 4 \cdot 4 \cdot 4$
12. $a \cdot a \cdot b \cdot b$
13. $t \cdot t \cdot t \cdot y \cdot y \cdot y \cdot y$

◇ EXERCISES ◇

Find the value.

1. 1^5 2. 3^3 3. 2^5 4. 7^1 5. 6^2
6. 8^2 7. 9^2 8. 8^3 9. 10^2 10. 11^2
11. 7^3 12. 10^3 13. 4^5 14. 9^3 15. 20^2

Write using exponents.

16. $9 \cdot 9 \cdot 9 \cdot 9$
17. $12 \cdot 12 \cdot 12$
18. $b \cdot b \cdot b \cdot b \cdot b \cdot b$
19. $7 \cdot 7 \cdot 7 \cdot 7$
20. $11 \cdot 11 \cdot 11$
21. $10 \cdot 10 \cdot 10 \cdot 10 \cdot 10 \cdot 10 \cdot 10$
22. $k \cdot k$
23. $m \cdot m \cdot m \cdot m$
24. $15 \cdot 15 \cdot 15$
25. $2 \cdot 2 \cdot 4 \cdot 4 \cdot 4$
26. $5 \cdot 5 \cdot 5 \cdot 5 \cdot 8 \cdot 8 \cdot 8$
27. $t \cdot t \cdot w \cdot w \cdot w \cdot w$
28. $7 \cdot 7 \cdot 6 \cdot 6 \cdot 6 \cdot 6$
29. $g \cdot g \cdot h$
30. $r \cdot r \cdot r \cdot k \cdot k$
31. $x \cdot x \cdot y \cdot y \cdot y$
32. $12 \cdot 12 \cdot 14$
33. $d \cdot d \cdot d \cdot d \cdot d \cdot y$

Evaluate.

34. y^3 if $y = 3$
35. x^4 if $x = 3$
36. t^2 if $t = 5$
37. w^5 if $w = 2$
38. r^4 if $r = 10$
39. t^6 if $t = 2$
40. y^3 if $y = 7$
41. m^4 if $m = 5$
42. a^8 if $a = 2$
43. $5x^2$ if $x = 4$
44. $3x^3$ if $x = 2$
45. $4b^2$ if $b = 5$
46. $6a^5$ if $a = 2$
47. $8n^3$ if $n = 3$
48. $9n^2$ if $n = 3$
49. $10y^2$ if $y = 5$
50. $9a^6$ if $a = 2$
51. $15b^3$ if $b = 1$

Evaluate.

★ 52. $4a^2b^3$ if $a = 3$ and $b = 2$
★ 53. $6x^4y$ if $x = 3$ and $y = 4$
★ 54. $5ab^4$ if $a = 7$ and $b = 2$
★ 55. $8t^4u^3$ if $t = 2$ and $u = 3$
★ 56. $7x^2y^3$ if $x = 3$ and $y = 2$
★ 57. $100x^5y^3$ if $x = 2$ and $y = 5$
★ 58. $11pt^5$ if $p = 9$ and $t = 1$
★ 59. $7a^4b^4$ if $a = 2$ and $b = 2$

FACTORS AND EXPONENTS

Exponents in Formulas

◆ **OBJECTIVE** ◆

To evaluate formulas containing exponents

◆ **RECALL** ◆

Evaluate x^3 if $x = 2$.

$$2^3 \rightarrow 2 \cdot 2 \cdot 2 = 8$$

Example 1

The formula for the area of a square is $A = s^2$. Find A if $s = 7$.

$A = s^2$

Substitute 7 for s. $A = 7^2 = 7 \cdot 7$ or 49

So, A is 49 if $s = 7$.

practice ▷ $A = s^2$. Find A for the given value of s.

1. 6   2. 8   3. 13   4. 19

Example 2

The formula for the area of a circle is $A = \pi r^2$. Find A in terms of π if $r = 9$.

$A = \pi r^2$

Substitute 9 for r. $A = \pi \cdot 9^2$
$9^2 = 9 \cdot 9 = 81$ $A = \pi \cdot 81$, or 81π

So, $A = 81\pi$ if $r = 9$.

practice ▷ $A = \pi r^2$. Find A in terms of π for the given values of r.

5. 4   6. 3   7. 10   8. 11

Example 3

The formula for the volume of a cylinder is $V = \pi r^2 h$. Find V in terms of π if $r = 4$ and $h = 10$.

$V = \pi r^2 h$

Substitute 4 for r and 10 for h. $V = \pi \cdot 4^2 \cdot 10$
$V = \pi \cdot 16 \cdot 10$
$160 \cdot \pi = 160\pi$ $V = \pi \cdot 160$ or 160π

So, $V = 160\pi$ if $r = 4$ and $h = 10$.

practice ▷ $V = \pi r^2 h$. Find V in terms of π for the given values of r and h.

    9. $r = 5, h = 8$ 10. $r = 3, h = 10$
    11. $r = 10, h = 4$ 12. $r = 7, h = 3$

Example 4   The formula for the surface area of a cube is $6e^2$ where e is the length of an edge. Find S if $e = 5$.

$S = 6e^2$

Substitute 5 for e. $S = 6 \cdot 5^2$
 $S = 6 \cdot 25$ or 150

$$\begin{array}{r} 25 \\ \times\ 6 \\ \hline 150 \end{array}$$

So, $S = 150$ if $e = 5$.

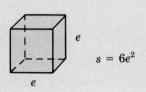

---

practice ▷ $S = 6e^2$. Find S for the given value of e.

    13. 3   14. 6   15. 30   16. 10

◇ EXERCISES ◇

$A = s^2$. Find A for the given value of s.

| 1. 2 | 2. 5 | 3. 3 | 4. 4 | 5. 18 |
| 6. 10 | 7. 16 | 8. 21 | 9. 40 | 10. 15 |

$A = \pi r^2$. Find A in terms of π for the given value of r.

| 11. 5 | 12. 6 | 13. 8 | 14. 12 | 15. 2 |
| 16. 14 | 17. 22 | 18. 20 | 19. 17 | 20. 30 |

$V = \pi r^2 h$. Find V in terms of π for the given values of r and h.

21. $r = 3, h = 3$ 22. $r = 6, h = 5$ 23. $r = 8, h = 7$
24. $r = 6, h = 8$ 25. $r = 5, h = 7$ 26. $r = 20, h = 11$

$S = 6e^2$. Find S for the given value of e.

| 27. 4 | 28. 7 | 29. 1 | 30. 9 | 31. 8 |
| 32. 12 | 33. 14 | 34. 19 | 35. 21 | 36. 42 |

The formula for the volume of a cube is $V = e^3$ where e is the length of a side. Find V for the given value of e.

| 37. 6 | 38. 2 | 39. 5 | 40. 7 | 41. 10 |
| 42. 15 | 43. 11 | 44. 18 | 45. 25 | 46. 13 |

The formula for the surface area of a cylinder is $S = 2\pi r^2 + 2\pi r h$. Find S in terms of π.

★ 47. $r = 4, h = 10$ ★ 48. $r = 8, h = 6$ ★ 49. $r = 10, h = 3$

EXPONENTS IN FORMULAS

Non-Routine Problems

There is no simple method used to solve these problems. Use whatever methods you can. Do not expect to solve these easily.

1. The Gordons have two children. The sum of their ages is 15 and the product is 36. What are their ages?

2. You have three piles of pennies. There are 10 pennies in the first pile, 18 pennies in the second pile, and a certain number in the third pile. If you double the number of pennies in each pile, you will have 70 pennies. How many are there in the third pile?

3. A farmer has a herd of cows and some chickens. Altogether there are 36 heads and 84 legs. How many cows and how many chickens are there?

4. There are books packed in a certain number of boxes. The math books are packed 12 to a box and science books 16 to a box. There are altogether 108 books. How many boxes contain math books, and how many boxes contain science books?

5. A third grader was given a problem of dividing one number by another. Instead of dividing, the student subtracted one number from the other. The teacher said that the answer was right. What numbers was the student asked to divide?

6. Lester has $23 in his savings account. Rhonda has $50 in hers. Beginning now, each of them will add $1 each week to their accounts. After how many weeks will Rhonda have twice as much as Lester?

7. A package of buttons sells for $3.10. There are silver and gold buttons in the package. Each silver button costs $0.08. Each gold button costs $0.10. How many silver and how many gold buttons are there in the package?

COMPUTER ACTIVITIES

You have used the **BASIC** operations of $+$, $-$, $*$, and $/$.
Many algebraic expressions also contain exponents.
But there is no way for a computer keyboard to "raise" the 3 in an expression like 4^3. Another device is used. 4^3 is written in **BASIC** as $4 \wedge 3$, where 3 is the exponent and the symbol $\wedge$ is used to show exponentiation.
The order of operations rule can now be extended as follows:

    1st: raise to a power; $\wedge$
    2nd: do multiplication or division; $*$ or $/$
    3rd: do addition or subtraction; $+$ or $-$

Example $A = 2 * 4 \wedge 3 - 6 \wedge 2$. Find the value of A by computation. Write a program to find A. Then type and **RUN** it.

             By computation: $A = 2 * 4 \wedge 3 - 6 \wedge 2$
  Do exponentiation first: $A = 2 * 64 - 36$
    $4 \wedge 3 = 4^3 = 64$ $A = 128 - 36$
    $6 \wedge 2 = 6^2 = 36$ $A = 92$

Type in this program.

```
10   LET A = 2 * 4 ^ 3 - 6 ^ 2
20   PRINT A
30   END
```
         RUN the program.   ]RUN
The screen displays the answer. 92

Exercises

Compute.

1. $7 \wedge 2$ **2.** $4 * 3 \wedge 2 - 4$ **3.** $19 + 4 \wedge 3 - 6 * 7$ **4.** $49 - 7 \wedge 2 + 8 * 6 - 5$

Find the value of each variable by computation. Then write a program to find the value. Type and **RUN** it.

5. $A = 8 * 4 \wedge 2$ **6.** $B = 6 * 3 \wedge 2 + 4 * 2 \wedge 5$ **7.** $T = 3 * 2 \wedge 4 + 8 \wedge 2$

You are now ready to learn more about the techniques of programming. The Computer Section beginning on page 420 will provide you with an introduction to the concepts of programming.

Chapter Review

Rewrite using the distributive property. [53]

1. $8(6 + 1)$
2. $(5 + 2)6$
3. $7(5 - 4)$

Rewrite using the distributive property. [53]

4. $8 \cdot 7 + 8 \cdot 13$
5. $3 \cdot 6 - 3 \cdot 2$
6. $17 \cdot 2 + 15 \cdot 2$

Simplify. [53, 56]

7. $6(5x + 2)$
8. $3(5m - 1)$
9. $8k + 5k$
10. $6w + 5y + 2 + 3y + 1 + 5w$
11. $5k + 2 + 4m$
12. $7a + 2 - 2a$

Simplify. Then evaluate for these values of the variables: $a = 5, b = 2, c = 6$. [58]

13. $3a + 5 + 8a$
14. $4b + 2 + 3b$
15. $9 + 2c + c$
16. $3a + b + 4a + b$
17. $c + 5a + c + 4 - 3a$
18. $8 + 2b + 3 + b + a + c + 3a + 1$

Solve each equation. Check the solution. [61]

19. $2h + 8 = 14$
20. $14 = 5m - 1$
21. $9y - 2 = 16$
22. $\frac{x}{3} + 1 = 10$

Write in mathematical form. [63]

23. 8 times k, decreased by 5
24. 9 less than a number divided by 2

Solve these problems. [65]

25. 1 less than 7 times a number is 13. Find the number.
26. 36 is the same as a number increased by twice the number. Find the number.

Find the value. [69]

27. 2^3
28. 1^9
29. 6^3

Evaluate. [69]

30. x^4 if $x = 7$
31. $7a^3$ if $a = 3$
32. $20y^2$ if $y = 4$

Find A in terms of π if $r = 6$. [72]

33. $A = \pi r^2$

Find V in terms of π if $r = 3$ and $h = 4$.

34. $V = \pi r^2 h$

Chapter Test

Rewrite using the distributive property.

1. $7(9 + 3)$
2. $(8 - 4)6$
3. $7 \cdot 9 - 7 \cdot 3$
4. $13 \cdot 5 + 11 \cdot 5$

Simplify.

5. $7(2k + 4)$
6. $8(5m - 2)$
7. $9t + 4t$
8. $c + 5 + 9d + 3 - 4d + 3c$

Simplify. Then evaluate for these values of the variables: $a = 5$, $b = 4$.

9. $7a + 2 + 3a$
10. $2a + 3b + 4$
11. $a + 3b + 6 + b + 4a + 1$

Solve each equation. Check the solution.

12. $2x + 10 = 14$
13. $14 = 3k - 7$
14. $8t - 3 = 21$
15. $\frac{x}{4} - 5 = 6$

Write in mathematical form.

16. a number divided by 5 is increased by 9
17. 13 less than twice a number

Solve these problems.

18. 5 less than 4 times a number is 15. Find the number.
19. 40 is the same as a number increased by 3 times the number. Find the number.

Find the value.

20. 3^4
21. 8^4

Evaluate.

22. x^4 if $x = 5$
23. $6a^2$ if $a = 4$

Find A in terms of π, if $r = 9$.

24. $A = \pi r^2$

Find V in terms of π if $r = 4$ and $h = 9$.

25. $V = \pi r^2 h$

Multiplying and Dividing Fractions

4

Bryce Canyon National Park in Utah covers 36,010 acres.

Introduction to Fractions

◇ OBJECTIVES ◇

To find numbers corresponding to points on a number line

To find what fractional part of a drawing is shaded

To find products like $6 \cdot \frac{1}{7}$

◇ RECALL ◇

Points on a number line correspond to numbers.

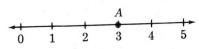

The number 3 corresponds to point A.

Example 1

What number corresponds to point A?

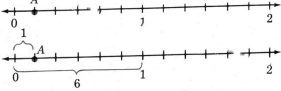

There are 6 equal subdivisions from 0 to 1.

1 of 6 equal parts
$\frac{1}{6}$

A corresponds to 1 of 6 equal parts.

So, $\frac{1}{6}$ corresponds to A.

practice ▷ What number corresponds to point A?

1.    2.

Example 2

What number corresponds to point A?

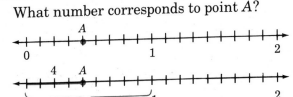

Count the number of equal subdivisions. There are 9 equal subdivisions from 0 to 1.

4 of 9 equal parts
$\frac{4}{9}$

A corresponds to 4 of 9 equal parts.

So, $\frac{4}{9}$ corresponds to A.

practice ▷ What number corresponds to point A?

3. 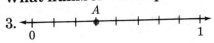   4.

Fractions such as $\frac{4}{9}$ or $\frac{1}{6}$ are called rational numbers.

4 is the numerator. ⟶ $\frac{4}{9}$
9 is the denominator. ⟶

A fraction like $\frac{6}{0}$ is meaningless.

Example 3

Think: Check by multiplication.

Show that $\frac{6}{0}$ is meaningless.

$\frac{6}{0}$ means $6 \div 0$.

Let $6 \div 0 = a$.
Then $0 \cdot a = 6$.
But, $0 \cdot a = 0$, **not** 6.

So, $\frac{6}{0}$ is meaningless.

The denominator of a fraction *cannot* be 0. Division by 0 is undefined.

practice ▷ Which of the following are meaningless?

5. $\frac{7}{9}$ 6. $\frac{0}{2}$ 7. $\frac{5}{0}$

Example 4

Use a number line to show that $5 \cdot \frac{1}{6} = \frac{5}{6}$.

Divide the number line into 6 equal parts from 0 to 1. Count 5 subdivisions from 0 to $\frac{5}{6}$.

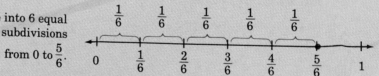

So, $5 \cdot \frac{1}{6} = \frac{5}{6}$.

Example 5

Multiply. $4 \cdot \frac{1}{7}$

$4 \cdot \frac{1}{7}$

$\frac{4 \cdot 1}{7} = \frac{4}{7}$

So, $4 \cdot \frac{1}{7} = \frac{4}{7}$ and $\frac{1}{5} \cdot 5 = 1$.

Multiply. $\frac{1}{5} \cdot 5$

$\frac{1}{5} \cdot 5$ means $5 \cdot \frac{1}{5}$.

$\frac{5 \cdot 1}{5} = \frac{5}{5}$, or 1

practice ▷ Multiply.

8. $3 \cdot \frac{1}{5}$ 9. $4 \cdot \frac{1}{9}$ 10. $\frac{1}{6} \cdot 5$ 11. $\frac{1}{8} \cdot 8$ 12. $\frac{1}{15} \cdot 15$

◇ ORAL EXERCISES ◇

Multiply.

1. $3 \cdot \frac{1}{4}$
2. $\frac{1}{3} \cdot 2$
3. $8 \cdot \frac{1}{8}$
4. $7 \cdot \frac{1}{11}$
5. $\frac{11}{2} \cdot 2$

Which of the following are meaningless?

6. $\frac{7}{8}$
7. $\frac{0}{4}$
8. $\frac{9}{0}$
9. $\frac{71}{10}$
10. $\frac{14}{0}$

◇ EXERCISES ◇

What number corresponds to point A?

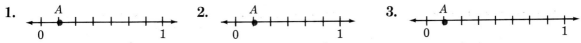

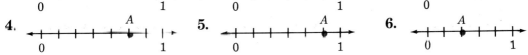 

Multiply.

7. $3 \cdot \frac{1}{7}$
8. $\frac{1}{5} \cdot 2$
9. $7 \cdot \frac{1}{9}$
10. $\frac{1}{4} \cdot 3$
11. $6 \cdot \frac{1}{6}$
12. $\frac{1}{15} \cdot 4$
13. $7 \cdot \frac{1}{8}$
14. $\frac{1}{9} \cdot 5$
15. $\frac{1}{8} \cdot 7$
16. $7 \cdot \frac{1}{11}$
17. $\frac{1}{4} \cdot 4$
18. $\frac{1}{3} \cdot 2$
19. $\frac{1}{13} \cdot 5$
20. $3 \cdot \frac{1}{3}$
21. $\frac{1}{7} \cdot 5$
22. $\frac{1}{12} \cdot 7$
23. $4 \cdot \frac{1}{15}$
24. $\frac{1}{12} \cdot 12$
25. $5 \cdot \frac{1}{8}$
26. $\frac{1}{10} \cdot 3$
27. $\frac{1}{7} \cdot 7$
28. $\frac{1}{12} \cdot 5$
★29. $3 \cdot 4 \cdot \frac{1}{17}$
★30. $7 \cdot \frac{1}{15} \cdot 2$
★31. $4 \cdot \frac{1}{9} \cdot 2$

Is it possible to make five squares of the same size by moving only three sticks? How?

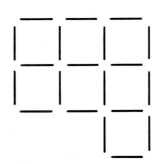

INTRODUCTION TO FRACTIONS

Multiplying by a Whole Number

◇ OBJECTIVES ◇

To multiply expressions like $12 \cdot \frac{2}{3}$ and $1\frac{1}{2} \cdot 3$

To change a mixed number to a fraction

◇ RECALL ◇

$2 \cdot \frac{1}{3} = \frac{2}{3}$

$\frac{3}{8} = 3 \cdot \frac{1}{8}$

Example 1

Multiply. $8 \cdot \frac{3}{4}$

$\frac{3}{4} = 3 \cdot \frac{1}{4}$

$8 \cdot \frac{3}{4}$

$8 \cdot 3 \cdot \frac{1}{4}$

$24 \cdot \frac{1}{4}$

$\frac{24 \cdot 1}{4} = \frac{24}{4}$, or 6

So, $8 \cdot \frac{3}{4} = \frac{24}{4}$, or 6.

Example 2

Multiply. $7 \cdot \frac{2}{5}$

$7 \cdot \frac{2}{5} = \frac{7 \cdot 2}{5}$

$\frac{14}{5}$ ← improper fraction

In $\frac{14}{5}$, the numerator is larger than the denominator. Write the remainder over the divisor as a fraction.

$\frac{14}{5} = 5\overline{)14}^{2}$, or $2\frac{4}{5}$ ← mixed number

divisor 4 ← remainder

So, $7 \cdot \frac{2}{5} = \frac{14}{5}$, or $2\frac{4}{5}$.

---

practice ▷ Multiply.

1. $6 \cdot \frac{2}{3}$
2. $10 \cdot \frac{2}{5}$
3. $5 \cdot \frac{5}{6}$
4. $8 \cdot \frac{2}{3}$
5. $4 \cdot \frac{3}{7}$

In Example 2, division was used to change $\frac{14}{5}$ to a mixed number. In the next example, the mixed number $4\frac{3}{5}$ is changed to a fraction.

Example 3

Change $4\frac{3}{5}$ to a fraction.

$$4\frac{3}{5} \qquad 4 \cdot \overset{3}{\underset{5}{}} +$$

Multiply 5 by 4. Then add 3.

$$\frac{(5 \cdot 4) + 3}{5}$$

$$\frac{20 + 3}{5} = \frac{23}{5}$$

So, $4\frac{3}{5} = \frac{23}{5}$.

---

practice ▷ Change to a fraction.

6. $3\frac{4}{5}$ 7. $5\frac{2}{3}$ 8. $4\frac{2}{5}$ 9. $6\frac{1}{3}$ 10. $4\frac{5}{6}$ 11. $9\frac{3}{4}$

Example 4

Multiply. $3\frac{2}{3} \cdot 2$

Change $3\frac{2}{3}$ to a fraction.

$$3\frac{2}{3} = \frac{(3 \cdot 3) + 2}{3} = \frac{9 + 2}{3} = \frac{11}{3}$$

Multiplication can be done in any order.

$3\frac{2}{3} \cdot 2$

$\frac{11}{3} \cdot 2$

$2 \cdot \frac{11}{3}$

$\frac{22}{3}$

$$\begin{array}{r} 7 \\ 3\overline{)22} \\ \underline{21} \\ 1 \end{array} = 7\frac{1}{3}$$

$7\frac{1}{3}$

So, $3\frac{2}{3} \cdot 2 = 7\frac{1}{3}$.

---

practice ▷ Multiply.

12. $3\frac{1}{3} \cdot 2$ 13. $2\frac{2}{3} \cdot 2$ 14. $1\frac{1}{5} \cdot 2$ 15. $5 \cdot 1\frac{1}{2}$ 16. $3 \cdot 1\frac{1}{4}$

MULTIPLYING BY A WHOLE NUMBER

◇ ORAL EXERCISES ◇

Multiply.

1. $3 \cdot \frac{2}{7}$
2. $4 \cdot \frac{3}{13}$
3. $2 \cdot \frac{4}{9}$
4. $3 \cdot \frac{2}{11}$
5. $4 \cdot \frac{2}{13}$
6. $2 \cdot \frac{4}{17}$
7. $2 \cdot \frac{5}{13}$
8. $5 \cdot \frac{3}{19}$
9. $2 \cdot \frac{6}{13}$
10. $\frac{5}{19} \cdot 2$
11. $\frac{2}{9} \cdot 4$
12. $\frac{4}{13} \cdot 3$

Change to a fraction.

13. $3\frac{1}{2}$
14. $2\frac{1}{3}$
15. $4\frac{1}{2}$
16. $3\frac{1}{3}$
17. $5\frac{1}{4}$
18. $2\frac{1}{8}$
19. $4\frac{1}{5}$
20. $2\frac{1}{6}$
21. $3\frac{1}{4}$
22. $6\frac{1}{2}$
23. $7\frac{1}{2}$
24. $4\frac{1}{7}$

◇ EXERCISES ◇

Multiply.

1. $10 \cdot \frac{3}{5}$
2. $4 \cdot \frac{3}{4}$
3. $3 \cdot \frac{2}{3}$
4. $9 \cdot \frac{2}{3}$
5. $14 \cdot \frac{2}{7}$
6. $15 \cdot \frac{2}{5}$
7. $12 \cdot \frac{3}{4}$
8. $10 \cdot \frac{2}{5}$
9. $14 \cdot \frac{1}{7}$
10. $15 \cdot \frac{2}{3}$
11. $9 \cdot \frac{2}{3}$
12. $18 \cdot \frac{2}{3}$
13. $2 \cdot \frac{3}{5}$
14. $4 \cdot \frac{3}{5}$
15. $2 \cdot \frac{2}{3}$
16. $5 \cdot \frac{3}{7}$
17. $3 \cdot \frac{4}{5}$
18. $2 \cdot \frac{4}{7}$
19. $7 \cdot \frac{2}{3}$
20. $5 \cdot \frac{4}{7}$
21. $8 \cdot \frac{2}{3}$
22. $7 \cdot \frac{2}{5}$
23. $6 \cdot \frac{3}{5}$
24. $3 \cdot \frac{5}{7}$

Change to a fraction.

25. $6\frac{2}{3}$
26. $4\frac{2}{3}$
27. $9\frac{3}{4}$
28. $6\frac{3}{5}$
29. $4\frac{3}{8}$
30. $7\frac{7}{8}$
31. $5\frac{5}{9}$
32. $4\frac{2}{7}$
33. $6\frac{3}{7}$
34. $5\frac{2}{9}$
35. $8\frac{5}{6}$
36. $6\frac{6}{7}$

Multiply.

37. $3\frac{1}{3} \cdot 2$
38. $1\frac{1}{3} \cdot 5$
39. $1\frac{1}{2} \cdot 7$
40. $3\frac{2}{3} \cdot 2$
41. $1\frac{1}{4} \cdot 3$
42. $2 \cdot 2\frac{1}{3}$
43. $5 \cdot 1\frac{1}{2}$
44. $2 \cdot 1\frac{1}{3}$
★ 45. $8 \cdot 2\frac{1}{2} \cdot 3$
★ 46. $6 \cdot 3\frac{1}{3} \cdot 15$

Solve these problems.

47. Jason works on the yearbook $\frac{3}{4}$ hour each day. How much time does he put in during a 5-day week?

48. Janet is a clerk at Walling Drugs. She works $2\frac{1}{2}$ h each night. How many hours does she work during a 5-day week?

Multiplying by a Fraction

◇ OBJECTIVE ◇

To multiply expressions like
$\frac{2}{5} \cdot \frac{3}{4}$ and $3\frac{1}{2} \cdot \frac{1}{8}$

◇ RECALL ◇

$$8 \cdot \frac{1}{8} = \frac{8 \cdot 1}{8} = \frac{8}{8} = 1$$

Since $8 \cdot \frac{1}{8} = 1$, 8 and $\frac{1}{8}$ are reciprocals of each other. Two numbers are *reciprocals* if their product is 1.

Example 1

Two numbers are reciprocals if their product is 1

Find the reciprocal of 3.
Think: 3 times what number is 1?
$$3 \cdot \frac{1}{3} = 1$$
So, $\frac{1}{3}$ is the reciprocal of 3.

Example 2

Show that $\frac{1}{3} \cdot \frac{1}{2} = \frac{1}{6}$.

First show that the reciprocal of 6 is $\frac{1}{3} \cdot \frac{1}{2}$.

$$6 \cdot \frac{1}{3} \cdot \frac{1}{2}$$

$3 \cdot 2 = 6$ $\qquad 3 \cdot 2 \cdot \frac{1}{3} \cdot \frac{1}{2}$

Regroup. $\qquad \left(3 \cdot \frac{1}{3}\right)\left(2 \cdot \frac{1}{2}\right)$

$3 \cdot \frac{1}{3} = 1 \quad | \quad 2 \cdot \frac{1}{2} = 1 \qquad \underbrace{\phantom{(3 \cdot \tfrac{1}{3})}}_{(1)} \; \underbrace{\phantom{(2 \cdot \tfrac{1}{2})}}_{(1)}$

$\qquad\qquad\qquad 1$

$$6 \cdot \frac{1}{3} \cdot \frac{1}{2} = 1$$

The reciprocal of 6 is $\frac{1}{3} \cdot \frac{1}{2}$. Since $6 \cdot \frac{1}{3} \cdot \frac{1}{2} = 1$ and $6 \cdot \frac{1}{6} = 1$,

The reciprocal of 6 is also $\frac{1}{6}$. $\qquad \frac{1}{3} \cdot \frac{1}{2} = \frac{1}{6}$.

A number has exactly one reciprocal. So, $\frac{1}{3} \cdot \frac{1}{2} = \frac{1}{6}$.

Example 3

Multiply. $\frac{3}{7} \cdot \frac{4}{5}$

$\frac{3}{7} = 3 \cdot \frac{1}{7}$ and $\frac{4}{5} = 4 \cdot \frac{1}{5}$

$\frac{3}{7} \cdot \frac{4}{5}$

$3 \cdot \frac{1}{7} \cdot 4 \cdot \frac{1}{5}$

Reorder. $3 \cdot 4 \cdot \frac{1}{7} \cdot \frac{1}{5}$

$\frac{1}{7} \cdot \frac{1}{5} = \frac{1}{35}$ $12 \cdot \frac{1}{35} = \frac{12}{35}$

So, $\frac{3}{7} \cdot \frac{4}{5} = \frac{3 \cdot 4}{7 \cdot 5}$, or $\frac{12}{35}$.

In Example 3, we found that $\frac{3}{7} \cdot \frac{4}{5} = \frac{3 \cdot 4}{7 \cdot 5}$.

Example 4

Multiply. $\frac{3}{5} \cdot \frac{7}{2}$

Multiply the numerators.
Multiply the denominators.

$\frac{3}{5} \cdot \frac{7}{2} = \frac{3 \cdot 7}{5 \cdot 2} = \frac{21}{10}$

$10\overline{)21} \atop \underline{20} \atop 1$ = $2\frac{1}{10}$ So, $\frac{3}{5} \cdot \frac{7}{2} = 2\frac{1}{10}$.

practice ▷ Multiply.

1. $\frac{2}{3} \cdot \frac{5}{3}$ 2. $\frac{5}{3} \cdot \frac{4}{3}$ 3. $\frac{3}{4} \cdot \frac{5}{7}$ 4. $\frac{1}{3} \cdot \frac{7}{2}$ 5. $\frac{3}{4} \cdot \frac{5}{2}$

Example 5

Multiply. $3\frac{1}{2} \cdot \frac{1}{8}$

$3\frac{1}{2} = \frac{(2 \cdot 3) + 1}{2} = \frac{6 + 1}{2} = \frac{7}{2}$

$3\frac{1}{2} \cdot \frac{1}{8} = \frac{7}{2} \cdot \frac{1}{8} = \frac{7 \cdot 1}{2 \cdot 8} = \frac{7}{16}$

So, $3\frac{1}{2} \cdot \frac{1}{8} = \frac{7}{16}$.

practice ▷ Multiply.

6. $2\frac{1}{2} \cdot \frac{1}{3}$ 7. $3\frac{1}{3} \cdot \frac{1}{9}$ 8. $2\frac{1}{4} \cdot \frac{1}{7}$ 9. $\frac{2}{5} \cdot 1\frac{1}{5}$ 10. $2\frac{2}{3} \cdot \frac{1}{3}$

◇ ORAL EXERCISES ◇

Multiply.

1. $\frac{3}{5} \cdot \frac{2}{5}$
2. $\frac{1}{7} \cdot \frac{2}{3}$
3. $\frac{3}{4} \cdot \frac{1}{2}$
4. $\frac{5}{8} \cdot \frac{1}{3}$
5. $\frac{1}{2} \cdot \frac{7}{9}$
6. $\frac{4}{7} \cdot \frac{2}{3}$
7. $\frac{3}{5} \cdot \frac{1}{5}$
8. $\frac{4}{9} \cdot \frac{2}{3}$
9. $\frac{2}{5} \cdot \frac{3}{7}$
10. $\frac{2}{9} \cdot \frac{1}{3}$

◇ EXERCISES ◇

Multiply.

1. $\frac{3}{5} \cdot \frac{1}{4}$
2. $\frac{1}{3} \cdot \frac{7}{2}$
3. $\frac{7}{2} \cdot \frac{1}{2}$
4. $\frac{3}{5} \cdot \frac{7}{2}$
5. $\frac{5}{2} \cdot \frac{5}{3}$
6. $\frac{3}{2} \cdot \frac{5}{8}$
7. $\frac{3}{4} \cdot \frac{3}{5}$
8. $\frac{2}{7} \cdot \frac{1}{3}$
9. $\frac{3}{2} \cdot \frac{3}{7}$
10. $\frac{4}{9} \cdot \frac{1}{5}$
11. $\frac{3}{2} \cdot \frac{7}{2}$
12. $\frac{2}{5} \cdot \frac{8}{3}$
13. $\frac{4}{3} \cdot \frac{5}{3}$
14. $4\frac{1}{2} \cdot \frac{1}{7}$
15. $1\frac{2}{3} \cdot \frac{5}{2}$
16. $\frac{2}{3} \cdot 2\frac{2}{3}$
★ 17. $\frac{2}{3} \cdot \frac{7}{3} \cdot \frac{5}{3}$
★ 18. $\frac{4}{5} \cdot \frac{8}{3} \cdot \frac{8}{5}$
★ 19. $3\frac{1}{2} \cdot 1\frac{1}{4} \cdot \frac{3}{4}$

Solve these problems.

20. John worked $\frac{1}{2}$ hour. Lucy worked $1\frac{1}{2}$ times as many hours. How many hours did Lucy work?

21. Maria walked $\frac{3}{4}$ kilometer. Joe walked $\frac{1}{2}$ as far. How far did Joe walk?

Algebra Maintenance

Simplify.

1. $8 \cdot 5k$
2. $7(3x + 4)$
3. $5(3j - 3)$
4. $4x + 3 + 7x + 4$
5. $2a + b + 3b + c + 4a + 4b$

Solve each equation. Check.

6. $3x + 4 = 22$
7. $15 = 7y - 6$
8. $12 = 4z - 8$

Write in mathematical form.

9. 7 times m, decreased by 4
10. 4 less than three times a number

MULTIPLYING BY A FRACTION

Prime Factorization

◇ OBJECTIVES ◇

To factor a number into primes
To determine whether a number is prime, composite, or neither

◇ RECALL ◇

$2 \cdot 2 \cdot 2 \cdot 2 = 2^4$
$3 \cdot 3 = 3^2$
$2 \cdot 2 \cdot 3 \cdot 3 = 2^2 \cdot 3^3$ exponents
$3 \cdot 4 = 12$
↑ ↑   ↑
factors   product

A *prime number* is a whole number greater than 1 whose only factors are itself and 1.
A *composite number* is a whole number that has factors other than itself and 1.
The number 1 is neither prime nor composite.

Example 1

5 and 3 are numbers whose product is 15; each is prime.

Factor 15 into primes.
$15 = 5 \cdot 3$ or $3 \cdot 5$
A number is factored into primes when each of its factors is prime. The order in which the factors is written does not matter.

---

practice ▷   Factor into primes.

1. 6   2. 10   3. 21   4. 35

A number may have more than one pair of factors. To factor the number into primes, you can begin by using any pair of its factors.

Example 2

6 is a composite number. Factor 6 into $3 \cdot 2$.

Factor 12 into primes.
You can begin in more than one way.
$12 = 6 \cdot 2$ $12 = 3 \cdot 4$
$ = 3 \cdot 2 \cdot 2$ $ = 3 \cdot 2 \cdot 2$
Each factorization gives the same prime factors. In each case, 3 is used as a factor once. 2 is used as a factor two times.
So, $12 = 2 \cdot 2 \cdot 3$ or $2^2 \cdot 3$.

Example 3

Factor 7 into primes, if possible.
The only factors of 7 are 7 and 1.
7 is prime. It can be factored no further.

practice Factor into primes, if possible.

5. 20   6. 13   7. 18   8. 8

Example 4

Factor 36 into primes.
Write using exponents.

First find two factors. $36 = 4 \cdot 9$ $36 = 3 \cdot 12$ $36 = 6 \cdot 6$

Then continue to factor any composite numbers. $= 2 \cdot 2 \cdot 3 \cdot 3$ $= 3 \cdot 4 \cdot 3$ $3 \cdot 2 \cdot 3 \cdot 2$

$= 3 \cdot 2 \cdot 2 \cdot 3$

The order of the factors does not matter. The three methods give the same result.
So, $36 = 3 \cdot 3 \cdot 2 \cdot 2$ or $3^2 \cdot 2^2$.

practice Factor into primes. Write using exponents.

9. 24   10. 40   11. 32   12. 54

◇ ORAL EXERCISES ◇

Tell whether each number is prime, composite, or neither.

1. 11   2. 22   3. 1   4. 80   5. 13

◇ EXERCISES ◇

Prime, composite, or neither?

1. 25   2. 36   3. 31   4. 75   5. 120   6. 15
7. 1   8. 48   9. 5   10. 32   11. 17   12. 26

Factor into primes, if possible.

13. 4   14. 22   15. 19   16. 9   17. 49   18. 39

Factor into primes, if possible. Write using exponents.

19. 24   20. 28   21. 27   22. 45   23. 30   24. 5
25. 44   26. 50   27. 48   28. 75   29. 99   30. 16
31. 60   32. 72   33. 81   34. 13   35. 100   36. 64
★ 37. 400   ★ 38. 120   ★ 39. 140   ★ 40. 225

PRIME FACTORIZATION

Greatest Common Factor (GCF)

◇ **OBJECTIVE** ◇

To find the greatest common factor (GCF) of two numbers

◇ **RECALL** ◇

Factor 20 into primes.
$20 = 5 \cdot 4$ or $20 = 10 \cdot 2$
$= 5 \cdot 2 \cdot 2$ $= 5 \cdot 2 \cdot 2$

Example 1

Find the greatest common factor (GCF) of 12 and 30.
Factor each number into primes.
$12 = 6 \cdot 2$ $30 = 10 \cdot 3$
$= 2 \cdot 3 \cdot 2$ $= 2 \cdot 5 \cdot 3$
There are at *most* one 2 and one 3 common to 12 and 30.
So, the greatest common factor is $2 \cdot 3$, or 6.

Example 2

Find the GCF of 10 and 21.
Factor each number into primes.
$10 = 5 \cdot 2$ $21 = 7 \cdot 3$

Note: 1 is a factor of every number.

There are no common *prime* factors.
So, the GCF of 10 and 21 is 1.

---

practice ▷ **Find the greatest common factor (GCF).**

1. 4 and 16   2. 9 and 21   3. 6 and 25   4. 6 and 14

Example 3

Find the GCF of 40 and 24.
First factor each number into primes.
$40 = 10 \cdot 4$ $24 = 12 \cdot 2$
$= 5 \cdot 2 \cdot 4$ $= 4 \cdot 3 \cdot 2$
$= 5 \cdot 2 \cdot 2 \cdot 2$ $= 2 \cdot 2 \cdot 3 \cdot 2$
There are at *most* three 2's common to 40 and 24.
So, the GCF of 40 and 24 is $2 \cdot 2 \cdot 2$, 2^3, or 8.

practice ▷ **Find the greatest common factor (GCF).**

5. 20 and 30   6. 18 and 45   7. 28 and 32   8. 12 and 18

◇ **ORAL EXERCISES** ◇

Find the greatest common factor (GCF), given the prime factorization.

1. $2 \cdot 3 \cdot 2 \cdot 5$ and $2 \cdot 7 \cdot 3$
2. $3 \cdot 3 \cdot 2 \cdot 5$ and $3 \cdot 5 \cdot 5 \cdot 3$
3. $2 \cdot 7 \cdot 7 \cdot 5 \cdot 5$ and $7 \cdot 7 \cdot 3$
4. $5 \cdot 5 \cdot 5 \cdot 3$ and $5 \cdot 5 \cdot 3 \cdot 2 \cdot 3$
5. $11 \cdot 7 \cdot 2 \cdot 3 \cdot 2$ and $11 \cdot 7 \cdot 7 \cdot 3$
6. $3 \cdot 5 \cdot 7 \cdot 2$ and $11 \cdot 13$

◇ **EXERCISES** ◇

Find the greatest common factor (GCF).

1. 6; 15
2. 12; 42
3. 10; 15
4. 35; 45
5. 28; 42
6. 30; 48
7. 18; 55
8. 27; 81
9. 14; 25
10. 44; 60
11. 25; 40
12. 16; 28
13. 36; 54
14. 15; 40
15. 14; 30
16. 15; 33
17. 24; 28
18. 32; 20
19. 16; 20
20. 63; 72
21. 44; 66
22. 32; 26
23. 27; 18
24. 26; 65
25. 36; 42
26. 32; 80
27. 64; 24
28. 28; 35
★ 29. 12; 18; 24
★ 30. 33; 55; 77

Calculator

The prime factors of a number can be found by a series of divisions by primes until the answer is 1. Below is a way to factor 84 into primes.

$84 \div 2 = 42$
$42 \div 2 = 21$
$21 \div 3 = 7$
$7 \div 7 = 1$

Divide by the least prime contained in the number.

The prime factors of 84 are 2, 2, 3, and 7.
So, $84 = 2 \cdot 2 \cdot 3 \cdot 7$ or $2^2 \cdot 3 \cdot 7$.

Factor into primes.

1. 90   2. 120   3. 220   4. 1,680

GREATEST COMMON FACTOR

Simplest Form

◆ OBJECTIVE ◆
To simplify fractions

◆ RECALL ◆
$\frac{4}{5} \cdot \frac{2}{3} = \frac{4 \cdot 2}{5 \cdot 3} = \frac{8}{15}$ ∣ $\frac{7}{7} = 1$

Example 1 Write $\frac{10}{14}$ in simplest form.

	One way	More compact form
	$\frac{10}{14}$	$\frac{10}{14}$
Factor 10 into primes. Factor 14 into primes.	$\frac{5 \cdot 2}{2 \cdot 7}$	$\frac{5 \cdot 2}{2 \cdot 7}$
Rearrange the like factors, the 2's, so that they are over each other.	$\frac{2 \cdot 5}{2 \cdot 7}$	$\frac{5 \cdot \cancel{2}^{1}}{\cancel{2}_{1} \cdot 7}$
	$\frac{2}{2} \cdot \frac{5}{7}$	$\frac{5 \cdot 1}{1 \cdot 7}$
$\frac{2}{2} = 1; 1 \cdot \frac{5}{7} = \frac{5}{7}$	$1 \cdot \frac{5}{7}$, or $\frac{5}{7}$	$\frac{5}{7}$

So, $\frac{10}{14} = \frac{5}{7}$.

Example 2 Write in simplest form. $\frac{6}{15}; \frac{10}{21}$

	$\frac{6}{15}$		$\frac{10}{21}$	
Factor 6 into primes. Factor 15 into primes.	$\frac{3 \cdot 2}{5 \cdot 3}$		$\frac{5 \cdot 2}{7 \cdot 3}$	Factor 10 into primes. Factor 21 into primes.
Divide out like factors.	$\frac{\cancel{3}^{1} \cdot 2}{5 \cdot \cancel{3}_{1}}$		There are no like factors, except 1. $\frac{10}{21}$ is in simplest form.	
	$\frac{1 \cdot 2}{5 \cdot 1}$, or $\frac{2}{5}$			

So, $\frac{6}{15} = \frac{2}{5}$; $\frac{10}{21}$ is in simplest form.

practice ▷ Write in simplest form.

1. $\frac{6}{10}$ 2. $\frac{9}{14}$ 3. $\frac{10}{15}$ 4. $\frac{14}{25}$ 5. $\frac{6}{9}$ 6. $\frac{10}{25}$

Example 3

Simplify $\frac{20}{6}$.

$20 = 2 \cdot 10$ $6 = 3 \cdot 2$
$\quad \searrow$
$2 \cdot 2 \cdot 5$

$$\frac{20}{6}$$

$$\frac{2 \cdot 2 \cdot 5}{3 \cdot 2}$$

Simplify means write in simplest form.

Divide out like factors. $\dfrac{\overset{1}{\cancel{2}} \cdot 2 \cdot 5}{3 \cdot \underset{1}{\cancel{2}}}$

$$\frac{1 \cdot 2 \cdot 5}{3 \cdot 1}$$

$3 \overline{)10}$, or $3\frac{1}{3}$ $\frac{10}{3}$, or $3\frac{1}{3}$
$\underline{9}$
1

So, $\frac{20}{6} = 3\frac{1}{3}$.

practice ▷ Simplify.

7. $\frac{12}{9}$ 8. $\frac{18}{10}$ 9. $\frac{15}{8}$ 10. $\frac{00}{20}$ 11. $\frac{21}{12}$ 12. $\frac{12}{10}$

Example 4

Simplify $\frac{18}{30}$.

Two ways

$\frac{18}{30}$ $\frac{18}{30}$

$18 = 6 \cdot 3$ $30 = 15 \cdot 2$ $\dfrac{6 \cdot 3}{15 \cdot 2} = \dfrac{3 \cdot 2 \cdot 3}{5 \cdot 3 \cdot 2}$ $\dfrac{9 \cdot 2}{10 \cdot 3} = \dfrac{3 \cdot 3 \cdot 2}{5 \cdot 2 \cdot 3}$
$\quad\searrow$ $\quad\searrow$
$3 \cdot 2 \cdot 3$ $5 \cdot 3 \cdot 2$

Divide out like factors. $\dfrac{3 \cdot \overset{1}{\cancel{2}} \cdot \overset{1}{\cancel{3}}}{5 \cdot \underset{1}{\cancel{3}} \cdot \underset{1}{\cancel{2}}}$ $\dfrac{\overset{1}{\cancel{3}} \cdot 3 \cdot \overset{1}{\cancel{2}}}{5 \cdot \underset{1}{\cancel{2}} \cdot \underset{1}{\cancel{3}}}$

$\dfrac{3 \cdot 1 \cdot 1}{5 \cdot 1 \cdot 1}$ $\dfrac{1 \cdot 3 \cdot 1}{5 \cdot 1 \cdot 1}$

$\dfrac{3}{5}$ ←——— same ———→ $\dfrac{3}{5}$

So, $\frac{18}{30} = \frac{3}{5}$.

practice ▷ Simplify.

13. $\frac{12}{20}$ 14. $\frac{12}{18}$ 15. $\frac{15}{20}$ 16. $\frac{8}{30}$ 17. $\frac{16}{27}$

SIMPLEST FORM

◇ ORAL EXERCISES ◇

Simplify.

1. $\dfrac{2 \cdot 3}{5 \cdot 3}$
2. $\dfrac{7 \cdot 5}{5 \cdot 11}$
3. $\dfrac{3 \cdot 5}{13 \cdot 3}$
4. $\dfrac{5 \cdot 11}{11 \cdot 7}$
5. $\dfrac{9 \cdot 2}{7 \cdot 9}$
6. $\dfrac{7 \cdot 2}{2 \cdot 11}$
7. $\dfrac{3 \cdot 5}{5 \cdot 5}$
8. $\dfrac{3 \cdot 11}{13 \cdot 3}$
9. $\dfrac{2 \cdot 2}{5 \cdot 2}$
10. $\dfrac{3 \cdot 7}{7 \cdot 17}$
11. $\dfrac{7 \cdot 2 \cdot 5}{2 \cdot 5 \cdot 11}$
12. $\dfrac{3 \cdot 2 \cdot 5}{2 \cdot 5 \cdot 7}$
13. $\dfrac{3 \cdot 7 \cdot 11}{11 \cdot 3 \cdot 13}$
14. $\dfrac{5 \cdot 2 \cdot 2}{2 \cdot 3 \cdot 5}$
15. $\dfrac{4 \cdot 5 \cdot 3}{5 \cdot 7 \cdot 4}$

◇ EXERCISES ◇

Simplify.

1. $\dfrac{4}{10}$
2. $\dfrac{2}{6}$
3. $\dfrac{4}{15}$
4. $\dfrac{6}{9}$
5. $\dfrac{5}{10}$
6. $\dfrac{6}{14}$
7. $\dfrac{14}{4}$
8. $\dfrac{8}{6}$
9. $\dfrac{12}{10}$
10. $\dfrac{18}{10}$
11. $\dfrac{20}{15}$
12. $\dfrac{27}{6}$
13. $\dfrac{18}{20}$
14. $\dfrac{8}{30}$
15. $\dfrac{8}{20}$
16. $\dfrac{18}{27}$
17. $\dfrac{12}{25}$
18. $\dfrac{12}{20}$
19. $\dfrac{10}{12}$
20. $\dfrac{14}{6}$
21. $\dfrac{8}{15}$
22. $\dfrac{10}{4}$
23. $\dfrac{14}{18}$
24. $\dfrac{20}{8}$
25. $\dfrac{15}{25}$
26. $\dfrac{12}{28}$
27. $\dfrac{28}{6}$
★ 28. $\dfrac{16}{50}$
★ 29. $\dfrac{36}{24}$
★ 30. $\dfrac{32}{48}$

Solve. Write answers in simplest form.

31. A basketball team won 14 out of 20 games. What fractional part of the games did they win?

32. A pizza was cut into 8 slices. The Yings ate 6 slices. What fractional part of the pizza did they eat?

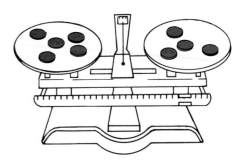

Bill has 9 quarters. They all appear to be genuine, but one is counterfeit. It weighs less than the others. Use a balance scale to find the counterfeit quarter in exactly two weighings.

Simplifying Products

◇ OBJECTIVE ◇
To simplify products

◇ RECALL ◇
2 cases of 6 bottles each
$2 \cdot 6 = 12$ ← *of* means *times*.

Example 1   Simplify. $\dfrac{3}{5} \cdot \dfrac{10}{9}$

$$\dfrac{3}{5} \cdot \dfrac{10}{9} = \dfrac{3 \cdot 10}{5 \cdot 9}$$

Factor 10: $10 = 5 \cdot 2$ $\dfrac{3 \cdot 5 \cdot 2}{5 \cdot 3 \cdot 3}$
Factor 9: $9 = 3 \cdot 3$

Divide out like factors. $\dfrac{\cancel{3} \cdot \cancel{5} \cdot 2}{\cancel{5} \cdot \cancel{3} \cdot 3}$

$$\dfrac{2}{3}$$

So, $\dfrac{3}{5} \cdot \dfrac{10}{9} = \dfrac{2}{3}$.

Example 2   Simplify. $2\dfrac{2}{5} \cdot \dfrac{1}{18}$

$2\dfrac{2}{5} = \dfrac{(5 \cdot 2) + 2}{5} = \dfrac{10 + 2}{5} = \dfrac{12}{5}$ $2\dfrac{2}{5} \cdot \dfrac{1}{18} = \dfrac{12}{5} \cdot \dfrac{1}{18} = \dfrac{12 \cdot 1}{5 \cdot 18}$

$12 = 2 \cdot 6$ $18 = 9 \cdot 2$ $\dfrac{2 \cdot 2 \cdot 3 \cdot 1}{5 \cdot 3 \cdot 3 \cdot 2}$
$2 \cdot 2 \cdot 3$ $3 \cdot 3 \cdot 2$

$\dfrac{\cancel{2} \cdot 2 \cdot \cancel{3} \cdot 1}{5 \cdot \cancel{3} \cdot 3 \cdot \cancel{2}}$

$1 \cdot 2 \cdot 1 \cdot 1 = 2$ $\dfrac{2}{15}$
$5 \cdot 1 \cdot 3 \cdot 1 = 15$

So, $2\dfrac{2}{5} \cdot \dfrac{1}{18} = \dfrac{2}{15}$.

---

practice ▷ Simplify.

1. $\dfrac{3}{5} \cdot \dfrac{5}{9}$ 2. $\dfrac{1}{3} \cdot \dfrac{6}{5}$ 3. $1\dfrac{1}{3} \cdot \dfrac{8}{12}$ 4. $3\dfrac{1}{2} \cdot \dfrac{1}{14}$

Example 3

Simplify. $2\frac{1}{7} \cdot 1\frac{1}{6}$

$2\frac{1}{7} = \frac{(7 \cdot 2) + 1}{7} = \frac{14 + 1}{7} = \frac{15}{7}$

$1\frac{1}{6} = \frac{(6 \cdot 1) + 1}{6} = \frac{6 + 1}{6} = \frac{7}{6}$

$2\frac{1}{7} \cdot 1\frac{1}{6} = \frac{15}{7} \cdot \frac{7}{6} = \frac{15 \cdot 7}{7 \cdot 6}$

$\frac{5 \cdot 3 \cdot 7}{7 \cdot 3 \cdot 2}$

$15 = 5 \cdot 3$
$6 = 3 \cdot 2$

Divide out like factors.

$\frac{5 \cdot \cancel{3} \cdot \cancel{7}}{\cancel{7} \cdot \cancel{3} \cdot 2}$

$\frac{5}{2}$, or $2\frac{1}{2}$

So, $2\frac{1}{7} \cdot 1\frac{1}{6} = 2\frac{1}{2}$.

practice ▷ Simplify.

5. $2\frac{1}{4} \cdot 1\frac{1}{3}$
6. $1\frac{3}{5} \cdot 1\frac{1}{4}$
7. $3\frac{1}{3} \cdot 2\frac{1}{4}$
8. $4\frac{1}{2} \cdot \frac{2}{3}$

Example 4

Simplify. $\frac{3}{4}$ of 18

of means *times*.

$\frac{3}{4}$ of 18 means $\frac{3}{4} \cdot 18$.

$18 = \frac{18}{1}$

$\frac{3}{4} \cdot \frac{18}{1} = \frac{3 \cdot 18}{4 \cdot 1}$

$18 = 3 \cdot 3 \cdot 2$
$4 = 2 \cdot 2$

$\frac{3 \cdot 3 \cdot 3 \cdot 2}{2 \cdot 2 \cdot 1}$

Divide out like factors.

$\frac{3 \cdot 3 \cdot 3 \cdot \cancel{2}}{2 \cdot \cancel{2} \cdot 1}$

$\frac{27}{2}$, or $13\frac{1}{2}$

So, $\frac{3}{4}$ of $18 = 13\frac{1}{2}$.

practice ▷ Simplify.

9. $\frac{3}{4}$ of 14
10. $\frac{5}{6}$ of 8
11. $\frac{3}{5}$ of 20
12. $\frac{2}{3}$ of 20

◇ ORAL EXERCISES ◇

Simplify.

1. $\frac{3}{2}$ 2. $\frac{7}{3}$ 3. $\frac{5}{2}$ 4. $\frac{7}{2}$ 5. $\frac{5}{3}$ 6. $\frac{7}{5}$ 7. $\frac{11}{3}$ 8. $\frac{8}{5}$
9. $\frac{5}{4}$ 10. $\frac{9}{5}$ 11. $\frac{9}{2}$ 12. $\frac{8}{3}$ 13. $\frac{9}{7}$ 14. $\frac{13}{5}$ 15. $\frac{7}{4}$ 16. $\frac{8}{7}$

◇ EXERCISES ◇

Simplify.

1. $\frac{3}{5} \cdot \frac{10}{9}$ 2. $\frac{2}{3} \cdot \frac{6}{5}$ 3. $\frac{5}{6} \cdot \frac{3}{10}$ 4. $\frac{3}{7} \cdot \frac{4}{15}$ 5. $\frac{2}{9} \cdot \frac{3}{4}$
6. $\frac{2}{7} \cdot \frac{14}{6}$ 7. $\frac{4}{21} \cdot \frac{7}{6}$ 8. $\frac{3}{14} \cdot \frac{10}{15}$ 9. $\frac{5}{8} \cdot \frac{6}{10}$ 10. $\frac{12}{15} \cdot \frac{3}{4}$
11. $1\frac{1}{5} \cdot \frac{5}{4}$ 12. $3\frac{1}{3} \cdot \frac{1}{2}$ 13. $\frac{7}{10} \cdot 2\frac{1}{2}$ 14. $1\frac{3}{4} \cdot \frac{1}{14}$ 15. $3\frac{1}{2} \cdot 2$
16. $1\frac{1}{3} \cdot \frac{9}{12}$ 17. $\frac{8}{15} \cdot 2\frac{1}{2}$ 18. $1\frac{1}{6} \cdot \frac{12}{14}$ 19. $\frac{2}{5} \cdot 3\frac{1}{3}$ 20. $2\frac{6}{7} \cdot \frac{14}{18}$
21. $2\frac{1}{2} \cdot 1\frac{2}{5}$ 22. $2\frac{1}{3} \cdot 1\frac{4}{7}$ 23. $1\frac{1}{5} \cdot 1\frac{1}{5}$ 24. $2\frac{1}{4} \cdot 1\frac{2}{3}$ 25. $2\frac{2}{3} \cdot 7\frac{1}{2}$
★ 26. $2\frac{1}{2} \cdot \frac{7}{15} \cdot \frac{3}{28}$ ★ 27. $1\frac{3}{5} \cdot 3\frac{1}{3} \cdot \frac{5}{12}$ ★ 28. $6\frac{2}{5} \cdot \frac{15}{24} \cdot 1\frac{1}{3}$
29. $\frac{5}{6}$ of 4 30. $\frac{3}{8}$ of 14 31. $\frac{4}{5}$ of 20 32. $\frac{2}{3}$ of 4 33. $\frac{3}{7}$ of 14

Solve these problems.

34. Mary bought $2\frac{1}{4}$ kilograms of meat. Her family ate $\frac{1}{3}$ of it. How much meat did they eat?

35. Bill rode his bike for $1\frac{1}{4}$ hours. He averaged $12\frac{4}{5}$ kilometers per hour. How far did he ride?

Calculator

Most calculators cannot display fractions. We can find products like $9 \cdot \frac{26}{6}$ on a calculator as shown below.

$9 \cdot 26 \div 6$ Press 9 ⊗ 26 ⊖ 6
$234 \div 6$ Display 39

Find the product.

1. $9 \cdot \frac{40}{6}$ 2. $12 \cdot \frac{28}{8}$ 3. $\frac{15}{6} \cdot 14$

SIMPLIFYING PRODUCTS

Problem Solving – Applications
Jobs for Teenagers

Use the chart to find the charge in Exercises 1–4.

HAPPY TIME CAMPGROUND	
Item	Rates per Hour
Bike riding	$ 1.50
Canoeing	$ 3.00
Archery	$ 2.00
Horseback riding	$15.00

1. $1\frac{1}{2}$ hours of bike riding

2. $1\frac{1}{4}$ hours of archery

3. $2\frac{1}{2}$ hours of canoeing

4. $3\frac{1}{2}$ hours of horseback riding

5. Bill rents umbrellas at the beach at $1.80 an hour. Find the charge for $\frac{3}{4}$ hour. Find the change from a $5 bill.

6. Carmine operates a horseback-riding concession. The rate is $16.00 an hour. Find the charge for $\frac{1}{2}$ hour. Find the change from a $10 bill.

7. José runs a canoe-renting concession. The rate is $2.80 an hour. Find the charge for $\frac{1}{2}$ hour. Find the change from a $20 bill.

8. Jim works in the office at a campground. The rate for a tent site is $4.50 per day. Find the charge for 3 day. Find the change from $20.

9. Gray Cloud works at a parking lot at an entertainment center. The rate is $1.40 an hour. Find the charge for $4\frac{1}{2}$ hours. Find the change from a $10 bill.

Dividing Fractions

◇ OBJECTIVE ◇
To divide fractions

◇ RECALL ◇

$$\frac{4 \cdot 5}{5 \cdot 4} = \frac{\overset{1}{\cancel{4}} \cdot \overset{1}{\cancel{5}}}{\underset{1}{\cancel{5}} \cdot \underset{1}{\cancel{4}}} = 1 \qquad 8 \div 2 \text{ means } \frac{8}{2}.$$

Example 1

Show that $\frac{3}{5} \div \frac{2}{7}$ means $\frac{3}{5} \cdot \frac{7}{2}$.

$$\frac{3}{5} \div \frac{2}{7} \text{ means } \frac{\frac{3}{5}}{\frac{2}{7}}.$$

We can multiply the numerator and the denominator by the same number, $\frac{7}{2}$.

$$\frac{2}{7} \cdot \frac{7}{2} = \frac{\overset{1}{\cancel{2}} \cdot \overset{1}{\cancel{7}}}{\underset{1}{\cancel{7}} \cdot \underset{1}{\cancel{2}}} = 1$$

$$\frac{\frac{3}{5} \cdot \frac{7}{2}}{\frac{2}{7} \cdot \frac{7}{2}}$$

$$\frac{\frac{3}{5} \cdot \frac{7}{2}}{1} = \frac{3}{5} \cdot \frac{7}{2}$$

So, $\frac{3}{5} \div \frac{2}{7} = \frac{3}{5} \cdot \frac{7}{2}$.

In Example 1, we found that $\frac{3}{5} \div \frac{2}{7}$ is the same as $\frac{3}{5} \cdot \frac{7}{2}$. $\frac{7}{2}$ is the reciprocal of $\frac{2}{7}$.

Example 2

Divide. $\frac{1}{12} \div \frac{7}{18}$

$$\frac{1}{12} \div \frac{7}{18}$$

To divide by $\frac{7}{18}$, multiply by its reciprocal, $\frac{18}{7}$.

$$\frac{1}{12} \cdot \frac{18}{7}$$

$$\frac{1 \cdot 18}{12 \cdot 7}$$

$18 = 9 \cdot 2 = 3 \cdot 3 \cdot 2$
$12 = 4 \cdot 3 = 2 \cdot 2 \cdot 3$
Divide out like factors.

$$\frac{1 \cdot 3 \cdot 3 \cdot 2}{2 \cdot 2 \cdot 3 \cdot 7} = \frac{1 \cdot \overset{1}{\cancel{3}} \cdot 3 \cdot \overset{1}{\cancel{2}}}{\underset{1}{\cancel{2}} \cdot 2 \cdot \underset{1}{\cancel{3}} \cdot 7} = \frac{3}{14}$$

So, $\frac{1}{12} \div \frac{7}{18} = \frac{3}{14}$.

practice ▷ Divide.

1. $\dfrac{6}{7} \div \dfrac{10}{7}$ 2. $\dfrac{3}{4} \div \dfrac{3}{2}$ 3. $\dfrac{7}{15} \div \dfrac{14}{10}$ 4. $\dfrac{3}{5} \div \dfrac{6}{5}$

Example 3   Divide.

$$4\tfrac{1}{2} \div 3 \qquad\qquad 5 \div 1\tfrac{1}{2}$$

$$3 = \tfrac{3}{1} \qquad 4\tfrac{1}{2} \div \tfrac{3}{1} \qquad\qquad \tfrac{5}{1} \div 1\tfrac{1}{2}$$

$$4\tfrac{1}{2} = \dfrac{(2 \cdot 4) + 1}{2} = \dfrac{8+1}{2} = \dfrac{9}{2} \qquad \dfrac{9}{2} \div \dfrac{3}{1} \qquad\qquad \dfrac{5}{1} \div \dfrac{3}{2}$$

To divide by $\tfrac{3}{1}$, multiply by $\tfrac{1}{3}$. $\qquad \dfrac{9}{2} \cdot \dfrac{1}{3} \qquad\qquad \dfrac{5}{1} \cdot \dfrac{2}{3}$

$$\dfrac{9 \cdot 1}{2 \cdot 3} \qquad\qquad \dfrac{5 \cdot 2}{1 \cdot 3}$$

$$\dfrac{\cancel{9}^{\,1} \cdot 3 \cdot 1}{2 \cdot \cancel{3}_{\,1}} = \dfrac{3}{2}, \text{ or } 1\tfrac{1}{2} \qquad \dfrac{10}{3}, \text{ or } 3\tfrac{1}{3}$$

So, $4\tfrac{1}{2} \div 3 = 1\tfrac{1}{2}$ and $5 \div 1\tfrac{1}{2} = 3\tfrac{1}{3}$.

---

practice ▷ Divide.

5. $2\tfrac{1}{4} \div 6$ 6. $14 \div 3\tfrac{1}{2}$ 7. $7\tfrac{1}{2} \div 3$ 8. $5\tfrac{1}{2} \div 22$

◇ **EXERCISES** ◇

Divide.

1. $\dfrac{3}{4} \div \dfrac{5}{6}$ 2. $\dfrac{2}{3} \div \dfrac{8}{9}$ 3. $\dfrac{5}{14} \div \dfrac{10}{7}$ 4. $\dfrac{1}{18} \div \dfrac{8}{12}$ 5. $\dfrac{3}{20} \div \dfrac{15}{8}$

6. $2\tfrac{1}{2} \div 5$ 7. $1\tfrac{1}{4} \div 5$ 8. $15 \div 3\tfrac{1}{3}$ 9. $18 \div 4\tfrac{1}{2}$ 10. $15 \div 2\tfrac{1}{2}$

11. $\dfrac{8}{5} \div \dfrac{4}{15}$ 12. $\dfrac{3}{20} \div \dfrac{6}{5}$ 13. $\dfrac{3}{5} \div \dfrac{9}{10}$ 14. $\dfrac{5}{7} \div \dfrac{15}{14}$ 15. $\dfrac{10}{8} \div \dfrac{15}{4}$

16. $\dfrac{4}{6} \div \dfrac{6}{21}$ 17. $1\tfrac{1}{3} \div \dfrac{8}{9}$ 18. $2\tfrac{6}{15} \div 1\tfrac{1}{3}$ 19. $7\tfrac{7}{12} \div 3\tfrac{1}{4}$ 20. $11\tfrac{1}{5} \div 1\tfrac{2}{5}$

★ 21. $\left(\dfrac{2}{3} \cdot \dfrac{9}{8}\right) \div 4\tfrac{1}{2}$ ★ 22. $3\tfrac{1}{3} \div \left(2\tfrac{1}{2} \cdot \dfrac{4}{5}\right)$ ★ 23. $\left(3\tfrac{1}{2} \cdot 1\tfrac{1}{3}\right) \div \left(\dfrac{8}{9} \cdot \dfrac{3}{4}\right)$

24. Donna mowed Mr. Weed's lawn for $7. It took $3\tfrac{1}{2}$ hours. How much did she make an hour?

25. José can run $3\tfrac{1}{2}$ laps in 7 minutes. How many laps can he run in 1 minute?

Evaluating Expressions

◇ OBJECTIVE ◇

To evaluate expressions like $\frac{4x-2}{12}$ if $x = 3$

◇ RECALL ◇

Evaluate $2x + 7$ if $x = 4$.

$2 \cdot 4 + 7$ Substitute
$8 + 7$ 4 for x.
15

Example 1

Evaluate $\frac{4x-2}{12}$ if $x = 3$.

Substitute 3 for x. → $\frac{4x-2}{12} = \frac{4 \cdot 3 - 2}{12}$

$\left. \begin{array}{c} \frac{4 \cdot 3 - 2}{12} \\ 12 - 2 \text{ or } 10 \end{array} \right\}$ → $= \frac{10}{12}$

Factor 10 into primes. $10 = 5 \cdot 2$ →
Factor 12 into primes. $12 = 2 \cdot 2 \cdot 3$ → $= \frac{5 \cdot 2}{2 \cdot 2 \cdot 3}$

Divide out like factors. → $= \frac{5 \cdot \cancel{2}}{2 \cdot \cancel{2} \cdot 3}$

$= \frac{5 \cdot 1}{2 \cdot 1 \cdot 3}$ or $\frac{5}{6}$

So, $\frac{4x-2}{12} = \frac{5}{6}$ if $x = 3$.

◇ EXERCISES ◇

Evaluate.

1. $\frac{3x-1}{10}$ if $x = 2$
2. $\frac{2x+3}{15}$ if $x = 3$
3. $\frac{5x+4}{21}$ if $x = 2$
4. $\frac{6+2x}{20}$ if $x = 2$
5. $\frac{5x-10}{35}$ if $x = 3$
6. $\frac{8+2x}{18}$ if $x = 4$
7. $\frac{3}{2x+5}$ if $x = 2$
8. $\frac{2}{2+4x}$ if $x = 2$
9. $\frac{10}{6x+3}$ if $x = 2$
★ 10. $\frac{6x-6}{7y-2}$ if $x = 5, y = 6$
★ 11. $\frac{3a+4}{2ab+a}$ if $a = 2, b = 3$

EVALUATING EXPRESSIONS

Equations

◇ OBJECTIVE ◇

To solve equations like
$\frac{3}{4}x = 10$ and $\frac{3}{5}x = \frac{2}{3}$

◇ RECALL ◇

$$\underset{\text{reciprocals}}{\frac{2}{3} \cdot \frac{3}{2}} = 1$$

$\frac{2}{3}$ times a number is 18.

Let x be the number. $\frac{2}{3} \cdot x = 18$, or $\frac{2}{3}x = 18$

Example 1   Find the value of x that makes $\frac{2}{3}x = 18$ true.

$\frac{2}{3}x$ means $\frac{2}{3} \cdot x$.

$$\frac{2}{3}x = 18$$

Multiply each side of the equation by $\frac{3}{2}$, the reciprocal of $\frac{2}{3}$.

$$\frac{3}{2} \cdot \frac{2}{3} \cdot x = \frac{3}{2} \cdot 18$$

$$1 \cdot x = \frac{3}{2} \cdot \frac{18}{1}$$

$$x = \frac{3 \cdot 3 \cdot 3 \cdot \cancel{2}}{\cancel{2} \cdot 1}$$

$$x = 27$$

Check.

$\frac{2}{3}x$	18
$\frac{2}{3} \cdot 27$	18
$\frac{2}{3} \cdot \frac{27}{1}$	
$\frac{2 \cdot \cancel{3} \cdot 3 \cdot 3}{\cancel{3} \cdot 1}$	
18	

Substitute 27 for x.

So, 27 is the value of x that makes $\frac{2}{3}x = 18$ true.

practice ▷   Solve.

1. $\frac{3}{4}x = 15$ 2. $\frac{3}{5}x = 9$ 3. $\frac{4}{7}x = 8$ 4. $\frac{5}{9}x = 20$

Example 2

Solve. $\frac{1}{5}x = \frac{8}{15}$

$\frac{1}{5}x$ means $\frac{1}{5} \cdot x$.

Multiply each side by the reciprocal of $\frac{1}{5}$.

$$\frac{1}{5}x = \frac{8}{15}$$

$$\frac{5}{1} \cdot \frac{1}{5} \cdot x = \frac{5}{1} \cdot \frac{8}{15}$$

$$1 \cdot x = \frac{5 \cdot 8}{15}$$

$$x = \frac{\overset{1}{\cancel{5}} \cdot 2 \cdot 2 \cdot 2}{\underset{1}{\cancel{5}} \cdot 3} = \frac{8}{3}$$

$$x = \frac{8}{3}, \text{ or } 2\frac{2}{3}.$$

So, $2\frac{2}{3}$ is the solution of $\frac{1}{5}x = \frac{8}{15}$.

practice ▷ Solve.

5. $\frac{1}{7}x = \frac{10}{21}$
6. $\frac{1}{10}x = \frac{5}{6}$
7. $\frac{1}{8}x = \frac{3}{14}$
8. $\frac{3}{7}x = \frac{9}{14}$

◇ ORAL EXERCISES ◇

By what number do you multiply each side of the equation to solve?

1. $\frac{3}{7}x = 15$
2. $\frac{14}{3}x = 7$
3. $\frac{5}{6}x = \frac{11}{15}$
4. $\frac{2}{3}x = 8$
5. $\frac{4}{5}x = \frac{6}{25}$
6. $\frac{3}{5}x = \frac{9}{10}$
7. $\frac{4}{5}x = \frac{2}{3}$
8. $\frac{3}{7}x = \frac{9}{14}$
9. $\frac{10}{3}x = \frac{5}{7}$
10. $\frac{4}{5}x = 12$

◇ EXERCISES ◇

Solve.

1. $\frac{2}{3}x = 6$
2. $\frac{4}{5}x = 12$
3. $\frac{5}{7}x = 10$
4. $\frac{3}{4}x = 15$
5. $\frac{3}{5}x = 9$
6. $\frac{1}{3}x = \frac{7}{15}$
7. $\frac{1}{2}x = \frac{9}{4}$
8. $\frac{3}{7}x = \frac{5}{14}$
9. $\frac{2}{5}x = \frac{7}{10}$
10. $\frac{2}{3}x = \frac{5}{6}$
11. $\frac{3}{7}x = 9$
12. $\frac{2}{3}x = 18$
13. $\frac{3}{4}x = 8$
14. $\frac{3}{2}x = 10$
15. $\frac{1}{5}x = 3$
16. $\frac{4}{5}x = 8$
17. $\frac{2}{3}x = 24$
18. $\frac{3}{5}x = 24$
19. $\frac{6}{5}x = 10$
20. $\frac{3}{4}x = 20$
21. $\frac{6}{5}x = \frac{2}{3}$
22. $\frac{4}{5}x = \frac{8}{3}$
23. $\frac{2}{3}x = \frac{15}{6}$
24. $\frac{4}{9}x = \frac{2}{3}$
25. $\frac{3}{4}x = \frac{15}{6}$
★ 26. $\frac{7}{2}x = \frac{14}{3}$
★ 27. $\frac{27}{2}x = \frac{41}{4}$
★ 28. $16 = \frac{2}{3}x$
★ 29. $\frac{3}{8} = \frac{1}{2}x$
★ 30. $\frac{25}{3} = \frac{5}{6}x$

EQUATIONS

Evaluating Formulas

◇ **OBJECTIVE** ◇

To evaluate formulas involving fractions

◇ **RECALL** ◇

6^2 means $6 \cdot 6$, or 36 $\quad|\quad$ $\frac{12}{3}$ means $12 \div 3$, or 4

Example 1

A formula for the area of a triangle is $A = \frac{1}{2}bh$.
Find A if b is 6 and h is $3\frac{1}{2}$.

Substitute 6 for b and $3\frac{1}{2}$ for h.

$6 = \frac{6}{1}$; $3\frac{1}{2} = \frac{(2 \cdot 3) + 1}{2} = \frac{6 + 1}{2} = \frac{7}{2}$

$$A = \frac{1}{2}bh$$
$$= \frac{1}{2} \cdot 6 \cdot 3\frac{1}{2}$$
$$= \frac{1}{2} \cdot \frac{6}{1} \cdot \frac{7}{2}$$
$$= \frac{1 \cdot 6 \cdot 7}{2 \cdot 1 \cdot 2} = \frac{1 \cdot 3 \cdot 2 \cdot 7}{2 \cdot 1 \cdot 2}$$
$$= \frac{1 \cdot 3 \cdot \cancel{2} \cdot 7}{\cancel{2} \cdot 1 \cdot 2} = \frac{21}{2}, \text{ or } 10\frac{1}{2}$$

So, A is $10\frac{1}{2}$ when b is 6 and h is $3\frac{1}{2}$.

Example 2

$\frac{22}{7} \doteq \pi$
↑
approximately equal to

r^2 means $r \cdot r$.

Substitute $2\frac{1}{3}$ for r.

$2\frac{1}{3} = \frac{(3 \cdot 2) + 1}{3} = \frac{6 + 1}{3} = \frac{7}{3}$

A formula for the area of a circle is $A = \pi r^2$.
Use $\frac{22}{7}$ for π.
Find A if r is $2\frac{1}{3}$.

$$A = \pi r^2 \doteq \frac{22}{7} \cdot r \cdot r$$
$$\doteq \frac{22}{7} \cdot 2\frac{1}{3} \cdot 2\frac{1}{3}$$
$$\doteq \frac{22}{7} \cdot \frac{7}{3} \cdot \frac{7}{3} = \frac{22 \cdot \cancel{7} \cdot 7}{\cancel{7} \cdot 3 \cdot 3}$$
$$\doteq \frac{154}{9}, \text{ or } 17\frac{1}{9}$$

So, A is about $17\frac{1}{9}$ when r is $2\frac{1}{3}$.

practice ▷ 1. $A = \frac{1}{2}bh$. Find A if b is 5 and h is $2\frac{2}{3}$.

2. $A \doteq \frac{22}{7}r^2$. Find A if r is $3\frac{1}{2}$.

Example 3

A formula for the volume of a cone is $V = \frac{1}{3}\pi r^2 h$. Find V in terms of π if r is $2\frac{1}{2}$ and h is 4.

$$V = \frac{1}{3}\pi r^2 h, \text{ or } \frac{1}{3}\pi \cdot r \cdot r \cdot h$$

$2\frac{1}{2} = \frac{(2 \cdot 2) + 1}{2} = \frac{4+1}{2} = \frac{5}{2}$

$\frac{1}{3}\pi \cdot 2\frac{1}{2} \cdot 2\frac{1}{2} \cdot 4 = \frac{1}{3} \cdot \frac{5}{2} \cdot \frac{5}{2} \cdot \frac{4}{1} \cdot \pi$

$\frac{1 \cdot 5 \cdot 5 \cdot 2 \cdot 2}{3 \cdot 2 \cdot 2 \cdot 1}\pi = \frac{1 \cdot 5 \cdot 5 \cdot \cancel{2} \cdot \cancel{2}}{3 \cdot \cancel{2} \cdot \cancel{2} \cdot 1}\pi, \text{ or } \frac{25}{3}\pi$

So, V is $8\frac{1}{3}\pi$ when r is $2\frac{1}{2}$ and h is 4.

practice ▷ $V = \frac{1}{3}\pi r^2 h$. Find V in terms of π.

3. r is $1\frac{1}{3}$, h is 3

4. r is $2\frac{1}{4}$, h is 8

◇ EXERCISES ◇

Find the value.

1. $A = \frac{1}{2}bh$ if b is 10, h is $\frac{5}{2}$
2. $A = \frac{1}{2}bh$ if b is 6, h is $\frac{10}{3}$
3. $M = \frac{2}{3}p^2$ if p is 6
4. $A \doteq \frac{22}{7}r^2$ if r is $1\frac{3}{4}$
5. $A \doteq \frac{22}{7}r^2$ if r is $1\frac{1}{6}$
6. $G = \frac{3}{4}a^2$ if a is $3\frac{1}{3}$
7. $R = \frac{2}{3}kb$ if k is $\frac{3}{5}$, b is 7
8. $M = \frac{3}{5}p^2$ if p is $2\frac{1}{2}$
9. $C \doteq \frac{22}{7}d$ if d is $1\frac{3}{4}$

Find the value. Leave answers in terms of π.

10. $V = \frac{1}{3}\pi r^2 h$ if r is $3\frac{1}{2}$, h is 6
11. $V = \frac{1}{3}\pi r^2 h$ if r is $1\frac{1}{2}$, h is 9

Find the value.

★ 12. $V = \frac{4}{3}\pi r^3$ if $\pi \doteq \frac{22}{7}$, r is $1\frac{1}{2}$

★ 13. $V = \frac{1}{3}\pi r^2 h$ if $\pi \doteq \frac{22}{7}$, r is $10\frac{1}{2}$, h is 6

EVALUATING FORMULAS

A fraction such as $\frac{28}{12}$ can be simplified by first finding the GCF of 12 and 28. The computer program below does this.

Example Type this program for simplifying any fraction of the form $\frac{A}{B}$, $B \neq 0$.

RUN the program to simplify $\frac{28}{12}$.

Clear the screen.	100	HOME
	110	INPUT "TYPE IN NUMERATOR ";A
	120	INPUT "TYPE IN DENOMINATOR ";B
Computer decides which is smaller, A or B. Trial divisors are decreased by 1.	130	IF A - B < 0 THEN LET X = A: GOTO 150
	140	LET X = B
	150	FOR I = X TO 1 STEP - 1
	160	LET N = A / I
	170	LET D = B / I
Computer tests whether I is a common divisor of A and B. Prints two lines of space above answer.	180	IF N = INT (N) AND D = INT (D) THEN 200
	190	NEXT I
	200	PRINT : PRINT
	210	PRINT A"/"B" = "N"/"D
	220	IF D = 1 THEN PRINT "OR "N
	230	END
		]RUN
You type 28.		TYPE IN NUMERATOR 28
You type 12.		TYPE IN DENOMINATOR 12
Computer displays the result.		28/12 = 7/3

See the Computer Section beginning on page 420 for more information.

Exercises

Simplify each fraction. Then **RUN** the program to check your answer.

1. $\frac{24}{72}$
2. $\frac{18}{12}$
3. $\frac{48}{24}$
4. $\frac{1,440}{2,160}$

5. Why does the computer seem to take so long to simplify $\frac{1,440}{2,160}$?

6. Change the program to find the product of any two fractions such as X/Y and Z/W. (Hint: A = X * Z, B = Y * W.) **RUN** the program to simplify (2 / 5) * (4 / 8).

Chapter Review

What number corresponds to point *A*? [79]

1.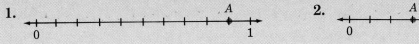
2.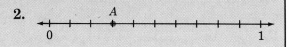

Multiply. [82]

3. $4 \cdot \frac{1}{9}$
4. $\frac{1}{10} \cdot 7$
5. $6 \cdot \frac{2}{5}$
6. $10 \cdot \frac{2}{3}$

Tell whether each number is prime, composite, or neither. [88]

7. 18
8. 24
9. 1
10. 19

Factor into primes. Write using exponents. [88]

11. 12
12. 28
13. 45
14. 42

Find the greatest common factor (GCF). [90]

15. 20; 24
16. 27; 18
17. 81; 108
18. 32; 98

Simplify. [92]

19. $\frac{6}{10}$
20. $\frac{20}{6}$
21. $\frac{14}{18}$
22. $\frac{18}{12}$

Simplify. [95]

23. $\frac{5}{14} \cdot \frac{7}{15}$
24. $1\frac{3}{8} \cdot \frac{1}{8}$
25. $7 \cdot \frac{3}{14}$
26. $\frac{3}{4}$ of 14

Divide. [99]

27. $\frac{2}{3} \div \frac{10}{9}$
28. $4\frac{1}{2} \div 6$
29. $15 \div 2\frac{1}{2}$
30. $2\frac{1}{3} \div \frac{4}{9}$

Evaluate. [101]

31. $\frac{4x - 2}{15}$ if $x = 2$
32. $\frac{4x + 3}{30}$ if $x = 3$
33. $\frac{2y + 5}{45}$ if $y = 5$

Solve. [102]

34. $\frac{3}{5}x = 21$
35. $\frac{7}{10}x = \frac{4}{15}$
36. $\frac{2}{3}x = \frac{5}{6}$

Find the value. [104]

37. $A = \frac{1}{2}bh$ if b is 8, h is $1\frac{1}{2}$
38. $A = \frac{22}{7}r^2$ if r is $1\frac{2}{3}$

39. $V = \frac{1}{3}\pi r^2 h$. Find V in terms of π if r is $2\frac{1}{3}$ and h is 9.

Chapter Test

What number corresponds to point A?

1.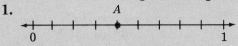
2.

Tell whether each number is prime, composite, or neither.

3. 15   4. 1   5. 37   6. 20   7. 39

Factor into primes. Write using exponents.

8. 24   9. 40   10. 18   11. 32   12. 48

Find the greatest common factor (GCF).

13. 12; 20   14. 18; 36   15. 24; 36

Simplify.

16. $\frac{4}{10}$ 17. $\frac{14}{6}$ 18. $\frac{15}{18}$

Simplify.

19. $\frac{4}{21} \cdot \frac{7}{6}$ 20. $1\frac{1}{2} \cdot \frac{1}{9}$ 21. $1\frac{1}{3} \cdot 4\frac{1}{2}$

Simplify.

22. $\frac{4}{5}$ of 15   23. $\frac{3}{5}$ of 7

Divide.

24. $\frac{3}{10} \div \frac{12}{5}$ 25. $3\frac{1}{2} \div 14$

Solve.

26. $\frac{2}{3}x = 8$ 27. $\frac{8}{5}x = \frac{4}{15}$

28. $A = \frac{1}{2}bh$. Find A if $b = 10$ and $h = 4\frac{1}{4}$.

29. $V = \frac{1}{3}\pi r^2 h$. Find V in terms of π if $r = 2\frac{3}{4}$ and $h = 8$.

Evaluate.

30. $\frac{7x - 1}{25}$ if $x = 3$ 31. $\frac{4x + 2}{24}$ if $x = 4$

Cumulative Review

Evaluate if $a = 5$, $x = 3$, and $t = 7$.

1. $a + 16$
2. $t + 4a - 19$
3. $3 + 2xa$
4. $2tx + 6$
5. x^4
6. $4a^3$

Simplify.

7. $3y + 6 + 8y$
8. $3(10m - 4)$
9. $3a + 4 - 2a + 5$
10. $\frac{6}{14}$
11. $\frac{6}{27} \cdot \frac{9}{21}$
12. $3\frac{1}{4} \cdot 2\frac{2}{3}$

Solve. Check.

13. $x + 5 = 13$
14. $y - 4 = 4$
15. $12 = z - 7$
16. $36 = 12m$
17. $\frac{t}{3} = 7$
18. $5 = \frac{n}{9}$
19. $4t + 3 = 11$
20. $5 = 3x - 13$
21. $\frac{3}{4}m = \frac{3}{5}$

Write in mathematical form.

22. 17 times n
23. x divided by 4
24. 9 divided by r
25. 7 times x, increased by 7
26. 3 less than 4 times a number

Which property is illustrated?

27. $41 + 0 = 41$
28. $(3 \cdot 5) \cdot 12 = 3 \cdot (5 \cdot 12)$
29. $32 \cdot 1 = 32$
30. $32 + 216 = 216 + 32$
31. $32 \cdot 9 = 9 \cdot 32$
32. $32 \cdot 9 + 32 \cdot 13 = 32(9 + 13)$

33. $P = 2l + 2w$. Find P if $l = 3\frac{1}{2}$ in. and $w = 1\frac{1}{4}$ in.
34. $A = \frac{bh}{2}$. Find A if $b = 7$ in. and $h = 5$ in.
35. $A = \pi r^2$. Find A in terms of π if $r = 9$.
36. $V = \pi r^2 h$. Find V in terms of π if $r = 5$ and $h = 4$.

Factor into primes.

37. 60
38. 42
39. 120

Solve these problems.

40. $35 less than Susan's salary is $145. What is Susan's salary?
41. The sum of a number and 26 is 42. What is the number?
42. Four times Alfonso's age is 64. How old is Alfonso?
43. Olga's score of 9 is the same as Tanya's score divided by 3. What is Tanya's score?

CUMULATIVE REVIEW

Adding and Subtracting Fractions

5

The Gateway Arch in St. Louis, Missouri, has a span and height of 630 ft.

Fractions with the Same Denominator

◇ OBJECTIVE ◇

To add and subtract fractions with the same denominator

◇ RECALL ◇

Show that $6 \cdot 5 + 3 \cdot 5 = (6 + 3)5$.

$6 \cdot 5 + 3 \cdot 5$	$(6 + 3)5$
$30 + 15$	$9 \cdot 5$
45	45

So, $6 \cdot 5 + 3 \cdot 5 = (6 + 3)5$.

Example 1 Show that $\frac{4}{7} + \frac{2}{7} = \frac{6}{7}$ and $\frac{5}{9} - \frac{4}{9} = \frac{1}{9}$.

$\frac{4}{7} + \frac{2}{7}$	$\frac{6}{7}$	$\frac{5}{9} - \frac{4}{9}$	$\frac{1}{9}$
$\frac{4}{7} = 4 \cdot \frac{1}{7} \quad \frac{2}{7} = 2 \cdot \frac{1}{7} \quad 4 \cdot \frac{1}{7} + 2 \cdot \frac{1}{7}$	$\frac{6}{7}$	$5 \cdot \frac{1}{9} - 4 \cdot \frac{1}{9}$	$\frac{1}{9}$
Distributive property $(4 + 2)\frac{1}{7}$		$(5 - 4)\frac{1}{9}$	
$6 \cdot \frac{1}{7}$		$1 \cdot \frac{1}{9}$	
$\frac{6}{7}$		$\frac{1}{9}$	

So, $\frac{4}{7} + \frac{2}{7} = \frac{6}{7}$ and $\frac{5}{9} - \frac{4}{9} = \frac{1}{9}$.

$\frac{4}{7} + \frac{2}{7} = \frac{4 + 2}{7} = \frac{6}{7}$ To add fractions with the same denominator: Add numerators. Keep the denominator.

$\frac{5}{9} - \frac{1}{9} = \frac{5 - 1}{9} = \frac{4}{9}$ To subtract fractions with the same denominator: Subtract numerators. Keep the denominator.

Example 2 Add. $\frac{2}{9} + \frac{5}{9}$ | Subtract. $\frac{5}{7} - \frac{2}{7}$

$\frac{2}{9} + \frac{5}{9} = \frac{2 + 5}{9} = \frac{7}{9}$ | $\frac{5}{7} - \frac{2}{7} = \frac{5 - 2}{7} = \frac{3}{7}$

So, $\frac{2}{9} + \frac{5}{9} = \frac{7}{9}$ and $\frac{5}{7} - \frac{2}{7} = \frac{3}{7}$.

practice ▷ Add or subtract.

1. $\dfrac{5}{7} + \dfrac{1}{7}$ 2. $\dfrac{10}{11} - \dfrac{3}{11}$ 3. $\dfrac{2}{5} + \dfrac{2}{5}$ 4. $\dfrac{8}{9} - \dfrac{4}{9}$

Example 3 Add or subtract. Simplify if possible.

$$\dfrac{5}{8} + \dfrac{1}{8} \qquad\qquad \dfrac{9}{10} - \dfrac{3}{10}$$

$$\dfrac{5}{8} + \dfrac{1}{8} = \dfrac{5+1}{8} = \dfrac{6}{8} \qquad \dfrac{9}{10} - \dfrac{3}{10} = \dfrac{9-3}{10} = \dfrac{6}{10}$$

Factor the numerator. $\dfrac{6}{8} = \dfrac{2 \cdot 3}{2 \cdot 2 \cdot 2} \qquad\qquad \dfrac{6}{10} = \dfrac{2 \cdot 3}{2 \cdot 5}$
Factor the denominator.

Divide out like factors. $\dfrac{\cancel{2} \cdot 3}{\cancel{2} \cdot 2 \cdot 2}$, or $\dfrac{3}{4} \qquad \dfrac{\cancel{2} \cdot 3}{\cancel{2} \cdot 5}$, or $\dfrac{3}{5}$

So, $\dfrac{5}{8} + \dfrac{1}{8} = \dfrac{3}{4}$ and $\dfrac{9}{10} - \dfrac{3}{10} = \dfrac{3}{5}$.

---

practice ▷ Add or subtract. Simplify if possible.

5. $\dfrac{5}{9} + \dfrac{1}{9}$ 6. $\dfrac{5}{6} - \dfrac{1}{6}$ 7. $\dfrac{1}{8} + \dfrac{3}{8}$ 8. $\dfrac{5}{12} - \dfrac{1}{12}$

Example 4 Add or subtract. Simplify if possible.

$$\begin{array}{c} \dfrac{5}{6} \\ +\dfrac{9}{6} \\ \hline \dfrac{14}{6} \end{array} \qquad\qquad \begin{array}{c} \dfrac{11}{8} \\ -\dfrac{1}{8} \\ \hline \dfrac{10}{8} \end{array}$$

$\dfrac{5}{6} + \dfrac{9}{6} = \dfrac{5+9}{6} = \dfrac{14}{6}$

Divide $6\overline{)14} = 2\dfrac{2}{6}$ $2\dfrac{2}{6}$ $1\dfrac{2}{8}$
$\underline{12}$
2 $2\dfrac{1}{3}$ $1\dfrac{1}{4}$

$\dfrac{2}{6} = \dfrac{\cancel{2}}{\cancel{2} \cdot 3} = \dfrac{1}{3}$ So, $\dfrac{5}{6} + \dfrac{9}{6} = 2\dfrac{1}{3}$ and $\dfrac{11}{8} - \dfrac{1}{8} = 1\dfrac{1}{4}$.

---

practice ▷ Add or subtract. Simplify if possible.

9. $\begin{array}{r} \dfrac{5}{4} \\ +\dfrac{1}{4} \\ \hline \end{array}$ 10. $\begin{array}{r} \dfrac{7}{9} \\ +\dfrac{5}{9} \\ \hline \end{array}$ 11. $\begin{array}{r} \dfrac{1}{3} \\ +\dfrac{7}{3} \\ \hline \end{array}$ 12. $\begin{array}{r} \dfrac{13}{10} \\ -\dfrac{1}{10} \\ \hline \end{array}$

◇ ORAL EXERCISES ◇

Add or subtract.

1. $\frac{2}{7} + \frac{1}{7}$
2. $\frac{3}{10} + \frac{6}{10}$
3. $\frac{1}{5} + \frac{3}{5}$
4. $\frac{7}{9} + \frac{1}{9}$
5. $\frac{3}{8} + \frac{4}{8}$
6. $\frac{4}{7} + \frac{2}{7}$
7. $\frac{5}{9} - \frac{4}{9}$
8. $\frac{7}{12} - \frac{6}{12}$
9. $\frac{4}{10} - \frac{1}{10}$
10. $\frac{5}{14} - \frac{2}{14}$
11. $\frac{7}{8} - \frac{2}{8}$
12. $\frac{5}{6} - \frac{4}{6}$

13. $\frac{5}{11} + \frac{3}{11}$
14. $\frac{7}{11} - \frac{3}{11}$
15. $\frac{2}{15} + \frac{5}{15}$
16. $\frac{5}{6} - \frac{4}{6}$
17. $\frac{5}{9} + \frac{2}{9}$
18. $\frac{8}{13} - \frac{6}{13}$

◇ EXERCISES ◇

Add or subtract. Simplify if possible.

1. $\frac{3}{9} + \frac{5}{9}$
2. $\frac{1}{11} + \frac{5}{11}$
3. $\frac{7}{10} + \frac{2}{10}$
4. $\frac{3}{8} + \frac{1}{8}$
5. $\frac{3}{6} + \frac{1}{6}$
6. $\frac{5}{9} - \frac{2}{9}$
7. $\frac{5}{9} - \frac{1}{9}$
8. $\frac{8}{12} - \frac{2}{12}$
9. $\frac{7}{10} - \frac{2}{10}$
10. $\frac{7}{8} - \frac{1}{8}$
11. $\frac{2}{9} + \frac{4}{9}$
12. $\frac{7}{12} + \frac{2}{12}$
13. $\frac{3}{8} + \frac{5}{8}$
14. $\frac{6}{5} + \frac{2}{5}$
15. $\frac{9}{8} + \frac{3}{8}$
16. $\frac{15}{8} - \frac{3}{8}$
17. $\frac{13}{12} - \frac{1}{12}$
18. $\frac{17}{10} - \frac{3}{10}$
19. $\frac{11}{6} - \frac{1}{6}$
20. $\frac{9}{4} - \frac{3}{4}$

21. $\frac{1}{9} + \frac{5}{9}$
22. $\frac{5}{8} + \frac{7}{8}$
23. $\frac{5}{12} + \frac{11}{12}$
24. $\frac{4}{6} + \frac{2}{6}$
25. $\frac{3}{4} + \frac{7}{4}$
26. $\frac{11}{4} + \frac{1}{4}$

27. $\frac{11}{12} - \frac{3}{12}$
28. $\frac{19}{6} - \frac{5}{6}$
29. $\frac{15}{4} - \frac{1}{4}$
30. $\frac{5}{18} - \frac{2}{18}$
31. $\frac{7}{12} - \frac{5}{12}$
32. $\frac{63}{15} - \frac{43}{15}$

★ 33. $\frac{8}{12} + \frac{7}{12} - \frac{13}{12}$
★ 34. $\frac{17}{15} - \frac{4}{15} + \frac{7}{15}$
★ 35. $\frac{20}{21} - \frac{9}{21} - \frac{4}{21}$

Solve these problems.

36. John spent $\frac{3}{4}$ hour before supper on homework and $\frac{1}{4}$ hour after supper on homework. How long did he work altogether?

37. Maria sold $\frac{3}{4}$ case of applesauce in the morning and $\frac{1}{4}$ case in the afternoon. How much more applesauce did she sell in the morning?

FRACTIONS WITH THE SAME DENOMINATOR

The Least Common Multiple (LCM)

◆ OBJECTIVE ◆

To find the least common multiple (LCM) of two or three numbers

◆ RECALL ◆

Factor 12 into primes.

$12 = 4 \cdot 3$
$= 2 \cdot 2 \cdot 3 \leftarrow$ all primes

A *multiple* of a number is the product of that number and any whole number.

Note: 0 is a multiple of every whole number. $n \cdot 0 = 0$ for any number n.

$2 \cdot 0 = 0$	0 is a multiple of 2.
$2 \cdot 1 = 2$	2 is a multiple of 2.
$2 \cdot 2 = 4$	4 is a multiple of 2.
$2 \cdot 3 = 6$	6 is a multiple of 2.

Definition of least common multiple (LCM)

The *least common multiple (LCM)* of two or more numbers is the least nonzero whole number that is a multiple of the numbers.

Example 1

Find the least common multiple (LCM) of 4 and 5.

Find some multiples of 4.
Do not include 0.

Multiples of 4: 4, 8, 12, 16, 20, 24, 28, 32, 36, 40, 44, 48, 52, 56, 60, . . .

Find some multiples of 5.
Do not include 0.

Multiples of 5: 5, 10, 15, 20, 25, 30, 35, 40, 45, 50, 55, 60, 65, . . .

Find some multiples that are the same (common) for both 4 and 5.

Common multiples: 20, 40, 60, . . .
↑
least common multiple

So, 20 is the least common multiple (LCM) of 4 and 5.

Writing several common multiples can be time consuming. A shorter way to find the LCM is shown in the next example.

Example 2

Find the LCM of 6 and 20. Use prime factorization.

Factor 20 into primes. $20 = 10 \cdot 2 = 5 \cdot 2 \cdot 2$
Factor 6 into primes. $6 = 3 \cdot 2$
At *most* one 5, one 3, two 2's occur in either 6 or 20.
The LCM is $5 \cdot 3 \cdot 2 \cdot 2$.
$5 \cdot 3 \cdot 2 \cdot 2 = 60$ So, the LCM of 6 and 20 is 60.

Example 3

Find the LCM of 2, 4, and 6.
Use prime factorization.

Two is prime. $\quad 2 = 2 \cdot 1 = 2 \qquad$ one 2
Factor 4 into primes. $\quad 4 = 2 \cdot 2 \qquad$ two 2's
Factor 6 into primes. $\quad 6 = 3 \cdot 2 \qquad$ one 3, one 2
At *most* two 2's, one 3 occur in either 2, 4, or 6.
So, the LCM of 2, 4, and 6 is $2 \cdot 2 \cdot 3$, or 12.

---

practice ▷ Find the LCM using prime factorization.

1. 3 and 4   2. 5 and 6   3. 4, 5, and 9   4. 5, 10, and 15

You will use the concept of LCM later in adding fractions with unlike denominators. You will have to find the LCM of the denominators. The LCM will then be referred to as the LCD (lowest common denominator).

◇ EXERCISES ◇

Find the first three multiples excluding zero.

1. 5   2. 10   3. 9   4. 8   5. 11   6. 12

Find the LCM. (Use prime factorization.)

7. 4; 7   8. 9; 5   9. 4; 6   10. 2; 3
11. 3; 8   12. 10; 20   13. 5; 15   14. 6; 9
15. 8; 16   16. 2; 7   17. 7; 8   18. 9; 4
19. 4; 6; 8   20. 12; 18; 30 21. 2; 3; 5   22. 4; 6; 10
23. 5; 6; 10   24. 20; 24; 32 ★ 25. 15; 24; 12   ★ 26. 36; 54; 81

THE LEAST COMMON MULTIPLE (LCM)

Fractions with Unlike Denominators

◇ OBJECTIVE ◇

To add and subtract fractions with unlike denominators

◇ RECALL ◇

$$\frac{4}{7} + \frac{2}{7} = \frac{4+2}{7}, \text{ or } \frac{6}{7}$$

$$\frac{4}{7} - \frac{2}{7} = \frac{4-2}{7}, \text{ or } \frac{2}{7}$$

Example 1   Add. Simplify if possible. $\frac{1}{5} + \frac{2}{15}$

The denominators are not the same. Factor 15 into primes. $15 = 5 \cdot 3$

$$\frac{1}{5} + \frac{2}{15} = \frac{1}{5} + \frac{2}{5 \cdot 3}$$

5 and 3 are the only factors present.

To have a common denominator, each denominator must have exactly the same factors.

$$\frac{1}{5} + \frac{2}{5 \cdot 3}$$

needs 3   has 5 and 3

LCD is the same as LCM.   The least common denominator (LCD) is $5 \cdot 3$, or 15.

Multiply the numerator and denominator of $\frac{1}{5}$ by 3.

$$\frac{1 \cdot 3}{5 \cdot 3} + \frac{2}{5 \cdot 3}$$

The denominators are now the same.

$$\frac{3}{15} + \frac{2}{15}$$

$$\frac{3}{15} + \frac{2}{15} = \frac{3+2}{15} = \frac{5}{15}$$

$$\frac{5}{15}$$

$$\frac{5}{15} = \frac{\cancel{5} \cdot 1}{\cancel{5} \cdot 3} = \frac{1}{3}$$

$$\frac{1}{3}$$

So, $\frac{1}{5} + \frac{2}{15} = \frac{1}{3}$.

---

practice ▷   Add or subtract. Simplify if possible.

1. $\frac{2}{7} + \frac{3}{14}$ 2. $\frac{1}{6} + \frac{2}{3}$ 3. $\frac{7}{15} - \frac{1}{5}$ 4. $\frac{1}{2} - \frac{3}{10}$

Example 2

Subtract. Simplify if possible. $\frac{13}{6} - \frac{2}{3}$

The denominators are not the same.
Factor 6 into primes. $6 = 3 \cdot 2$.

$$\begin{array}{r}\frac{13}{6}\\-\frac{2}{3}\\\hline\end{array} \qquad \begin{array}{r}\frac{13}{3\cdot 2}\\-\frac{2}{3}\\\hline\end{array} \begin{array}{l}\leftarrow \text{ has } 3\cdot 2\\ \\ \leftarrow \text{ needs } 2\end{array}$$

The least common denominator (LCD) is $3 \cdot 2$, or 6.

Multiply the numerator and denominator of $\frac{2}{3}$ by 2.

$$\begin{array}{r}\frac{13}{3\cdot 2}\\-\frac{2\cdot 2}{3\cdot 2}\\\hline\end{array} \qquad \begin{array}{r}\frac{13}{6}\\-\frac{4}{6}\\\hline\frac{9}{6}\end{array} = 1\frac{3}{6}, \text{ or } 1\frac{1}{2}$$

So, $\frac{13}{6} - \frac{2}{3} = 1\frac{1}{2}$.

practice ▷ Add or subtract. Simplify if possible.

5. $\frac{7}{6} + \frac{1}{3}$
6. $\frac{4}{5} + \frac{8}{15}$
7. $\frac{17}{10} - \frac{1}{2}$
8. $\frac{13}{6} - \frac{1}{2}$

Example 3

Add or subtract. Simplify if possible.

$$\frac{1}{3} + \frac{1}{2} \qquad\qquad \frac{1}{2} - \frac{1}{5}$$

Each denominator is prime. Neither 3 nor 2 can be factored.

$\frac{1}{3} + \frac{1}{2}$ $\qquad$ $\frac{1}{2} - \frac{1}{5}$

needs 2   needs 3 $\qquad$ needs 5   needs 2

The LCD is $3 \cdot 2$, or 6.

$$\frac{1\cdot 2}{3\cdot 2} + \frac{1\cdot 3}{2\cdot 3} \qquad\qquad \frac{1\cdot 5}{2\cdot 5} - \frac{1\cdot 2}{5\cdot 2}$$

$$\frac{2}{6} + \frac{3}{6} \qquad\qquad \frac{5}{10} - \frac{2}{10}$$

$$\frac{5}{6} \qquad\qquad \frac{3}{10}$$

So, $\frac{1}{3} + \frac{1}{2} = \frac{5}{6}$ and $\frac{1}{2} - \frac{1}{5} = \frac{3}{10}$.

practice ▷ Add or subtract. Simplify if possible.

9. $\frac{1}{5} + \frac{1}{3}$
10. $\frac{1}{3} + \frac{3}{7}$
11. $\frac{1}{2} - \frac{1}{7}$
12. $\frac{1}{2} - \frac{2}{5}$

◇ ORAL EXERCISES ◇

Substitute for n the factor needed to make the denominators the same.

1. $\dfrac{1n}{3n} + \dfrac{5}{3 \cdot 2}$
2. $\dfrac{1}{7 \cdot 2} + \dfrac{3n}{7n}$
3. $\dfrac{1n}{5n} + \dfrac{2}{5 \cdot 3}$
4. $\dfrac{1}{7 \cdot 2} + \dfrac{3n}{7n}$
5. $\dfrac{4n}{2n} - \dfrac{3}{2 \cdot 5}$
6. $\dfrac{15}{7 \cdot 3} - \dfrac{2n}{7n}$
7. $\dfrac{13}{5 \cdot 7} - \dfrac{4n}{5n}$
8. $\dfrac{3n}{5n} - \dfrac{4}{5 \cdot 7}$
9. $\dfrac{4}{5 \cdot 3} + \dfrac{9n}{5n}$
10. $\dfrac{6n}{2n} + \dfrac{7}{2 \cdot 3}$
11. $\dfrac{5}{7 \cdot 3} + \dfrac{6n}{7n}$
12. $\dfrac{8}{5 \cdot 5} + \dfrac{7n}{5n}$

◇ EXERCISES ◇

Add or subtract. Simplify if possible.

1. $\dfrac{1}{6} + \dfrac{1}{2}$
2. $\dfrac{3}{7} + \dfrac{1}{14}$
3. $\dfrac{2}{9} + \dfrac{1}{3}$
4. $\dfrac{1}{3} + \dfrac{7}{15}$
5. $\dfrac{3}{14} + \dfrac{2}{7}$

6. $\dfrac{7}{10} - \dfrac{1}{2}$
7. $\dfrac{2}{3} - \dfrac{1}{6}$
8. $\dfrac{11}{15} - \dfrac{2}{5}$
9. $\dfrac{7}{9} - \dfrac{2}{3}$
10. $\dfrac{5}{4} - \dfrac{1}{2}$

11. $\dfrac{5}{6} + \dfrac{1}{2}$
12. $\dfrac{8}{9} + \dfrac{2}{3}$
13. $\dfrac{7}{5} + \dfrac{3}{10}$
14. $\dfrac{19}{10} - \dfrac{1}{2}$
15. $\dfrac{27}{14} - \dfrac{3}{7}$
16. $\dfrac{15}{4} - \dfrac{1}{2}$

17. $\dfrac{1}{7} + \dfrac{1}{5}$
18. $\dfrac{1}{3} + \dfrac{1}{7}$
19. $\dfrac{2}{3} + \dfrac{1}{2}$
20. $\dfrac{1}{3} + \dfrac{2}{5}$
21. $\dfrac{2}{3} + \dfrac{3}{5}$

22. $\dfrac{1}{2} - \dfrac{2}{5}$
23. $\dfrac{2}{3} - \dfrac{1}{2}$
24. $\dfrac{4}{5} - \dfrac{1}{3}$
25. $\dfrac{8}{5} - \dfrac{1}{6}$
26. $\dfrac{5}{2} - \dfrac{3}{7}$

27. $\dfrac{1}{3} + \dfrac{2}{15}$
28. $\dfrac{6}{7} + \dfrac{5}{14}$
29. $\dfrac{19}{14} - \dfrac{1}{2}$
30. $\dfrac{3}{5} + \dfrac{1}{2}$
★ 31. $\dfrac{13}{38} + \dfrac{3}{19}$
★ 32. $\dfrac{22}{35} - \dfrac{3}{7}$

Solve these problems.

33. David practices the flute $\dfrac{1}{2}$ h in the morning and $\dfrac{3}{4}$ h in the evening. How much longer does he practice in the evening?

34. Myra read $\dfrac{3}{10}$ of a book on Saturday and $\dfrac{1}{5}$ of the book on Sunday. How much of the book did she read?

Addition and Subtraction

◇ OBJECTIVE ◇
To add and subtract fractions with no denominator prime

◇ RECALL ◇
The LCD for $\frac{1}{2 \cdot 3} + \frac{1}{3}$ is $2 \cdot 3$.

Example 1

Add. Simplify if possible.
$$\frac{5}{6} + \frac{3}{4}$$

$\frac{5}{6} + \frac{3}{4}$

$6 = 3 \cdot 2 \qquad 4 = 2 \cdot 2 \qquad \frac{5}{3 \cdot 2} + \frac{3}{2 \cdot 2}$

At most there are one 3 and two 2's in any denominator. The LCD is $2 \cdot 2 \cdot 3$.

Each denominator needs $2 \cdot 2 \cdot 3$ to be the same.

$$\frac{5}{3 \cdot 2} + \frac{3}{2 \cdot 2}$$
$\qquad \uparrow \qquad\quad \uparrow$
needs 2   needs 3

$$\frac{5 \cdot 2}{3 \cdot 2 \cdot 2} + \frac{3 \cdot 3}{2 \cdot 2 \cdot 3}$$

$$\frac{10}{12} + \frac{9}{12}$$

$\frac{10}{12} + \frac{9}{12} = \frac{10 + 9}{12} = \frac{19}{12}$ $\qquad \frac{19}{12}$

$\qquad\qquad\qquad\qquad\qquad\qquad\qquad 1\frac{7}{12}$

$\frac{7}{12}$ is in simplest form. So, $\frac{5}{6} + \frac{3}{4} = 1\frac{7}{12}$.

---

practice ▷ Add. Simplify if possible.

1. $\frac{5}{6}$
   $+\frac{2}{9}$

2. $\frac{3}{10}$
   $+\frac{3}{4}$

3. $\frac{1}{6}$
   $+\frac{1}{4}$

4. $\frac{4}{9}$
   $+\frac{5}{6}$

Example 2

Subtract. Simplify if possible. $\frac{7}{10} - \frac{1}{4}$

$$\frac{7}{10} - \frac{1}{4}$$

$10 = 5 \cdot 2 \qquad 4 = 2 \cdot 2$

$$\frac{7}{5 \cdot 2} - \frac{1}{2 \cdot 2}$$

Each denominator needs $2 \cdot 2 \cdot 5$ to be the same.   needs 2   needs 5

$$\frac{7 \cdot 2}{5 \cdot 2 \cdot 2} - \frac{1 \cdot 5}{2 \cdot 2 \cdot 5}$$

$$\frac{14}{20} - \frac{5}{20}$$

$\frac{9}{20}$ is in simplest form. $\frac{9}{20}$

So, $\frac{7}{10} - \frac{1}{4} = \frac{9}{20}$.

practice ▷ Subtract. Simplify if possible.

5. $\frac{5}{6} - \frac{1}{4}$ 6. $\frac{9}{10} - \frac{1}{4}$ 7. $\frac{5}{9} - \frac{1}{6}$ 8. $\frac{5}{6} - \frac{1}{10}$

Example 3

Add. Simplify if possible. $\frac{5}{6} + \frac{7}{12} + \frac{1}{3}$

$6 = 3 \cdot 2$
$12 = 2 \cdot 2 \cdot 3 \qquad$ 3 is prime.

At most there are two 2's and one 3 in any denominator. The LCD is $2 \cdot 2 \cdot 3$.

$$\frac{5}{6} + \frac{7}{12} + \frac{1}{3}$$

$$\frac{5}{3 \cdot 2} + \frac{7}{2 \cdot 2 \cdot 3} + \frac{1}{3}$$

needs 2   has $2 \cdot 2 \cdot 3$ needs $2 \cdot 2$

$$\frac{5 \cdot 2}{3 \cdot 2 \cdot 2} + \frac{7}{3 \cdot 2 \cdot 2} + \frac{1 \cdot 2 \cdot 2}{3 \cdot 2 \cdot 2}$$

$$\frac{10}{12} + \frac{7}{12} + \frac{4}{12}$$

$\frac{10}{12} + \frac{7}{12} + \frac{4}{12} = \frac{10 + 7 + 4}{12} = \frac{21}{12}$

$\frac{21}{12} = 1\frac{9}{12} = 1\frac{3}{4}$

$\frac{9}{12} = \frac{3 \cdot \cancel{3}}{2 \cdot 2 \cdot \cancel{3}} = \frac{3}{4}$ So, $\frac{5}{6} + \frac{7}{12} + \frac{1}{3} = 1\frac{3}{4}$.

practice ▷ Add. Simplify if possible.

9. $\frac{5}{12} + \frac{3}{4} + \frac{1}{2}$ 10. $\frac{1}{6} + \frac{3}{4} + \frac{1}{2}$ 11. $\frac{1}{2} + \frac{1}{4} + \frac{9}{20}$

◇ EXERCISES ◇

Add or subtract. Simplify if possible.

1. $\dfrac{5}{6} + \dfrac{1}{4}$
2. $\dfrac{1}{6} + \dfrac{4}{9}$
3. $\dfrac{1}{4} + \dfrac{9}{10}$
4. $\dfrac{3}{10} + \dfrac{5}{6}$
5. $\dfrac{7}{15} + \dfrac{1}{6}$
6. $\dfrac{7}{10} + \dfrac{2}{15}$

7. $\dfrac{3}{4} - \dfrac{1}{6}$
8. $\dfrac{7}{9} - \dfrac{1}{6}$
9. $\dfrac{5}{6} - \dfrac{3}{4}$
10. $\dfrac{3}{10} - \dfrac{1}{4}$
11. $\dfrac{17}{18} - \dfrac{5}{6}$
12. $\dfrac{7}{12} - \dfrac{1}{4}$

13. $\dfrac{2}{3} + \dfrac{5}{12} + \dfrac{1}{6}$
14. $\dfrac{3}{4} + \dfrac{1}{6} + \dfrac{7}{12}$
15. $\dfrac{5}{6} + \dfrac{3}{4} + \dfrac{1}{2}$
16. $\dfrac{5}{8} + \dfrac{3}{4} + \dfrac{1}{2}$

17. $\dfrac{7}{12} + \dfrac{1}{6} + \dfrac{1}{2}$
18. $\dfrac{7}{20} + \dfrac{1}{2} + \dfrac{3}{4}$
19. $\dfrac{5}{8} + \dfrac{1}{12} + \dfrac{3}{4}$
20. $\dfrac{8}{15} + \dfrac{3}{10} + \dfrac{1}{4}$

Solve these problems.

21. Tom babysat for $\dfrac{3}{4}$ h on Friday, $\dfrac{1}{2}$ h on Saturday, and $\dfrac{3}{4}$ h on Sunday. For how many hours did he babysit?

22. A pharmacist wants to reduce a compound of $\dfrac{7}{8}$ kg by $\dfrac{1}{4}$ kg. How much will be left in the compound?

Reading in Math

In each group of four mathematical terms below, one does not belong. Identify the term that does not belong.

1. variable — constant — denominator — expression
2. open sentence — equation — fraction — expression
3. multiply — sum — add — subtract
4. commutative — add — associative — distributive
5. multiply — formula — variable — value
6. LCM — subtract — LCD — multiple

ADDITION AND SUBTRACTION

Mixed Numbers

◇ **OBJECTIVE** ◇

To add and subtract mixed numbers

◇ **RECALL** ◇

$\frac{10}{6} = 1\frac{4}{6}$, or $1\frac{2}{3}$

Example 1

The denominators are the same.

Add or subtract. Simplify if possible.

$3\frac{1}{8} + 6\frac{5}{8}$ $\qquad$ $9\frac{5}{6} - 7\frac{1}{6}$

Add fractions.	Subtract fractions.
Add whole numbers.	Subtract whole numbers.

$\frac{6}{8} = \frac{3 \cdot \cancel{2}}{\cancel{2} \cdot 2 \cdot 2} = \frac{3}{4}$ $\qquad$ $\frac{4}{6} = \frac{2 \cdot 2}{\cancel{2} \cdot 3} = \frac{2}{3}$

$\begin{array}{r} 3\frac{1}{8} \\ +6\frac{5}{8} \\ \hline \frac{6}{8} \end{array}$ $\quad$ $\begin{array}{r} 3\frac{1}{8} \\ +6\frac{5}{8} \\ \hline 9\frac{6}{8} \end{array}$, or $9\frac{3}{4}$

$\begin{array}{r} 9\frac{5}{6} \\ -7\frac{1}{6} \\ \hline \frac{4}{6} \end{array}$ $\quad$ $\begin{array}{r} 9\frac{5}{6} \\ -7\frac{1}{6} \\ \hline 2\frac{4}{6} \end{array}$, or $2\frac{2}{3}$

So, $3\frac{1}{8} + 6\frac{5}{8} = 9\frac{3}{4}$ and $9\frac{5}{6} - 7\frac{1}{6} = 2\frac{2}{3}$.

---

practice ▷ Add or subtract. Simplify if possible.

1. $5\frac{1}{10} + 3\frac{1}{10}$
2. $6\frac{1}{8} + 7\frac{3}{8}$
3. $9\frac{11}{12} - 4\frac{5}{12}$
4. $8\frac{7}{8} - 1\frac{1}{8}$

Example 2

Subtract. Simplify if possible. $8\frac{3}{4} - 2\frac{1}{6}$

$\begin{array}{r} 8\frac{3}{4} \\ -2\frac{1}{6} \end{array}$ $\quad$ $\begin{array}{r} 8\frac{3}{2 \cdot 2} \\ -2\frac{1}{3 \cdot 2} \end{array}$

Each denominator needs $2 \cdot 2 \cdot 3$ to be the same.

The LCD is $2 \cdot 2 \cdot 3$.

$\begin{array}{r} 8\frac{3}{2 \cdot 2} \leftarrow \text{needs } 3 \\ -2\frac{1}{3 \cdot 2} \leftarrow \text{needs } 2 \end{array}$ $\quad$ $\begin{array}{r} 8\frac{3 \cdot 3}{2 \cdot 2 \cdot 3} \\ -2\frac{1 \cdot 2}{3 \cdot 2 \cdot 2} \end{array}$ $\quad$ $\begin{array}{r} 8\frac{9}{12} \\ -2\frac{2}{12} \\ \hline 6\frac{7}{12} \end{array}$

So, $8\frac{3}{4} - 2\frac{1}{6} = 6\frac{7}{12}$.

practice Add or subtract. Simplify if possible.

5. $3\frac{1}{4} + 2\frac{1}{6}$ 6. $6\frac{1}{5} + 3\frac{1}{15}$ 7. $7\frac{5}{6} - 3\frac{3}{4}$ 8. $2\frac{3}{4} - \frac{1}{8}$

Example 3   Add. $6 + 4\frac{1}{2}$ Subtract. $7\frac{5}{6} - 4$

$6 + 4\frac{1}{2} = (6 + 4) + \frac{1}{2} = 10 + \frac{1}{2}$, or $10\frac{1}{2}$ $6 + 4\frac{1}{2} = 10\frac{1}{2}$ $7\frac{5}{6} - 4 = 3\frac{5}{6}$

$7\frac{5}{6} - 4 = 7 + \frac{5}{6} - 4 = 3 + \frac{5}{6}$, or $3\frac{5}{6}$ So, $6 + 4\frac{1}{2} = 10\frac{1}{2}$ and $7\frac{5}{6} - 4 = 3\frac{5}{6}$.

---

practice Add or subtract.

9. $4\frac{3}{4} + 7$ 10. $3 + 5\frac{7}{8}$ 11. $9\frac{3}{5} - 6$ 12. $8\frac{9}{10} - 1$

Example 4   Add and simplify. $2\frac{1}{3} + 3\frac{5}{6} + 4\frac{1}{2}$

Factor the denominators. $2\frac{1}{3}$ $2\frac{1}{3}$
3 is prime.

$6 = 3 \cdot 2$ $3\frac{5}{6}$ $3\frac{5}{3 \cdot 2}$ Each denominator needs $3 \cdot 2$ to be the same. The LCD is $3 \cdot 2$.

2 is prime. $+ 4\frac{1}{2}$ $+ 4\frac{1}{2}$

Multiply the numerator and denominator by 2. $2\frac{1}{3}$ ← needs 2 $2\frac{1 \cdot 2}{3 \cdot 2}$ $2\frac{2}{6}$

$3\frac{5}{3 \cdot 2}$ ← has $3 \cdot 2$ $3\frac{5}{3 \cdot 2}$ $3\frac{5}{6}$

Multiply the numerator and denominator by 3. $+ 4\frac{1}{2}$ ← needs 3 $+ 4\frac{1 \cdot 3}{2 \cdot 3}$ $+ 4\frac{3}{6}$

$9\frac{10}{6}$

$9\frac{10}{6}$ means $9 + \frac{10}{6} = 9 + 1\frac{4}{6} = 10\frac{4}{6}$, or $10\frac{2}{3}$.

So, $2\frac{1}{3} + 3\frac{5}{6} + 4\frac{1}{2} = 10\frac{2}{3}$.

---

practice Add. Simplify if possible.

13. $3\frac{1}{5} + 2\frac{7}{10} + 5\frac{1}{2}$ 14. $\frac{2}{5} + 2\frac{7}{15} + 6\frac{1}{3}$ 15. $4\frac{1}{2} + 7\frac{1}{8} + 1\frac{3}{4}$

MIXED NUMBERS

◇ EXERCISES ◇

Add or subtract. Simplify if possible.

1. $7\frac{1}{10} + 3\frac{3}{10}$
2. $8\frac{1}{9} + 6\frac{5}{9}$
3. $7\frac{1}{8} + 6\frac{3}{8}$
4. $9\frac{7}{15} - 4\frac{2}{15}$
5. $8\frac{3}{4} - 7\frac{1}{4}$
6. $8\frac{3}{10} + \frac{1}{2}$
7. $8\frac{7}{10} + 4\frac{1}{4}$
8. $4\frac{3}{5} + \frac{1}{3}$
9. $3\frac{3}{8} + 2\frac{1}{4}$
10. $7\frac{1}{6} + 3\frac{5}{9}$
11. $7\frac{1}{2} - 2\frac{1}{4}$
12. $8\frac{2}{3} - 5\frac{1}{6}$
13. $7\frac{7}{8} - \frac{3}{4}$
14. $9\frac{4}{5} - 3\frac{1}{2}$
15. $4\frac{5}{6} - 3\frac{1}{4}$
16. $9 + 3\frac{1}{2}$
17. $7\frac{1}{4} - 2$
18. $6\frac{5}{8} + 4$
19. $9 + 6\frac{1}{2}$
20. $7\frac{3}{4} - 2$
21. $2\frac{3}{5} + 5\frac{9}{10}$
22. $4\frac{1}{2} + 7\frac{5}{6}$
23. $7\frac{1}{4} + 9\frac{2}{3}$
24. $5\frac{2}{3} + 4\frac{5}{9}$
25. $8\frac{1}{2} + 6\frac{7}{8}$
26. $8\frac{2}{3} + 3\frac{1}{6} + 4\frac{1}{2}$
27. $\frac{3}{10} + 8\frac{1}{2} + 9\frac{2}{5}$
28. $6\frac{2}{3} + 5\frac{1}{2} + 3\frac{5}{6}$
29. $3\frac{13}{15} + 2\frac{1}{5} + 5\frac{1}{3}$
30. $7\frac{1}{8} + 6\frac{3}{4} + \frac{1}{2}$
31. $7\frac{9}{10} + 4\frac{3}{5} + 1\frac{1}{2}$
32. $3\frac{3}{4} + 2\frac{1}{3} + 4\frac{5}{6}$
33. $2\frac{5}{8} + 8\frac{2}{3} + 6\frac{1}{4}$

Solve these problems.

34. At birth, a baby weighed $3\frac{3}{10}$ kilograms. Two months later, it weighed $4\frac{3}{4}$ kilograms. How much weight did it gain?

35. Marni's punchbowl holds $4\frac{5}{8}$ liters of punch. Gerard's punchbowl holds $1\frac{1}{4}$ liters more. How much punch does Gerard's bowl hold?

36. A grocer ordered $3\frac{1}{2}$ cases of soup on Monday and $7\frac{3}{4}$ cases of soup on Tuesday. How many cases of soup were ordered?

37. Abdul ran 3 kilometers in $9\frac{3}{4}$ minutes last week. This week, he ran 3 kilometers in $8\frac{1}{2}$ minutes. How many minutes less is this?

Challenge

Find the value of n in each case.

1. $\frac{1}{2} + \frac{1}{3} = \frac{35}{n}$
2. $1\frac{2}{3} + 3\frac{3}{4} = 4\frac{n}{24}$
3. $\frac{5}{6} - \frac{1}{2} = \frac{n}{15}$
4. $7\frac{1}{2} - 2\frac{3}{4} = 3\frac{28}{n}$

Problem Solving – Applications
Jobs for Teenagers

1. Kim has a new job as a lifeguard. To get in shape, she swam $8\frac{3}{4}$ laps on Monday. She swam $7\frac{3}{4}$ laps on Tuesday. Find the total laps she swam.

2. José's swimming coach told him to swim 65 laps this weekend. He swam $15\frac{1}{4}$ laps on Friday and $40\frac{1}{4}$ laps on Saturday. How many laps must he swim on Sunday to make his goal?

3. Sue's softball coach told her to run 30 laps this weekend. She ran $14\frac{1}{2}$ laps on Saturday morning and $10\frac{1}{4}$ laps on Saturday afternoon. How many laps must she run on Sunday to make the 30-lap goal?

4. Rudy is manager of the track team. He needed to lose weight. He lost $1\frac{1}{2}$ kg the first week, 2 kg the second week, and $1\frac{1}{4}$ kg the third week. He now weighs $71\frac{1}{4}$ kg. How much did he weigh before?

5. Martha weighed 60 kg. Her soccer coach wanted her to reduce her weight. She lost $1\frac{1}{2}$ kg each week for 3 weeks. How much does she weigh now?

6. Peter is supposed to practice 10 h this weekend for a tennis match. He practiced $2\frac{1}{2}$ h on Saturday morning, $1\frac{1}{4}$ h that afternoon, and 3 h on Sunday morning. How many more hours must he practice on Sunday?

PROBLEM SOLVING

Renaming in Subtraction

◇ **OBJECTIVE** ◇

To subtract mixed numbers using renaming

◇ **RECALL** ◇

Regroup 1 ten as 10 ones.

$$483 \atop -219$$

$$\overset{7\ 13}{4\cancel{8}\cancel{3}} \atop {-219 \over 264}$$

Example 1

Rename $6\frac{3}{10}$ as $5\frac{n}{10}$ by replacing n.

$$6\frac{3}{10} = 6 + \frac{3}{10}$$

Rename 6 as 5 + 1. $= 5 + 1 + \frac{3}{10}$

Rename 1 as $\frac{10}{10}$. $= 5 + \frac{10}{10} + \frac{3}{10}$

$\frac{10}{10} + \frac{3}{10} = \frac{13}{10}$ $= 5 + \frac{13}{10}$

So, $6\frac{3}{10} = 5\frac{13}{10}$.

---

practice ▷ **Rename by replacing n.**

1. $8\frac{1}{3} = 7\frac{n}{3}$

2. $7\frac{4}{5} = 6\frac{n}{5}$

3. $9\frac{1}{6} = 8\frac{n}{6}$

Example 2

Subtract. $9\frac{2}{5} - 2\frac{4}{5}$

Write in vertical form.

$$9\frac{2}{5} \atop -2\frac{4}{5}$$

Think: Cannot subtract $\frac{4}{5}$ from $\frac{2}{5}$; Rename $9\frac{2}{5}$.

$$8\frac{7}{5} \atop -2\frac{4}{5} \atop 6\frac{3}{5}$$

$9\frac{2}{5} = 8 + 1 + \frac{2}{5} = 8 + \frac{5}{5} + \frac{2}{5} = 8 + \frac{7}{5} = 8\frac{7}{5}$

So, $9\frac{2}{5} - 2\frac{4}{5} = 6\frac{3}{5}$.

CHAPTER FIVE

Example 3

Subtract and simplify. $6\frac{1}{3} - 2\frac{5}{6}$

Factor denominators; 3 is prime.
The LCD is $3 \cdot 2$.

$$6\frac{1}{3}$$
$$-2\frac{5}{6}$$

Step 1
$$6\frac{1}{3}$$
$$-2\frac{5}{3 \cdot 2}$$

Step 2
$$6\frac{1 \cdot 2}{3 \cdot 2}$$
$$-2\frac{5}{3 \cdot 2}$$

$6\frac{2}{6} = 6 + \frac{2}{6}$

$5 + 1 + \frac{2}{6}$

$5 + \frac{6}{6} + \frac{2}{6} = 5\frac{8}{6}$

Step 3
$$6\frac{2}{6}$$
$$-2\frac{5}{6}$$

Step 4
$$5\frac{8}{6}$$
$$-2\frac{5}{6}$$
$$3\frac{3}{6}, \text{ or } 3\frac{1}{2}$$

So, $6\frac{1}{3} - 2\frac{5}{6} = 3\frac{1}{2}$

practice ▷ Subtract. Simplify if possible.

4. $7\frac{1}{10} - 3\frac{3}{5}$ 5. $4\frac{1}{3} - 2\frac{5}{6}$ 6. $7\frac{1}{4} - 4\frac{1}{2}$ 7. $5\frac{3}{5} - 2\frac{7}{10}$

Example 4

Subtract. $8 - 5\frac{2}{3}$

$$8$$
$$-5\frac{2}{3}$$

Think: Rename 8 as $7\frac{3}{3}$.

$$7\frac{3}{3}$$
$$-5\frac{2}{3}$$
$$2\frac{1}{3}$$

So, $8 - 5\frac{2}{3} = 2\frac{1}{3}$.

practice ▷ Subtract.

8. $7 - 3\frac{1}{5}$ 9. $9 - 4\frac{3}{4}$ 10. $7 - 6\frac{1}{3}$ 11. $8 - 3\frac{1}{2}$

RENAMING IN SUBTRACTION

◇ ORAL EXERCISES ◇

Rename by replacing *n*.

1. $4\frac{1}{5} = 3\frac{n}{5}$
2. $5\frac{1}{8} = 4\frac{n}{8}$
3. $9\frac{2}{3} = 8\frac{n}{3}$
4. $6\frac{1}{2} = 5\frac{n}{2}$
5. $7\frac{3}{4} = 6\frac{n}{4}$
6. $9\frac{1}{6} = 8\frac{n}{6}$
7. $8\frac{1}{4} = 7\frac{n}{4}$
8. $5\frac{1}{3} = 4\frac{n}{3}$
9. $9\frac{3}{4} = 8\frac{n}{4}$
10. $7\frac{1}{9} = 6\frac{n}{9}$

◇ EXERCISES ◇

Subtract. Simplify if possible.

1. $3\frac{1}{4} - 1\frac{3}{4}$
2. $5\frac{1}{6} - 2\frac{5}{6}$
3. $7\frac{1}{9} - 3\frac{4}{9}$
4. $6\frac{1}{10} - 4\frac{7}{10}$
5. $3\frac{3}{8} - 1\frac{5}{8}$
6. $8\frac{1}{12} - 3\frac{5}{12}$

7. $8\frac{1}{6} - 4\frac{1}{2}$
8. $5\frac{3}{5} - 2\frac{9}{10}$
9. $7\frac{1}{6} - 4\frac{2}{3}$
10. $9\frac{7}{10} - 3\frac{4}{5}$
11. $8\frac{1}{4} - 3\frac{1}{2}$

12. $5\frac{1}{6} - 2\frac{1}{3}$
13. $3\frac{1}{8} - 2\frac{3}{4}$
14. $7\frac{1}{9} - 3\frac{2}{3}$
15. $8\frac{1}{5} - \frac{3}{10}$
16. $8\frac{1}{3} - 2\frac{5}{6}$

17. $6 - 2\frac{1}{3}$
18. $6 - 3\frac{2}{5}$
19. $8 - 7\frac{1}{2}$
20. $6 - 5\frac{1}{4}$
21. $9 - 4\frac{5}{6}$

Solve these problems.

22. Josh is on a diet. One Monday he weighed 70 kg. The following Monday, he weighed $68\frac{1}{4}$ kg. How much weight did he lose?

23. A pair of slacks is $80\frac{1}{4}$ cm long. They must be shortened by $8\frac{1}{2}$ cm. How long will they be then?

24. Elke had 5 kg of potatoes. She cooked $2\frac{3}{8}$ kg of potatoes. How many kg of potatoes are left?

25. The Wilsons are on a trip of $85\frac{3}{10}$ km. They have gone $67\frac{1}{2}$ km. How much farther must they go?

Calculator

Estimate the number of digits in the answer. Check with a calculator.

1. $4{,}628 \div 52$
2. $102{,}764 \div 46$
3. $29{,}880 \div 72$

CHAPTER FIVE

Evaluating Expressions

◆ OBJECTIVE ◆

To evaluate expressions like
$\frac{3x - 5}{5} + \frac{2}{15}$ if $x = 2$

◆ RECALL ◆

Evaluate $3a + 4$ if $a = 5$.
$3 \cdot 5 + 4$ Substitute 5 for a.
$15 + 4$ or 19

Example 1 Evaluate $\frac{2a}{9} + \frac{2}{9}$ if $a = 3$.

Substitute 3 for a. $\frac{2 \cdot 3}{9} + \frac{2}{9}$

$\frac{6}{9} + \frac{2}{9}$

Denominators are the same; add the numerators. $\frac{8}{9}$

So, the value of $\frac{2a}{9} + \frac{2}{9}$ if $a = 3$ is $\frac{8}{9}$.

Example 2 Evaluate $\frac{2x + 1}{21} + \frac{5}{21}$ if $x = 4$. Simplify if possible.

Substitute 4 for x. $\frac{2 \cdot 4 + 1}{21} + \frac{5}{21}$

$2 \cdot 4 + 1 = 9$ $\frac{9}{21} + \frac{5}{21}$

Denominators are the same; add the numerators. $\frac{14}{21}$

Factor 14 into primes.
Factor 21 into primes. $\frac{7 \cdot 2}{7 \cdot 3}$

Divide out the like factors. $\frac{\cancel{7} \cdot 2}{\cancel{7} \cdot 3}$ or $\frac{2}{3}$

So, the value of $\frac{2x + 1}{21} + \frac{5}{21}$ if $x = 4$ is $\frac{2}{3}$.

---

practice ▷ Evaluate. Simplify if possible.

1. $\frac{2a}{11} + \frac{4}{11}$ if $a = 3$
2. $\frac{5x}{20} + \frac{1}{20}$ if $x = 3$
3. $\frac{1}{30} + \frac{3m + 5}{30}$ if $m = 2$

Example 3 Evaluate $\frac{2}{15} + \frac{3x-5}{5}$ if $x = 4$. Simplify if possible.

Substitute 4 for x. $\frac{2}{15} + \frac{3 \cdot 4 - 5}{5}$

$3 \cdot 4 - 5 = 12 - 5 = 7$. $\frac{2}{15} + \frac{7}{5}$

Factor 15 into primes. $\frac{2}{5 \cdot 3} + \frac{7}{5}$

Multiply numerator and denominator of $\frac{7}{5}$ by 3; the LCD is $5 \cdot 3$. $\frac{2}{5 \cdot 3} + \frac{7 \cdot 3}{5 \cdot 3}$

$\frac{2}{15} + \frac{21}{15}$

$15 \overline{)23} \quad 1\frac{8}{15}$

$\frac{23}{15} = 1\frac{8}{15}$

So, the value of $\frac{2}{15} + \frac{3x-5}{5}$ if $x = 4$ is $1\frac{8}{15}$.

practice Evaluate. Simplify if possible.

4. $\frac{1}{4} + \frac{2x+1}{12}$ if $x = 2$

5. $\frac{2x-5}{24} + \frac{3}{8}$ if $x = 5$

♦ ORAL EXERCISES ♦

Evaluate.

1. $\frac{3x}{4}$ if $x = 4$

2. $\frac{3y}{5}$ if $y = 10$

3. $\frac{4n}{7}$ if $n = 14$

♦ EXERCISES ♦

Evaluate. Simplify if possible.

1. $\frac{4x}{13} + \frac{5}{13}$ if $x = 1$

2. $\frac{7}{15} + \frac{2a}{15}$ if $a = 3$

3. $\frac{2a-1}{11} + \frac{1}{11}$ if $a = 4$

4. $\frac{2x-3}{20} + \frac{3}{20}$ if $x = 5$

5. $\frac{7y-10}{9} + \frac{2}{9}$ if $y = 2$

6. $\frac{7}{30} + \frac{5x-14}{30}$ if $x = 3$

7. $\frac{7t-13}{6} + \frac{5}{24}$ if $t = 2$

8. $\frac{7}{9} + \frac{2k-11}{3}$ if $k = 6$

9. $\frac{7}{18} + \frac{3y-14}{9}$ if $y = 5$

10. $\frac{1}{2} + \frac{7a-25}{10}$ if $a = 4$

11. $\frac{3x-4}{3} + \frac{2}{15}$ if $x = 2$

12. $\frac{7}{30} + \frac{4a-23}{10}$ if $a = 6$

★ 13. $\frac{3x+1}{40} + \frac{5x-9}{5}$ if $x = 2$

★ 14. $\frac{7m-9}{32} + \frac{9m-26}{8}$ if $m = 3$

Problem Solving – Careers
Bookkeepers

A bookkeeper prepares the payroll. Find the total number of hours the bookkeeper should record.

1. John Hendrickson

Mon.	Tues.	Wed.	Thurs.	Fri.
$8\frac{1}{2}$	$9\frac{1}{4}$	$7\frac{3}{4}$	—	—

2. Joanne Romano

Mon.	Tues.	Wed.	Thurs.	Fri.
$8\frac{1}{2}$	$8\frac{3}{4}$	9	$8\frac{1}{4}$	$8\frac{1}{2}$

A bookkeeper checks for mistakes in bills sent to a business. Is the total correct?

3. From: Hogg's Meats
   To: Jan's Deli
   For: Purchase of frankfurters

July 15	Aug. 1	Aug. 15	Total
$17\frac{1}{2}$ kg	$24\frac{1}{4}$ kg	$15\frac{3}{4}$ kg	$60\frac{1}{2}$ kg

4. From: ABC Textiles
   To: Fabric Yard
   For: Purchase of cotton fabric

Blue	Orange	Aqua	Total
$10\frac{1}{2}$ m	$20\frac{3}{8}$ m	$17\frac{1}{4}$ m	$48\frac{1}{8}$ m

A firm pays overtime for all hours over 40 hours per week. How many hours overtime should Mary record for each person?

5. Bill O'Neill

Mon.	Tues.	Wed.	Thurs.	Fri.
$9\frac{3}{4}$	$8\frac{1}{2}$	$7\frac{3}{4}$	$9\frac{1}{2}$	8

6. Margarita Juarez

Mon.	Tues.	Wed.	Thurs.	Fri.
$9\frac{1}{4}$	8	$8\frac{1}{2}$	$8\frac{3}{4}$	$9\frac{1}{2}$

PROBLEM SOLVING

Equations

◇ OBJECTIVES ◇
To solve equations whose solutions are fractions or mixed numbers

To solve equations that contain fractions

◇ RECALL ◇
Solve	$x + 3 = 9.$
Subtract 3 from each side.	$x + 3 - 3 = 9 - 3$
So, the solution is 6.	$x = 6$

Example 1

Solve $3x + 5 = 12$. Check the solution.

Subtract 5 from each side. $\quad 3x + 5 - 5 = 12 - 5$
$$3x = 7$$

Divide each side by 3. $\quad \dfrac{3x}{3} = \dfrac{7}{3}$

$$x = \dfrac{7}{3}, \text{ or } 2\dfrac{1}{3}.$$

Check: $\quad 3x + 5 = 12$

$\dfrac{7}{3}$ is a convenient form for checking. $\quad 3 \cdot \dfrac{7}{3} + 5 \;\Big|\; 12$

$\qquad\qquad\qquad\qquad\qquad\qquad 7 + 5 \;\Big|\;$

$\qquad\qquad\qquad\qquad\qquad\qquad\qquad 12 = 12$

So, $\dfrac{7}{3}$, or $2\dfrac{1}{3}$ is the solution of $3x + 5 = 12$.

Example 2

Solve $5y - 6 = 5$. Check the solution.

Add 6 to each side. $\quad 5y - 6 + 6 = 5 + 6$
$$5y = 11$$

Divide each side by 5. $\quad \dfrac{5y}{5} = \dfrac{11}{5}$

$$y = \dfrac{11}{5}, \text{ or } 2\dfrac{1}{5}$$

Check: $\quad 5y - 6 = 5$

$\dfrac{11}{5}$ is a convenient form for checking. $\quad 5 \cdot \dfrac{11}{5} - 6 \;\Big|\; 5$

$\qquad\qquad\qquad\qquad\qquad\qquad 11 - 6 \;\Big|\;$

$\qquad\qquad\qquad\qquad\qquad\qquad\qquad 5 = 5$

So, $\dfrac{11}{5}$, or $2\dfrac{1}{5}$ is the solution of $5y - 6 = 5$.

practice ▷ Solve. Check the solution.

1. $2x + 7 = 16$
2. $5y + 9 = 11$
3. $3z + 4 = 12$
4. $3x - 2 = 8$
5. $6y - 3 = 12$
6. $4z - 5 = 13$

Example 3

Solve $\frac{2}{3}x + 6 = 15$. Check the solution.

Subtract 6 from each side of the equation.
$$\frac{2}{3}x + 6 - 6 = 15 - 6$$
$$\frac{2}{3}x = 9$$

Multiply by the reciprocal of the coefficient of x.
$$\frac{3}{2} \cdot \frac{2}{3} = 1$$
$$\frac{3}{2} \cdot \frac{2}{3}x = \frac{3}{2} \cdot 9$$
$$x = \frac{27}{2}, \text{ or } 13\frac{1}{2}$$

$\frac{27}{2}$ is a convenient form for checking.

Check: $\frac{2}{3}x + 6 = 15$

$\frac{2}{3} \cdot \frac{27}{2} + 6 \;|\; 15$

$9 + 6$

$15 = 15$

So, $\frac{27}{2}$, or $13\frac{1}{2}$ is the solution of $\frac{2}{3}x + 6 = 15$.

---

practice ▷ Solve. Check the solution.

7. $\frac{4}{5}x + 7 = 9$
8. $\frac{3}{4}y + 1 = 8$
9. $\frac{1}{5}z + 7 = 11$

Example 4

Solve $\frac{4}{7}x - 2 = 9$. Check the solution.

Add 2 to each side.
$$\frac{4}{7}x - 2 + 2 = 9 + 2$$
$$\frac{4}{7}x = 11$$

Multiply each side by $\frac{7}{4}$.
$$\frac{7}{4} \cdot \frac{4}{7}x = \frac{7}{4} \cdot 11$$
$$x = \frac{77}{4}, \text{ or } 19\frac{1}{4}$$

$\frac{77}{4}$ is a convenient form for checking.

Check: $\frac{4}{7}x - 2 = 9$

$\frac{4}{7} \cdot \frac{77}{4} - 2 \;|\; 9$

$11 - 2$

$9 = 9$

So, $\frac{77}{4}$, or $19\frac{1}{4}$ is the solution of $\frac{4}{7}x - 2 = 9$.

EQUATIONS

practice ▷ Solve. Check the solution.

10. $\frac{2}{3}x - 1 = 2$ 11. $\frac{5}{6}y - 3 = 7$ 12. $\frac{3}{5}z - 5 = 2$

◇ ORAL EXERCISES ◇

Solve.

1. $x + \frac{3}{5} = \frac{4}{5}$
2. $x + \frac{1}{7} = \frac{6}{7}$
3. $x - \frac{1}{9} = \frac{4}{9}$
4. $x - \frac{1}{7} = \frac{2}{7}$
5. $x + \frac{1}{11} = \frac{7}{11}$
6. $x - \frac{1}{5} = \frac{2}{5}$
7. $x + \frac{2}{9} = \frac{7}{9}$
8. $x - \frac{1}{13} = \frac{3}{13}$
9. $x - \frac{1}{15} = \frac{7}{15}$
10. $x + \frac{1}{9} = \frac{8}{9}$
11. $x - \frac{3}{7} = \frac{2}{7}$
12. $x + \frac{2}{11} = \frac{9}{11}$

◇ EXERCISES ◇

Solve. Check the solution.

1. $4x + 7 = 9$
2. $3x + 6 = 7$
3. $5x + 9 = 15$
4. $4x + 1 = 21$
5. $5x + 3 = 12$
6. $9x + 2 = 4$
7. $6x - 2 = 1$
8. $7x - 4 = 2$
9. $8x - 5 = 12$
10. $9x - 3 = 10$
11. $5x - 2 = 4$
12. $12x - 4 = 7$
13. $\frac{1}{2}x + 5 = 8$
14. $\frac{3}{5}x + 9 = 14$
15. $\frac{7}{2}x + 7 = 9$
16. $\frac{4}{7}x - 1 = 2$
17. $\frac{9}{2}x - 3 = 15$
18. $\frac{3}{7}x - 6 = 11$
★ 19. $2x + 9\frac{1}{3} = 17\frac{2}{3}$
★ 20. $\frac{2}{3}x - 2\frac{3}{4} = 1\frac{1}{2}$
★ 21. $1\frac{1}{2}x + 4\frac{1}{3} = 9\frac{1}{4}$

To what single fraction does this simplify?

$$\cfrac{1}{1+\cfrac{1}{2+\cfrac{1}{3+\cfrac{1}{4+\cfrac{1}{5+\frac{1}{6}}}}}}$$

(*Hint:* Start at the bottom.)

$$\cfrac{1}{5+\cfrac{1}{\frac{1}{6}}} = \cfrac{1}{5\frac{1}{6}} = \cfrac{1}{\frac{31}{6}} = 1 \div \frac{31}{6} = \frac{1}{1} \cdot \frac{6}{31} = \frac{6}{31}$$

134 CHAPTER FIVE

Non-Routine Problems

There is no simple method used to solve these problems. Use whatever methods you can. Do not expect to solve these easily.

1. Your friend challenges you to use four 4's to make numbers, and she allows you to use any operations you wish. She gives you this beginning to get you going: $1 = \frac{4+4}{4+4}$, $2 = \frac{4+4}{\sqrt{4} \cdot \sqrt{4}}$. Recall that $\sqrt{4} = 2$ because $2 \cdot 2 = 4$. Now complete the work writing numbers 3 through 10. You must use *exactly* four 4's, no more no less.

2. Now you have the challenge of writing numbers using exactly three 9's. To make 10, you can write $9 + \frac{9}{9}$. To make 0, you can write $\frac{9-9}{9}$. Write the following numbers using exactly three 9's.
   a. 6
   b. 18
   c. 81
   d. 27
   e. 4
   f. 24
   g. 12
   h. 10

3. Write the following numbers using exactly three 6's. For example, to make 6, you can write $\frac{6 \cdot 6}{6}$. To make 2, you can write $\frac{6+6}{6}$.
   a. 0
   b. 6
   c. 60
   d. 18
   e. 11
   f. 30
   g. 7
   h. 5

4. You made the following transaction with your friend: You bought a pen from him for $4, sold it for $5, then bought it back for $6, and finally sold it for $7. Did you make any money? Did you lose?

5. The Carters have two children. The product of their ages is 32. The sum is 12. What are their ages?

6. I am thinking of a number. If the number is multiplied by 6, then the product divided by 9, and this result increased by 7, the final result is 15. What number am I thinking of?

7. Sandy has $8 in her savings account. Maria has $12. From now on, Sandy will add $1 to her account each week, and Maria will add $3. After how many weeks will Maria's account be twice as big as Sandy's?

COMPUTER ACTIVITIES

The first step in writing a program to add any two fractions is to come up with a general formula. This formula must tell the computer how to add two fractions, given their numerators and denominators.

Example 1 Write a formula for the sum of any two fractions $\frac{X}{Y}$ and $\frac{Z}{W}$.

$$S = \frac{X}{Y} + \frac{Z}{W}$$

First, find the LCD: $Y \cdot W$ $\qquad S = \frac{X \cdot W}{Y \cdot W} + \frac{Z \cdot Y}{W \cdot Y} \qquad S = \frac{X \cdot W + Z \cdot Y}{Y \cdot W}$

This formula for finding the sum of any two fractions is used in the next example to write a program for adding fractions.

Example 2 Write a program to add any two fractions $\frac{X}{Y}$ and $\frac{Z}{W}$, $Z \neq 0$, $W \neq 0$. Then **RUN** the program to find the sum $\frac{2}{3} + \frac{1}{10}$.

A is the numerator and B is the denominator of the sum in the formula from **Example 1**.

```
10  HOME
20  INPUT "WRITE 1ST NUMERATOR ";X
30  INPUT "WRITE 1ST DENOMINATOR ";Y
40  INPUT "WRITE 2ND NUMERATOR ";Z
50  INPUT "WRITE 2ND DENOMINATOR ";W
60  LET A = X * W + Y * Z
70  LET B = Y * W
80  PRINT : PRINT
90  PRINT X"/"Y" + "Z"/"W" = "A"/"B
100 END
```

See the Computer Section beginning on page 420 for more information.

Exercises

Find each sum on your own. Then use the program above to check.

1. $\frac{4}{7} + \frac{1}{3}$ 2. $\frac{2}{9} + \frac{3}{5}$ 3. $\frac{4}{5} + \frac{1}{3}$ 4. $\frac{3}{8} + \frac{1}{2}$

Notice that in Exercise 4 above, the result is not simplified. Modify the program in Example 1 above by adding the program from Chapter 4 on page 106. Use this new program to add each of the following and simplify the results.

5. $\frac{3}{8} + \frac{1}{2}$ 6. $\frac{1}{10} + \frac{1}{15}$ 7. $\frac{5}{12} + \frac{1}{4}$ 8. $\frac{3}{20} + \frac{3}{5}$

Chapter Review

Find the least common multiple (LCM). [114]

1. 6; 15
2. 4; 10
3. 4; 6; 8
4. 3; 5; 30

Add or subtract. Simplify if possible.

5. [111] $\frac{3}{10} + \frac{1}{10}$
6. [111] $\frac{7}{12} - \frac{5}{12}$
7. [116] $\frac{3}{5} + \frac{5}{15}$
8. [116] $\frac{3}{5} - \frac{1}{3}$

9. [116] $\frac{11}{6} - \frac{2}{3}$
10. [116] $\frac{3}{10} + \frac{4}{5}$
11. [116] $\frac{17}{6} - \frac{1}{2}$
12. [116] $\frac{5}{14} + \frac{1}{7}$

13. [119] $\frac{5}{6} + \frac{5}{9}$
14. [119] $\frac{7}{10} + \frac{3}{4}$
15. [119] $\frac{7}{10} - \frac{1}{4}$
16. [119] $\frac{5}{6} - \frac{3}{4}$

17. [122] $7\frac{3}{10} + 3\frac{1}{10}$
18. [122] $6\frac{3}{4} - 4\frac{1}{6}$
19. [122] $7\frac{1}{8} - 5$
20. [122] $9 + 4\frac{1}{3}$

21. [122] $9\frac{1}{3} - 6$
22. [122] $7\frac{2}{5} + 1\frac{3}{10} + 4\frac{1}{2}$
23. [126] $9\frac{1}{10} - 4\frac{3}{5}$

Evaluate. Simplify if possible. [129]

24. $\frac{5}{12} + \frac{6x-8}{12}$ if $x = 2$
25. $\frac{2t+1}{20} + \frac{1}{5}$ if $t = 5$

Solve. Simplify if possible. [132]

26. $\frac{3}{2}x + 5 = 9$
27. $4x - 6 = 12$
28. $\frac{4}{9}x - 5 = 1$

Solve these problems. [125]

29. John jogs $5\frac{1}{2}$ laps, rests, then jogs $3\frac{1}{4}$ laps. How many laps does he jog altogether?

30. Mary weighs $70\frac{1}{2}$ kg. How many kilograms must she lose in order to weigh $67\frac{1}{4}$ kg?

Chapter Test

Find the least common multiple (LCM).

1. 4; 6
2. 3; 5
3. 6; 10; 15

Add or subtract. Simplify if possible.

4. $\dfrac{5}{8} - \dfrac{1}{8}$
5. $\dfrac{4}{9} + \dfrac{5}{9}$
6. $\dfrac{1}{5} + \dfrac{2}{3}$

7. $\dfrac{11}{6} - \dfrac{1}{3}$
8. $\dfrac{3}{14} + \dfrac{1}{2}$
9. $\dfrac{1}{6} + \dfrac{2}{9}$
10. $\dfrac{3}{4} - \dfrac{1}{6}$

11. $\dfrac{7}{12} + \dfrac{2}{3} + \dfrac{1}{2}$
12. $\dfrac{1}{10} + \dfrac{1}{5} + \dfrac{1}{2}$

13. $9\dfrac{7}{10} + 2\dfrac{1}{10}$
14. $6\dfrac{3}{4} - 5\dfrac{1}{6}$
15. $8\dfrac{2}{3} - 5$
16. $7 + 5\dfrac{3}{4}$

17. $8\dfrac{1}{2} + 1\dfrac{7}{10} + 6\dfrac{1}{5}$
18. $8\dfrac{2}{5} + 4\dfrac{7}{15} + 7\dfrac{1}{3}$

Evaluate. Simplify if possible.

19. $\dfrac{2x+1}{7} + \dfrac{3}{7}$ if $x = 4$
20. $\dfrac{2m-1}{28} + \dfrac{1}{4}$ if $m = 3$

Solve. Simplify if possible.

21. $\dfrac{3}{4}x + 2 = 8$
22. $\dfrac{4}{5}x - 2 = 3$
23. $5x - 2 = 7$

Solve these problems.

24. Mona jogs $4\dfrac{1}{4}$ laps, rests, then jogs $8\dfrac{1}{2}$ laps. How many laps does she jog altogether?

25. Martin weighs $75\dfrac{3}{4}$ kg. How much will he weigh if he loses $9\dfrac{1}{2}$ kg?

Organizing Data

6

The Statue of Freedom tops the 287 ft Capitol Dome in Washington, D.C.

Reading and Making Tables

◇ OBJECTIVES ◇

To read a table
To make a table

◇ RECALL ◇

TEST SCORE	
Eric	97
Mort	88
Dawn	99 ← Highest Test Score

Example 1

What tax would you pay if your taxable income was between $22,250 and $22,300 and you are single? Between $22,700 and $22,750 and are married filing jointly?

Single

Taxable Income: $22,250–$22,300

1. Find $22,250–$22,300 in the left column.
2. Find *Single* at the top.
3. Find the intersection of the horizontal line from the left column and the vertical line from the top.
4. Your tax is $4,889.

Married filing jointly

Taxable Income: $22,700–$22,750

1. Find $22,700–$22,750 in the left column.
2. Find *Married filing jointly* at the top.
3. Find the intersection of the horizontal and vertical lines.
4. Your tax is $3,930.

TAX TABLE

If line 34 (taxable income) is—		And you are—			
At least	But less than	Single	Married filing jointly	Married filing separately	Head of a household
		Your tax is—			
22,000	22,050	4,805	3,737	5,909	4,438
22,050	22,100	4,821	3,751	5,930	4,453
22,100	22,150	4,838	3,764	5,951	4,468
22,150	22,200	4,855	3,778	5,973	4,483
22,200	22,250	4,872	3,792	5,994	4,499
22,250	22,300	4,889	3,806	6,015	4,514
22,300	22,350	4,905	3,820	6,036	4,529
22,350	22,400	4,922	3,833	6,058	4,545
22,400	22,450	4,939	3,847	6,079	4,560
22,450	22,500	4,956	3,861	6,100	4,575
22,500	22,550	4,973	3,875	6,121	4,591
22,550	22,600	4,989	3,889	6,142	4,606
22,600	22,650	5,006	3,903	6,164	4,621
22,650	22,700	5,023	3,916	6,185	4,637
22,700	22,750	5,040	3,930	6,206	4,652
22,750	22,800	5,056	3,944	6,227	4,667
22,800	22,850	5,073	3,958	6,249	4,682
22,850	22,900	5,090	3,972	6,270	4,698
22,900	22,950	5,107	3,986	6,293	4,713
22,950	23,000	5,124	3,999	6,317	4,728

So, the tax on taxable income between $22,250 and $22,300 and Single is $4,889. The tax on taxable income between $22,700 and $22,750 and Married filing jointly is $3,930.

practice ▷ Determine the tax using the tax table in Example 1.

1. Single
   Taxable income:
   $22,500–$22,550

2. Married filing separately
   Taxable income:
   $22,850–$22,900

It is often useful to make a table to display information.

Example 2

Make a table for the following information. Assume each fruit contains 100 edible grams. An apple contains 80 calories, 20 grams of carbohydrates, and 120 units of vitamin A. A banana contains 100 calories, 26 grams of carbohydrates, and 230 units of vitamin A. A $\frac{1}{2}$ cantaloupe contains 80 calories, 20 grams of carbohydrates, and 9,240 units of vitamin A.

1. Label the vertical scale with the names of the fruits.
2. Label the horizontal scale with the composition of the fruits.
3. Draw vertical and horizontal lines separating fruits and their composition.
4. Put data in the appropriate boxes.
5. Give the table a title.

COMPOSITION OF FOODS

	Calories	Carbohydrates (grams)	Vitamin A (units)
Apple	80	20	120
Banana	100	26	230
$\frac{1}{2}$ Cantaloupe	80	20	9,240

practice ▷ Make a table from the following information.

3. At 5% interest, $1,000 would grow to $1,051 at the end of 1 year; $1,105 in 2 years, $1,161 in 3 years; and $1,221 in 4 years. At 6%, it would grow to $1,061; $1,127; $1,197; and $1,271 respectively.

READING AND MAKING TABLES

◇ ORAL EXERCISES ◇

Answer the exercises below. Use the bus schedule.

1. When will the 6:20 Court St. bus arrive at Morgan Rd. and Ross Park?
2. When will the 7:50 Court St. bus arrive back at Court St.?
3. A passenger leaves Telegraph St. and Conklin Ave. at 7:25. When will he get to Vestal Ave. and Park Ave.?
4. A passenger leaves James St. and Washington St. at 6:03. When will she get to Court St.?

Leave Court St.	Telegraph St. and Conklin Ave.	James St. and Washington St.	Arrive Morgan Rd. and Ross Park	Leave Morgan Rd. and Ross Park	Vestal Ave. and Park Ave.	Arrive Court St.
5:55	5:59	6:03	6:07	6:07	6:15	6:20
6:20	6:25	6:29	6:34	6:35	6:40	6:50
6:50	6:55	6:59	7:04	7:05	7:10	7:20
7:20	7:25	7:29	7:34	7:35	7:40	7:50
7:50	7:55	7:59	8:04	8:05	8:10	8:20

◇ EXERCISES ◇

Answer Exercises 1–4 below. Use the table.

1. What tax will you pay if your taxable income is $19,225 and you are single?

2. What tax will you pay if your taxable income is $19,455 and you are married filing jointly?

3. What tax will you pay if your taxable income is $19,048 and you are the head of a household?

4. What tax will you pay if your taxable income is $19,275 and you are married filing separately?

If line 34 (taxable income) is—		And you are—			
At least	But less than	Single	Married filing jointly	Married filing separately	Head of a household
		Your tax is—			
19,000					
19,000	19,050	3,797	2,954	4,635	3,519
19,050	19,100	3,814	2,965	4,656	3,535
19,100	19,150	3,831	2,977	4,678	3,550
19,150	19,200	3,848	2,989	4,699	3,565
19,200	19,250	3,865	3,001	4,720	3,580
19,250	19,300	3,881	3,013	4,741	3,596
19,300	19,350	3,898	3,025	4,762	3,611
19,350	19,400	3,915	3,037	4,784	3,626
19,400	19,450	3,932	3,048	4,805	3,642
19,450	19,500	3,949	3,060	4,826	3,657

Algebra Maintenance

Solve.

1. $3x = 48$
2. $x - 12 = 7$
3. $24 = 5z - 1$
4. $x + \frac{1}{2} = \frac{3}{4}$
5. $3 = k - 2\frac{1}{2}$
6. $3x - 1 = 4$
7. $\frac{3}{4}r + 4 = 5$
8. $\frac{5}{6}t - 4 = 1$
9. $5k + 9 = 17$

Pictographs

◇ OBJECTIVE ◇
To read and make pictographs

◇ RECALL ◇
$4\frac{1}{2} \cdot 60 = \frac{9}{2} \cdot \frac{60}{1}$, or 270

Example 1

Pictures are used to show student enrollment.

The pictograph shows student enrollment in grades 10–12. How many are in each grade?

ENROLLMENT IN MADISON HIGH SCHOOL

Grade	Number of Students
10	🚹 🚹 🚹 🚹 🚹
11	🚹 🚹 🚹 🚹 🚹
12	🚹 🚹 🚹 🚹

Key: Each 🚹 represents 100 students.

Grade 10: 5 🚹 means 5 · 100, or 500

$4\frac{1}{2} \cdot 100 = \frac{9}{2} \cdot 100$, or 450   Grade 11: $4\frac{1}{2}$ 🚹 means $4\frac{1}{2} \cdot 100$, or 450

Grade 12: 4 🚹 means 4 · 100, or 400

So, there are 500 students in grade 10; 450 in grade 11; and 400 in grade 12.

Example 2

Make a pictograph for the following table.

NEW HOUSES

Jan.–Mar.	Apr.–June	July–Sept.	Oct.–Dec.
60	45	55	30

Title the pictograph.
Put heads on the columns.

Choose a symbol to represent houses.
Use 🏠 to represent 10 new houses.

For 60 houses, use 6 🏠.
For 45 houses, use $4\frac{1}{2}$ 🏠.
For 55 houses, use $5\frac{1}{2}$ 🏠.
For 30 houses, use 3 🏠.
Determine key.

NEW HOUSES

Months	Number of New Houses
Jan.–Mar.	🏠 🏠 🏠 🏠 🏠 🏠
Apr.–June	🏠 🏠 🏠 🏠 🏠
July–Sept.	🏠 🏠 🏠 🏠 🏠 🏠
Oct.–Dec.	🏠 🏠 🏠

Key: Each 🏠 represents 10 new houses.

PICTOGRAPHS

practice ▷ Make a pictograph.

ATTENDANCE AT SOCCER GAMES

Game	Number of Fans
1	6,000
2	7,500
3	7,000
4	4,000

◇ ORAL EXERCISES ◇

♁ represents 100 people. How many ♁ would you need to show each?

1. 400 people
2. 350 people
3. 1,000 people
4. 150 people

◇ EXERCISES ◇

1. How many birthdays in Jan.?
2. How many birthdays in Feb.?
3. How many birthdays in Mar.?
4. How many birthdays in Apr.?
5. What was the total number of birthdays for the 4 months?

6. How many boxes sold by the ninth grade?
7. How many boxes sold by the tenth grade?
8. How many boxes sold by the eleventh grade?
9. How many boxes sold by the twelfth grade?

BIRTHDAYS IN CHEROKEE GRADE SCHOOL

Key: Each 🎂 represents 10 birthdays.

SCHOOL STATIONERY SALE

Key: Each 📦 represents 20 boxes.

Make a pictograph.

10. **WORLD SERIES ATTENDANCE**

Day	Number of Fans
Friday	50,000
Saturday	65,000
Monday	70,000
Tuesday	60,000
Wednesday	75,000

★ 11. **STAMP COLLECTIONS**

Name	Number of Stamps
Josh	175
Randi	225
Mario	150
Carol	100

Problem Solving – Applications
Jobs for Teenagers

Example 1

Jason worked 2 hours cutting a lawn. He charged $3.75 per hour. His expenses were $0.50 per hour for gas and oil and $0.90 for equipment costs and repairs. How much was his net profit?

Net Profit = Amount Charged − Expenses

$2 \cdot \$3.75 = \$7.50 \qquad 2 \cdot \$0.50 + \$0.90 = \$1.90$

Net Profit = $7.50 − $1.90 = $5.60

$$\begin{array}{r} \$7.50 \\ -1.90 \\ \hline \$5.60 \end{array}$$

So, Jason's net profit was $5.60.

Example 2

One year, Wanda used 6 bags of fertilizer, $1\frac{1}{2}$ bags of seed, and 4 bags of limestone. The prices follow: fertilizer $8.95 per bag, seed $8.50 per bag, limestone $1.25 per bag. What was the total cost for the actual amounts used?

Total cost = fertilizer + seed + limestone.

Fertilizer $8.95 × 6 = $53.70

Seed $8.50 × $1\frac{1}{2}$ = $8.50 \times \frac{3}{2} = \frac{25.50}{2}$, or $12.75

Limestone $1.25 × 4 = $5.00

$53.70 + $12.75 + $5.00 = $71.45

$$\begin{array}{r} \$53.70 \\ 12.75 \\ +5.00 \\ \hline \$71.45 \end{array}$$

So, the total cost for the amounts used was $71.45.

PROBLEM SOLVING

Solve these problems.

1. Gale took 3 hours to cut the O'Neals' lawn. She charged $3.90 per hour. Her expenses were 60 cents per hour for gas and oil and 50 cents per hour for equipment costs and repairs. How much did she net?

2. Keith took 4 hours to cut the Issas' lawn. He charged $2.60 per hour. He used their lawnmower and gas. How much did he earn?

3. Carol purchased fertilizer at $9 per bag, limestone at $2 per bag, seed at $6.50 per bag, and weed killer at $4 per bottle. In one year she used 3 bags of fertilizer, $\frac{1}{2}$ bag of seed, $2\frac{1}{2}$ bags of limestone, and $\frac{1}{4}$ bottle of weed killer. What was the total cost for the actual amounts used?

4. Miguel purchased fertilizer at $12 per bag, seed at $8.50 per bag, limestone at $2.50 per bag, and weed killer at $4.20 per bottle. In one year he used bags of fertilizer, 1 bag of seed, $4\frac{1}{2}$ bags of limestone, and $\frac{1}{4}$ bottle of weed killer. What was the total cost for the actual amounts used?

5. Randy spent 8 hours working with a rototiller. He charged $6.50 per hour. His expenses were $20 for renting the rototiller and 50 cents per hour for gas. How much was his net profit?

6. Reiko worked 6 hours digging gardens with a rototiller. She charged $6.50 per hour. Her expenses were $4 per hour for the rototiller and 40 cents per hour for gas. How much was her net profit?

7. Karen took 3 hours to cut a lawn. She charged $4.10 per hour. Her expenses were 50 cents per hour for gas and oil and 40 cents per hour for equipment costs and repairs. How much did she net?

8. Ralph took $1\frac{1}{2}$ hours to cut the Burns' lawn. He charged $3 per hour. He had no expenses. How much did he earn?

9. Wally purchased fertilizer at $9.50 per bag, limestone at $2.25 per bag, seed at $6 per bag, and weed killer at $4.50 per bottle. In one year he used $5\frac{1}{2}$ bags of fertilizer, $1\frac{1}{2}$ bags of seed, 3 bags of limestone, and $\frac{1}{2}$ bottle of weed killer. Find the total cost for the actual amounts used.

10. Roz purchased fertilizer at $10 per bag, limestone at $1.75 per bag, seed at $7.25 per bag, and weed killer at $3.50 per bottle. In one year she used $4\frac{1}{2}$ bags of fertilizer, 1 bag of seed, 2 bags of limestone, and $\frac{1}{5}$ bottle of weed killer. What was the total cost of the actual amounts used?

11. Carmen spent 4 hours working with a rototiller. She charged $6 per hour. Her expenses were $10 for renting the rototiller and 40 cents per hour for gas. How much was her net profit?

12. Michael worked $7\frac{1}{2}$ hours landscaping the Paisley's lawn. He charged $10.50 per hour, plus bus fare. If his total bus fare was $2.95, how much must the Paisley's pay him?

Bar Graphs

◇ OBJECTIVES ◇

To read and make vertical bar graphs

To read and make horizontal bar graphs

◇ RECALL ◇

25,000 = 25 thousand
110,000,000 = 110 million

Example 1

How many babies were born each month at Northside Hospital?

The graph is a vertical bar graph.

Number of babies born:
in Sept.: 40
Oct.: 30
Nov.: 15
Dec.: 35

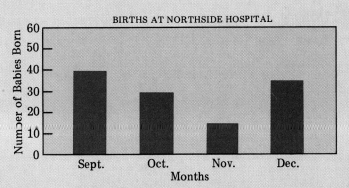

So, there were 40 babies born in Sept., 30 in Oct., 15 in Nov., and 35 in Dec.

Example 2

How many yearbooks were sold by each grade?

The graph is a horizontal bar graph.

Number of yearbooks sold:
grade 10: 80
grade 11: 130
grade 12: 200

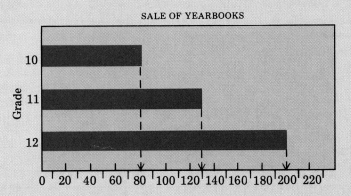

So, 80 yearbooks were sold by the tenth grade, 130 by the eleventh grade, and 200 by the twelfth grade.

practice

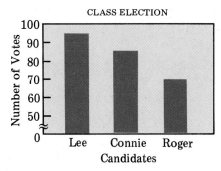

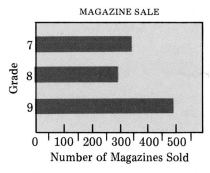

1. How many votes did each candidate get?

2. How many magazines were sold by each grade?

Example 3   Make a bar graph to show the approximate U.S. population.
1940, 130,000,000; 1950, 150,000,000;
1960, 180,000,000; 1970, 200,000,000

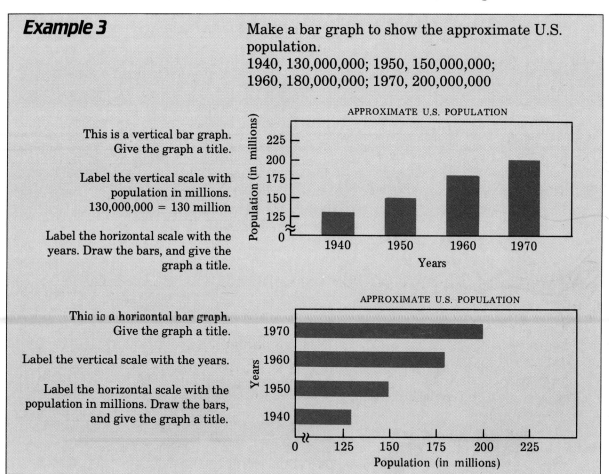

This is a vertical bar graph. Give the graph a title.

Label the vertical scale with population in millions. 130,000,000 = 130 million

Label the horizontal scale with the years. Draw the bars, and give the graph a title.

This is a horizontal bar graph. Give the graph a title.

Label the vertical scale with the years.

Label the horizontal scale with the population in millions. Draw the bars, and give the graph a title.

practice   3. Make a vertical bar graph and a horizontal bar graph to show the average yearly rainfall.
Atlanta, 130 cm; Los Angeles, 30 cm; Philadelphia, 100 cm; Minneapolis, 50 cm

◇ ORAL EXERCISES ◇

How many fans attended the games on each day?

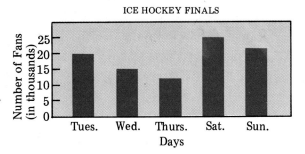
ICE HOCKEY FINALS

1. on Tues.
2. on Wed.
3. on Thurs.
4. on Sat.
5. on Sun.
6. On which day did the greatest number of fans attend?
7. On which day did the fewest number of fans attend?

◇ EXERCISES ◇

What is the life expectancy of each?

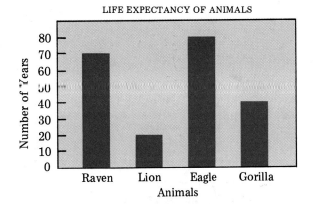
LIFE EXPECTANCY OF ANIMALS

1. the raven
2. the lion
3. the eagle
4. the gorilla
5. Which animal has the longest life expectancy?
6. Which animal has the shortest life expectancy?

Answer the exercises below. Use the graph at the right.

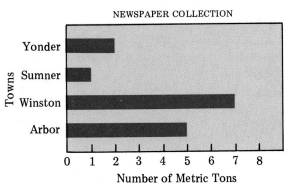
NEWSPAPER COLLECTION

7. How much newspaper did each town collect?
8. Which town collected the most newspaper?
9. Which town collected the least newspaper?
10. How much did all four towns collect together?
11. Make a vertical bar graph to show these basketball scores.
    Game 1, 90 points; Game 2, 85 points; Game 3, 95 points; Game 4, 80 points
12. Make a horizontal bar graph to show the average yearly rainfall.
    Buffalo, 100 cm; Salt Lake City, 40 cm; Miami, 125 cm; Juneau, 160 cm

BAR GRAPHS

Line Graphs

◇ OBJECTIVE ◇

To read and make line graphs

◇ RECALL ◇

Date	Withdrawal	Deposit	Balance	
Jan. 7			$12.50	
Jan. 15		$2.25	$14.75	}increase
Jan. 24	$1.25		$13.50	}decrease
Jan. 30		$2.50	$16.00	}increase

Example 1 Answer the exercises below. Use the line graph.

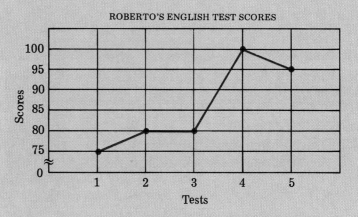

To find the score on a test, read up from the test number to the dot, then read across to the score.

Highest score: 100
Lowest score: − 75
Difference: 25

Test 3: 80 }increase
Test 4: 100

1. On which test did Roberto get the highest score?
   Roberto scored the highest on Test 4.
2. What is the difference between the highest and lowest scores?
   The difference between the highest and lowest scores is 25.
3. Did the scores increase, decrease, or stay the same from Test 3 to Test 4?
   The score increased from Test 3 to Test 4.

practice ▷ Use the graph in Example 1. Did the scores increase, decrease, or stay the same?

   1. From Test 1 to Test 2
   2. From Test 2 to Test 3
   3. From Test 4 to Test 5

Example 2

Make a line graph to show the average daily high temperatures. Mon., 12°; Tues., 15°; Wed., 12°; Thurs., 14°; Fri., 16°

Use graph paper or make a grid.
Give the graph a title.

Label the vertical scale with temperature in degrees.
Label the horizontal scale with days.
Draw dots to show the daily high temperatures.

Connect the dots.

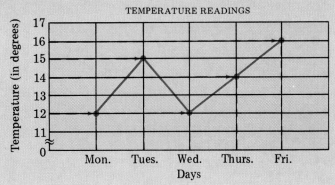

practice 4. Make a line graph to show how many books were sold each day. Mon., 25; Tues., 20; Wed., 15; Thurs., 20; Fri., 30

◇ ORAL EXERCISES ◇

What was the time for the men's 100-meter dash for each year?

1. 1964
2. 1968
3. 1972
4. 1976
5. 1980

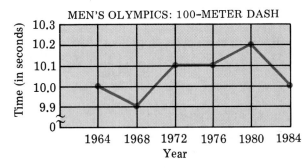

◇ EXERCISES ◇

Answer the exercises below. Use the line graph at the right.

1. In what month did Marcy read the greatest number of books?

2. In what month did Marcy read the least number of books?

3. In what months did Marcy read the same number of books?

4. How many books did Marcy read during the five months?

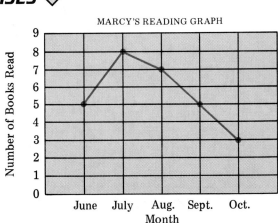

LINE GRAPHS

What was the approximate time for women's downhill skiing for each year?

5. 1952
6. 1956
7. 1960
8. 1964
9. 1968
10. 1972
11. 1976
12. 1980
13. What is the difference in the times of 1964 and 1972?
14. Make a line graph to show the average monthly rainfall.
   July, 18 cm; Aug., $16\frac{1}{2}$ cm; Sept., 23 cm; Oct., 20 cm; Nov., 8 cm; Dec., 4 cm

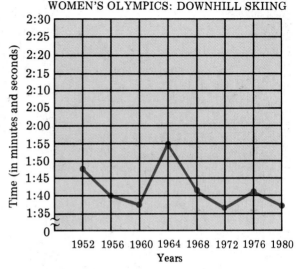

WOMEN'S OLYMPICS: DOWNHILL SKIING

Does the amount of rainfall increase or decrease? Use the graph you made in Exercise 14.

15. From July to Aug.?
16. From Aug. to Sept.?
17. From Oct. to Nov.?

18. Make a line graph to show the price of a stock during a one-week period.
   Mon., $4\frac{1}{2}$; Tues., $4\frac{3}{8}$; Wed., $4\frac{5}{8}$; Thurs., $5\frac{1}{8}$; Fri., $4\frac{3}{4}$

Did the price of the stock increase or decrease? Use the graph you made in Exercise 18.

19. From Mon. to Tues.?
20. From Wed. to Thurs.?
21. From Thurs. to Fri.?

Challenge

$1 \cdot 91 = 91$
$2 \cdot 91 = 182$
$3 \cdot 91 = 273$
$4 \cdot 91 = 364$

Do you see a pattern? Use the pattern to find these products.

$5 \cdot 91 = $ _____
$6 \cdot 91 = $ _____
$7 \cdot 91 = $ _____
$8 \cdot 91 = $ _____
$9 \cdot 91 = $ _____

Problem Solving – Careers
Appliance-Repair Specialists

1. Marco's Appliance Repair specializes in fixing small appliances. Last year, Marco's repaired 4,721 appliances. About how many appliances were repaired per week?
2. Chin's Major Appliance Service repairs refrigerators, washers, and dryers. Lucy Chin charges $25 for a house call and $16 per hour for labor. It took her 45 minutes to repair the Sragows' washer. How much did she charge if the new part cost $8.39?
3. Ira Kaplan repairs small appliances. He repairs about 50 appliances per week. He makes $250 profit for each appliance. About how much does he earn in a year?

Answer the exercises below. Use the paycheck stub.

Fitzpatrick Appliance Repair Co.													
Payroll statement for George M. Manitzas													Weekly Statement
1.	25 h Regular Hours	2.	2 h 15 min Overtime Hours	3.	293.92 Gross Pay	4.	28.46 Withholding Tax	5.	10.34 State Tax	6.	4.83 City Tax	7.	18.02 Social Security
8.	Bonds Savings 6.50	9.	Credit Union Savings 2.00	10.	Dues 4.00	11.	United Fund 2.00	12.	Insurance 2.40	13.	Net Pay	14.	End of Pay Period 10/24/86
Detach stub before cashing.													Retain this record for tax purposes.

4. What is the net pay for the pay period ending on 10/24/86?
5. What is the approximate net pay for the year?
6. What are the credit union savings for the year?
7. What are the approximate taxes for the year?

PROBLEM SOLVING

Using Graphs

◇ **OBJECTIVE** ◇

To solve problems by using graphs

◇ **RECALL** ◇

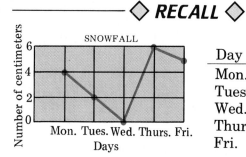

Day	Number of Centimeters
Mon.	4
Tues.	2
Wed.	0
Thurs.	6
Fri.	5

Example 1

Copy and complete the table below. Use the bar graph.

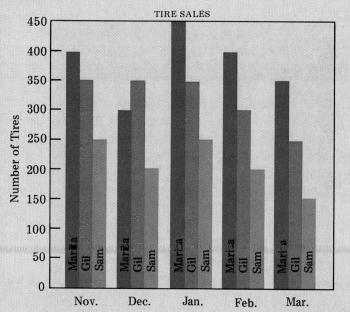

Add across to find the number of tires sold by each.
Add down to find the number of tires sold each month.

	Nov.	Dec.	Jan.	Feb.	Mar.	Total
Marita	400	300	450	400	350	1,900
Gil	350	350	350	300	250	1,600
Sam	250	200	250	200	150	1,050
Total	1,000	850	1,050	900	750	4,550

Example 2

Answer the questions. Use the table in Example 1.

1. Who sold the most tires?
   Marita sold 1,900 tires.   Marita sold the most tires.
2. Who sold the fewest tires?
   Sam sold 1,050 tires.   Sam sold the fewest tires.
3. In which month were the most tires sold?
   1,050 tires were sold in Jan.   The most tires were sold in January.
4. Who sold the most tires in December?
   Gil sold 350 tires in Dec.   Gil sold the most tires in December.

practice ▷ Use the table in Example 1.

1. Who sold the most tires in Mar.?
2. Who sold the fewest tires in Jan.?

Example 3

Answer the exercises below. Use the graph.

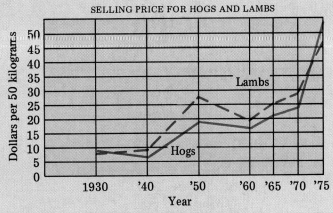

Key: Hogs ——— Lambs — — — —

1. In what year did the farmer get the best price for hogs and lambs?

   Look for highest point on the graph.   The farmer got the best price for hogs and lambs in 1975.
2. What was the selling price for hogs in 1930?

   Look at the solid line graph.   The selling price for hogs in 1930 was about $9 per 50 kilograms.

practice ▷ Use the graph in Example 3.

3. In what year did the farmer get the poorest price for hogs and lambs?
4. What was the selling price for lambs in 1950?

◇ EXERCISES ◇

Use the bar graph for Exercises 1–7.

1. Copy and complete the table.
2. Which store had the greatest sales?
3. Which store had the least sales?
4. In which year were sales the greatest?
5. In which year were sales the worst?
6. Which store's sales increased each year?
7. What were the total sales for all three stores?

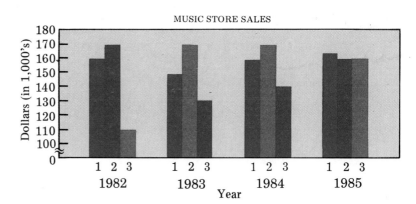

MUSIC STORE SALES

	1982	1983	1984	1985	Total
Store 1	$160,000				
Store 2	$170,000				
Store 3	$110,000				
Total	$440,000				

Answer the exercises below. Use the graph.

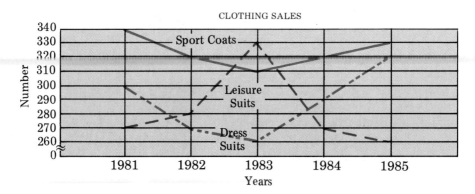

8. In what year were the most sport coats sold?
9. In what year were the least number of leisure suits sold?
10. What was the number of dress suits sold in 1984?
11. What was the total number of sport coats, dress suits, and leisure suits sold in 1982?

Applying Equations

 **OBJECTIVE**

To solve perimeter problems using equations

 **RECALL**

Perimeter means the distance around.

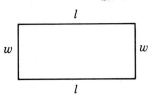

$P = l + w + l + w$
$P = 2l + 2w$

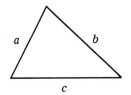

$P = a + b + c$

Example 1

The length of a rectangle is 19 cm and the perimeter is 62 cm. Find the width.

	$P = 2l + 2w$
Substitute 62 for P, 19 for l.	$62 = 2 \cdot 19 + 2w$
	$62 = 38 + 2w$
Subtract 38 from each side.	$62 - 38 = 38 - 38 + 2w$
	$24 = 2w$
Divide each side by 2.	$\dfrac{24}{2} = \dfrac{2w}{2}$
	$12 = w$, or $w = 12$

So, the width is 12 cm.

practice

1. The length of a rectangle is 15 cm and the perimeter is 40 cm. Find the width.

2. The width of a rectangle is 8 cm and the perimeter is 58 cm. Find the length.

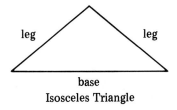

Isosceles Triangle

A triangle is *isosceles* if two of the sides have the same length (are congruent). The two congruent sides are called *legs*, and the third side is called the *base*.

Example 2

The length of each leg of an isosceles triangle is 10 cm and the perimeter is 25 cm. Find the length of the base, b.

$$P = a + b + c$$

Substitute 10 for a, 10 for c, and 25 for P.
$$25 = 10 + b + 10$$
$$25 = 20 + b$$

Subtract 20 from each side.
$$25 - 20 = 20 - 20 + b$$
$$5 = b, \text{ or } b = 5$$

So, the length of the base is 5 cm.

practice 3. The length of each leg of an isosceles triangle is 12 cm, and the perimeter is 30 cm. Find the length of the base.

4. The base of an isosceles triangle is 18 cm, and the perimeter is 42 cm. Find the length of each leg.

◇ ORAL EXERCISES ◇

Find the perimeter.

1.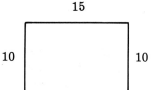
2.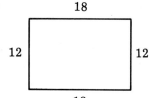
3.
4.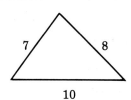
5.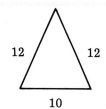
6.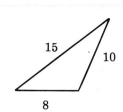

◇ EXERCISES ◇

Find the missing side.

1. $46 = 2 \cdot 15 + 2w$
2. $62 = 2l + 2 \cdot 10$
3. $108 = 2l + 2 \cdot 22$
4. $180 = 2 \cdot 61 + 2w$

Solve these problems.

5. The width of a rectangle is 14 cm, and the perimeter is 98 cm. Find the length.

6. The length of a rectangle is 17 cm, and the perimeter is 38 cm. Find the width.

7. The length of each leg of an isosceles triangle is 12 cm, and the perimeter is 34 cm. Find the base, b.

8. The base of an isosceles triangle is 20 cm, and the perimeter is 42 cm. Find the length of each leg.

9. The base of an isosceles triangle is 15 cm, and the perimeter is 35 cm. Find the length of each leg.

10. The length of each leg of an isosceles triangle is 17 cm, and the perimeter is 62 cm. Find the base.

11. The length of two sides of a triangle are 16 cm and 14 cm, and the perimeter is 42 cm. Find the length of the third side.

12. The length of two sides of a triangle are 8 cm and 7 cm, and the perimeter is 19 cm. Find the length of the third side.

★ 13. The perimeter of an equilateral triangle (3 congruent sides) is 33 cm. Find the length of each side.

★ 14. The perimeter of an equilateral triangle is 72 m. Find the length of each side.

★ 15. The length of a rectangle is 4 m more than 3 times the width. The perimeter is 32 m. Find the length and width.

★ 16. The width of a rectangle is one half the length. The perimeter is 24 m. Find the length and width.

★ 17. The first side of a triangle is 2 cm longer than the second side. The third side is 3 cm shorter than twice the second side. The perimeter is 15 cm. How long is each side?

★ 18. One side of a triangle is 4 cm shorter then the second side. The remaining side is twice the first side. The perimeter is 48 cm. Find the length of each side.

Calculator

Find the length of a rectangle. Round to the nearest tenth.

$P = 56.9$ cm
$w = 12.3$ cm

Formula $P = 2l + 2w$

$l = \dfrac{P - 2w}{2}$

PRESS $2 \times 12.3 \ominus 24.6$
$56.9 \ominus 24.6 \ominus 32.3 \oplus 2 \ominus$

DISPLAY   16.15

So, the length is 16.2 cm.

1. $P = 72.3$ cm and $w = 11.7$ cm
2. $P = 63.8$ m and $w = 9.8$ m

APPLYING EQUATIONS

Mathematics Aptitude Test

This test will prepare you for taking standardized tests. Choose the best answer for each question.

1. Which of the following is greater than $\frac{1}{3}$?
   a. $\left(\frac{1}{3}\right)^2$
   b. 0.03
   c. $\frac{1}{4}$
   d. None of the above.

2. If it is snowing at the rate of 3 centimeters per hour, then how many centimeters of snow will fall in n minutes?
   a. $\frac{n}{20}$
   b. $\frac{20}{n}$
   c. $20n$
   d. $3n$

3. There are 880 students in a college. One half of the enrollment is women. One fourth of the women are studying psychology. One fifth of those women studying psychology are studying statistics. One half of those women studying statistics are studying computers. How many women are studying computers?
   a. 10
   b. 11
   c. 44
   d. 396

4. The sum of two numbers is 10. Their difference is 4. What is the whole number quotient of the two numbers?
   a. 2
   b. 4
   c. 6
   d. 8

5. How many one fourths are there in $\frac{1}{2} + 1\frac{1}{4} + 2\frac{3}{4}$?
   a. $4\frac{1}{2}$
   b. 7
   c. 9
   d. 18

6. If the same number is added to both the numerator and denominator of a proper fraction, then the value of the new fraction is
   a. less than the value of the original fraction.
   b. greater than the value of the original fraction.
   c. the same as the value of the original fraction.
   d. cannot be determined.

7. If 4 times a number A is 12, and 5 times a number B is 11, then 10 times AB is?
   a. 66
   b. 75
   c. 132
   d. 230

8. How many nickels are there in q quarters?
   a. $\frac{q}{5}$
   b. $25q$
   c. $4q$
   d. $5q$

9. How many numbers between 1 and 100 are exactly divisible by 2 or 3 but not both?
   a. 67
   b. 76
   c. 77
   d. 83

COMPUTER ACTIVITIES

Another way to enter data into a computer is by using the **READ** and **DATA** statements. These statements are advantageous when you know ahead of time all the data you need to enter.

Example Listed are the wages and weekly hours of four workers. Write a program to show the hours, wages, and resulting pay for each of the four workers: $5.75/h, 35 h; $6.70/h, 40 h; $5.75/h, 39 h; $7.00/h, 35 h.

First write a formula to find the pay.

Pay = Wage times Hours
 P = W   *   H

Prints heading for 3 columns.	10	PRINT "WAGE","HOURS","PAY"
There are 4 sets of **DATA**.	20	FOR K = 1 TO 4
Two memory locations are reserved for wage, hours.	30	READ W,H
Formula for finding pay.	40	LET P = W * H
Four sets of **DATA**.	50	DATA 5.75,35, 6.70,40, 5.75,39, 7,35
Compute **PRINTS** wage, hours, and resulting pay.	60	PRINT W,H,P
	70	NEXT K
	80	END

See the Computer Section beginning on page 420 for more information.

EXERCISES

1. Modify the program above to compute the pay for the following 3 sets of **DATA**: $9.75/h, 40 h; $5.89/h, 35 h; $4.85/h, 38 h. Compute the pay for each. Then **RUN** the program to check your answers.

2. Write and **RUN** a program to find P in the formula P = 2L + 2W (perimeter of a rectangle). Use the **DATA** L = 5, W = 8; L = 19, W = 23; L = 42, W = 17.

Chapter Review

Answer Exercises 1–3 below using the tax table. [140]

1. What tax will you pay if your taxable income is $30,190 and you are single?

2. What tax will you pay if your taxable income is $30,220 and you are married filing jointly?

3. What tax will you pay if your taxable income is $30,075 and you are married filing separately?

4. Make a pictograph for attendance at four tennis matches. Match 1: 800 fans; Match 2: 750 fans; Match 3: 950 fans; Match 4: 900 fans [143]

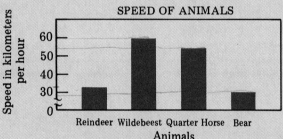

If line 34 (taxable income) is—		And you are—		
At least	But less than	Single	Married filing jointly	Married filing separately
		Your tax is—		
30,000	30,050	7,873	6,169	9,729
30,050	30,100	7,895	6,187	9,756
30,100	30,150	7,917	6,206	9,783
30,150	30,200	7,939	6,224	9,809
30,200	30,250	7,960	6,242	9,836

Use the bar graph to find the approximate speed for each. [147]

5. the reindeer
6. the wildebeest
7. the quarter horse
8. the bear

9. Make a line graph to show the average monthly temperature. Jan., 2°C; Mar., 7°C; May, 18°C; July, 26°C; Sept., 22°C; Nov., 9°C [150]

Use the graph to answer Exercises 10–11.

10. In what year did Brand A have the greatest sales? [154]

11. Who sold the most jeans in 1985?

Solve these problems. [157]

12. The length of each leg of an isosceles triangle is 11 cm and the perimeter is 38 cm. Find the length of the base.

13. The base of an isosceles triangle is 16 cm and the perimeter is 58 cm. Find the length of each side.

14. The length of a rectangle is 14 cm and the perimeter is 52 cm. Find the width.

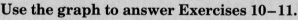

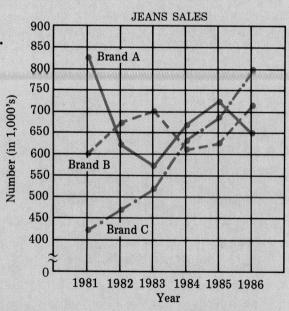

Chapter Test

Complete Exercises 1–3 below using the tax table.

1. What tax will you pay if your taxable income is $41,240 and you are single?

2. What tax will you pay if your taxable income is $41,125 and you are married filing jointly?

3. What tax will you pay if your taxable income is $41,080 and you are married filing separately?

If line 37 (taxable income) is—		And you are—		
At least	But less than	Single	Married filing jointly	Married filing separately
			Your tax is—	
41,000	41,050	11,859	9,595	14,255
41,050	41,100	11,881	9,614	14,279
41,100	41,150	11,903	9,634	14,304
41,150	41,200	11,925	9,653	14,328
41,200	41,250	11,947	9,673	14,353

Use the pictograph to answer these questions.

4. How many cars were sold by Dealer 1?

5. Who sold the most cars?

6. How many cars were sold in November?

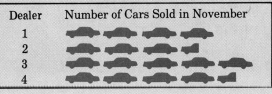

7. Make a vertical bar graph to show the approximate life expectancy of these animals.
   cow, 18 years; horse, 27 years; mouse, 4 years; monkey, 7 years

8. Make a line graph to show Leah's scores on 5 French tests.
   Test 1, 75; Test 2, 90; Test 3, 80; Test 4, 70; Test 5, 85

Use the bar graph to complete Exercises 9 and 10.

9. In 1986 which class had the best attendance?

10. In 1983 which class had the poorest attendance?

Solve these problems.

11. The length of each leg of an isosceles triangle is 15 cm and the perimeter is 41 cm. Find the length of the base.

12. The length of a rectangle is 16 cm and the perimeter is 58 cm. Find the width.

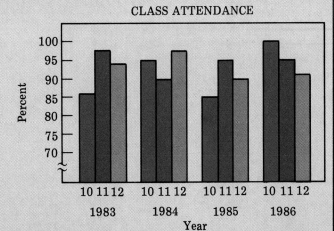

Decimals

7

In the spring, Niagara Falls doubles its flow to 100,000 ft³/s.

Introduction to Decimals

◇ OBJECTIVES ◇

To read and write numerals for word names from millions to millionths

To change fractions like $\frac{3}{10}$, $\frac{75}{100}$, and $\frac{4}{1,000}$ to decimals

To round decimals to the nearest tenth or hundredth

◇ RECALL ◇

Read $\frac{5}{10}$ as 5 tenths.

Read $\frac{7}{100}$ as 7 hundredths.

Read $\frac{95}{1,000}$ as 95 thousandths.

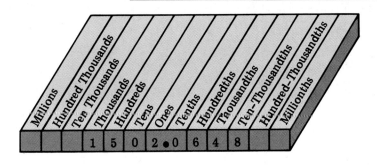

This number is read as follows:
One thousand, five hundred two, and six hundred forty-eight ten-thousandths.

Example 1

Study the table below. Then cover the right hand column, and give the word name for each decimal. Cover the left hand column, and give the decimal for each word name.

DECIMAL	WORD NAME
12.9	twelve and nine tenths
128.60	one hundred twenty-eight and sixty hundredths
0.481	four hundred eighty-one thousandths
5,063,198.0006	five million, sixty-three thousand, one hundred ninety-eight and six ten-thousandths
0.000741	seven hundred forty-one millionths

practice ▷

Write a decimal for each word name.
1. six and eight tenths
2. four hundred three millionths

Write a word name for each decimal.
3. 17.4
4. 0.0659

INTRODUCTION TO DECIMALS

We can use decimals to represent fractions.

2 tenths	25 hundredths	365 thousandths
$\frac{2}{10}$ or 0.2	$\frac{25}{100}$ or 0.25	$\frac{356}{1{,}000}$ or 0.356
one 0 — one decimal place	two 0's — two decimal places	three 0's — three decimal places

Example 2

Change $\frac{4}{100}$ to a decimal.

Write 4.   4.

0.04   Move the decimal point 2 places to the left.

So, $\frac{4}{100}$ = 0.04.

Example 3

Change $\frac{66}{1{,}000}$ to a decimal.

Write 66.   66.

0.066   Move the decimal point 3 places to the left.

So, $\frac{66}{1{,}000}$ = 0.066.

practice ▷ Change to decimals.

5. $\frac{7}{10}$ 6. $\frac{65}{100}$ 7. $\frac{46}{1{,}000}$ 8. $\frac{9}{100}$

Example 4

Round to the nearest tenth.
48.32, 48.35, and 48.37

Look at the hundredths digit.   48.3 **2**   48.3 **5**   48.3 **7**

If the hundredths digit is less than 5, round down.
tenths place — less than 5 — 48.3
5 or more — 48.4
5 or more — 48.4

If the hundredths digit is 5 or greater, round up.

So, 48.32 is 48.3, 48.35 is 48.4, and 48.37 is 48.4 to the nearest tenth.

Example 5

Round 0.3459 to the nearest hundredth.

Look at the thousandths digit.   0.34 **5** 9

The thousandths digit is 5; round up.   So, 0.3459 is 0.35 to the nearest hundredth.

practice ▷ Round to the nearest tenth and to the nearest hundredth.

9. 0.627 10. 0.6559 11. 0.683 12. 24.178

◇ ORAL EXERCISES ◇

Read each numeral.

1. 2.4
2. 36.21
3. 42.07
4. 0.05
5. 271.632
6. 306.0145
7. 79.006
8. 9.24863
9. 1,352.060
10. 4,628,000.9
11. 0.000592
12. 160.00083

◇ EXERCISES ◇

Write a numeral for each word name.

1. nine and five tenths
2. sixty-three hundredths
3. eight hundred twenty-three thousandths
4. one hundred five and eight thousandths
5. eight hundred thirteen and two hundred thirty-six millionths
6. two million, seven hundred twenty-four thousand and nine tenths

Change to decimals.

7. $\frac{76}{100}$ 8. $\frac{8}{10}$ 9. $\frac{6}{100}$ 10. $\frac{5}{1,000}$ 11. $\frac{67}{100}$ 12. $\frac{495}{1,000}$
13. $\frac{763}{1,000}$ 14. $\frac{98}{100}$ 15. $\frac{1}{100}$ 16. $\frac{3}{1,000}$ 17. $\frac{13}{1,000}$ 18. $\frac{4}{10}$
19. $\frac{54}{1,000}$ 20. $\frac{2}{100}$ 21. $\frac{16}{1,000}$ 22. $\frac{40}{1,000}$ 23. $\frac{11}{1,000}$ 24. $\frac{8}{100}$

Round to the nearest tenth.

25. 0.617 26. 0.657 27. 0.6874 28. 57.454 29. 116.93 30. 17.181
31. 14.683 32. 14.6237 33. 14.653 34. 114.08 35. 72.115 36. 112.0586

Round to the nearest hundredth.

37. 0.638 38. 0.635 39. 0.6328 40. 2.788 41. 3.169 42. 4.2215
43. 16.486 44. 16.483 45. 16.485 46. 29.057 47. 46.118 48. 192.7394

★ 49. Change $\frac{768}{1,000}$ to a decimal and round to the nearest hundredth.

★ 50. Change $\frac{2,651}{1,000}$ to a decimal and round to the nearest hundredth.

INTRODUCTION TO DECIMALS

Changing Fractions to Decimals

◇ OBJECTIVES ◇

To change fractions to decimals

To change a fraction like $\frac{5}{12}$ to a decimal to the nearest hundredth

◇ RECALL ◇

$\frac{15}{5}$ means $5\overline{)15}$ with quotient 3, 15, remainder 0.

Example 1 Change $\frac{2}{5}$ to a decimal.

Divide. $\frac{2}{5}$ means $5\overline{)2}$.

Write a 0 after the decimal point since 20 is divisible by 5.

$5\overline{)2.0} \rightarrow 5\overline{)2.0}$ with quotient 0.4, $2\,0$, remainder 0.

So, $\frac{2}{5} = 0.4$.

Example 2 Change $\frac{3}{8}$ to a decimal.

$\frac{3}{8}$ means $8\overline{)3}$.

Divide.
Write enough 0's after the decimal point so that the remainder will be 0.

Write one 0.
$8\overline{)3.0}$ quotient 0.3, $2\,4$, 6 ← remainder

Write two 0's.
$8\overline{)3.00}$ quotient 0.37, $2\,4$, 60, 56, 4 ← remainder

Write three 0's.
$8\overline{)3.000}$ quotient 0.375, $2\,4$, 60, 56, 40, 40, 0 ← remainder

So, $\frac{3}{8} = 0.375$.

---

practice ▷ Change to decimals.

1. $\frac{3}{5}$
2. $\frac{1}{4}$
3. $\frac{5}{8}$
4. $\frac{3}{20}$

Example 3

Change $\frac{4}{9}$ to a decimal to the nearest tenth.

$\frac{4}{9}$ means $9\overline{)4}$.

Divide.
To find the answer to the nearest tenth, carry out the division to hundredths.

```
  0.44
9)4.00
  3 6
    40
    36
     4
```

0.4 4 ←— less than 5
Round to 0.4.

So, $\frac{4}{9} = 0.4$ to the nearest tenth.

Notice that $\frac{4}{9} = 0.4444\ldots$ The 4's repeat forever. $0.4444\ldots$ is called a *repeating decimal*. It is written as $0.4\overline{4}$.

practice ▷ Change to a decimal to the nearest tenth.

5. $\frac{2}{3}$ 6. $\frac{7}{9}$ 7. $\frac{5}{6}$ 8. $\frac{7}{12}$

Example 4

Change $\frac{5}{12}$ to a decimal to the nearest hundredth.

Divide.
To find the answer to the nearest hundredth, carry out the division to thousandths.

```
   0.416
12)5.000
   4 8
     20
     12
      80
      72
       8
```

0.41 6 ←— 5 or more
Round to 0.42.

So, $\frac{5}{12} = 0.42$ to the nearest hundredth.

practice ▷ Change to a decimal to the nearest hundredth.

9. $\frac{4}{7}$ 10. $\frac{5}{9}$ 11. $\frac{11}{12}$ 12. $\frac{1}{6}$

◆ **EXERCISES** ◆

Change to decimals.

1. $\frac{1}{5}$
2. $\frac{4}{5}$
3. $\frac{3}{4}$
4. $\frac{1}{8}$
5. $\frac{7}{8}$
6. $\frac{1}{2}$
7. $\frac{7}{20}$
8. $\frac{9}{25}$
9. $\frac{3}{50}$
10. $\frac{11}{40}$
★ 11. $\frac{5}{16}$
★ 12. $\frac{13}{32}$

Change to a decimal to the nearest tenth.

13. $\frac{5}{7}$
14. $\frac{2}{3}$
15. $\frac{5}{9}$
16. $\frac{6}{7}$
17. $\frac{7}{9}$
18. $\frac{4}{7}$
19. $\frac{1}{3}$
20. $\frac{1}{7}$
21. $\frac{2}{9}$
22. $\frac{5}{11}$
23. $\frac{3}{14}$
24. $\frac{4}{11}$

Change to a decimal to the nearest hundredth.

25. $\frac{8}{9}$
26. $\frac{1}{12}$
27. $\frac{1}{9}$
28. $\frac{6}{7}$
29. $\frac{1}{7}$
30. $\frac{7}{12}$
31. $\frac{5}{11}$
32. $\frac{3}{14}$
33. $\frac{5}{6}$
34. $\frac{5}{14}$
★ 35. $\frac{5}{29}$
★ 36. $\frac{3}{41}$

37. Anthony saves $\frac{1}{5}$ of his salary. Write the part he saves as a decimal.

38. The gas tank of Edna's car is $\frac{3}{4}$ full. Write the part of the tank that is full as a decimal.

Algebra Maintenance

Simplify.

1. $3 \cdot \frac{4}{3}m$
2. $5\left(\frac{3}{5}x - 6\right)$
3. $7\left(\frac{3}{14}y + \frac{4}{7}\right)$
4. $4a + 2 - 7b + 3a - 4b$
5. $2x - 5 + \frac{3}{4}x + 5y + 8 - \frac{4}{5}y$

Solve each equation. Check.

6. $\frac{1}{2}y = 4$
7. $\frac{3}{4}t - 4 = 8$
8. $8k - \frac{3}{4} = \frac{1}{2}$

Solve.

9. One-half of a number, increased by 2, is 13. What is the number?

10. Larry's age is 5 more than one half Frank's age. Frank is 14 years old. How old is Larry?

Adding and Subtracting Decimals

◇ OBJECTIVES ◇
To add decimals
To subtract decimals

◇ RECALL ◇
Add. 48 + 3 + 116

```
   48
    3
 +116
  167
```

Example 1

Write in vertical form with decimal points in line with each other.

The answer has its decimal point in line with the others.

Add. 11.02 + 0.1 + 100

```
  11.02              11.02
   0.1       →        0.10
+100.            +  100.00
                     111.12
```

So, the sum is 111.12.

Example 2

Write in vertical form with decimal points in line with each other.

The answer has its decimal point in line with the others.

Add. 92.68 + 0.003 + 9.378

```
  92.68             92.680
   0.003             0.003
+  9.378         +   9.378
                   102.061
```

So, the sum is 102.061.

practice ▷ Add.

1. 149.68 + 0.48 + 91.32   2. 87.65 + 0.004 + 9.478

Example 3

Decimal points are lined up. The answer has its decimal point in line with the others.

Subtract. 146.38 − 28.99

```
  146.38            146.38
 − 28.99          −  28.99
                    117.39
```

So, the difference is 117.39.

practice ▷ Subtract.

3. 38.65 − 25.43   4. 17.146 − 9.38

Example 4

Evaluate $x + 5.8$ if $x = 0.932$.

Substitute 0.932 for x. Write in vertical form. Insert zeros. Then add.

$$\begin{array}{r} 0.932 \\ +5.800 \\ \hline 6.732 \end{array}$$

So, the sum is 6.732.

practice ▷ Evaluate for the given value of the variable.

5. $x + 9.8$ if $x = 12.05$ 6. $9.3 - y$ if $y = 0.46$

◇ EXERCISES ◇

Add.

1. $13.06 + 0.8 + 400.0$
2. $24.35 + 0.046 + 3.047$
3. $\$25.43 + \$170 + \$0.69$
4. $\$112.45 + \$1.04 + \$0.89$
5. $81.43 + 2.089 + 0.6 + 20.098$
6. $0.4308 + 2.85 + 300 + 47.1$

Subtract.

7. $48.75 - 26.32$
8. $16.146 - 8.236$
9. $\$412.42 - \$59.86$
10. $466 - 65.89$
11. $17.003 - 3.68$
12. $18.5 - 9.836$

Evaluate for the given value of the variable.

13. $x + 4.3$ if $x = 9.8$
14. $62.3 - x$ if $x = 5.9$
15. $y - 2.3$ if $y = 17.1$
16. $7.46 + z$ if $z = 8.25$

★ 17. Subtract 4.12 from the sum of 112.46, 2.18, and 1.06.

Calculator

1. Add. $249.6182 + 0.0083 + 12.49 + 0.0349$
2. Subtract. 241.346 from 3,568
3. Add. $0.0046 + 1.0049 + 21.1678 + 1119.32$
4. Subtract. 0.00584 from 12.001578
5. Subtract the sum of $115.45, $240.65, and $19.45 from the sum of $255.69, $189.39, and $1,156.95.

Multiplying Decimals

◇ OBJECTIVES ◇

To multiply decimals
To multiply whole numbers and decimals

◇ RECALL ◇

$0.03 = \frac{3}{100}$ $0.1 = \frac{1}{10}$

Example 1

Write each as a fraction and multiply.
Multiply the numerators.
Multiply the denominators.

$\frac{6}{1,000} = 0.006$

Show that $(0.03)(0.2) = 0.006$.
$(0.03)(0.2)$
$\frac{3}{100} \cdot \frac{2}{10}$
$\frac{6}{1,000}$
0.006 So, $(0.03)(0.2) = 0.006$.

Example 1 suggests that we count the decimal places.

```
  0.03   ← 2 decimal places
× 0.2    ← 1 decimal place
  0.006  ← 2 + 1, or 3 decimal places
```

Example 2

Multiply. $(3.45)(0.014)$
```
     3.45   ← 2 decimal places
  × 0.014   ← 3 decimal places
    1380
     345
   0.04830  ← 2 + 3, or 5 decimal places
```
So, $(3.45)(0.014) = 0.04830$, or 0.0483.

practice ▷ **Multiply.**

1. $(62.3)(2.4)$ 2. $(3.72)(0.017)$ 3. $(0.243)(0.076)$

Example 3

Substitute 0.8 for x and write in vertical form. Then multiply.
Three decimal places are in the product.

Evaluate $1.36x$ if $x = 0.8$.
```
   1.36
 × 0.8
  1.088
```
So, the product is 1.088.

practice ▷ Evaluate for the given value of the variable.

4. 7.52x if x = 0.6 **5.** 0.42y if y = 1.5

◇ ORAL EXERCISES ◇

Where should the decimal point be placed?

| 1. | 3.48
×0.2
696 | 2. | 12.16
×0.02
2432 | 3. | 0.24
×0.3
072 | 4. | 2.86
×0.014
04004 | 5. | 1.46
×0.028
04088 | 6. | 0.0867
×0.05
004335 |

◇ EXERCISES ◇

Multiply.

1. (2.16)(0.015)   2. (72.4)(3.2)   3. (1.06)(0.32)   4. (0.26)(0.04)
5. (1.1)(0.038)   6. (21.6)(0.042)   7. (2.17)(0.43)   8. (0.512)(3.4)

| 9. | 215
×0.034 | 10. | 26
×0.13 | 11. | 0.047
×35 | 12. | 0.0082
×23 | 13. | 21.3
×17 | 14. | 249
×0.046 |

| 15. | 145.32
×0.25 | 16. | 72.89
×0.06 | 17. | 15.45
×0.32 | 18. | 212.19
×0.013 | 19. | 11.43
×0.013 | 20. | 15.3
×0.0015 |

★ **21.** (7.1)(0.04)(1.13)   ★ **22.** (0.014)(0.13)(16.23)   ★ **23.** (30.62)(0.25)(4.6)

Evaluate for the given value of the variable.

24. 6.32y if y = 0.4   **25.** 17.8x if x = 0.12
26. 23.5z if z = 0.08   **27.** 1.53y if y = 1.09

Solve these problems.

28. Joanna saves $1.75 each week. How much has she saved after 48 weeks?

29. Plums cost $1.40 per kilogram. What is the cost of 4.5 kilograms?

See if you can move the disks to one of the other pegs. Follow these rules.

1. Move no more than one disk at a time.
2. Do not put a disk on top of one smaller than itself.
What is the least possible number of moves?

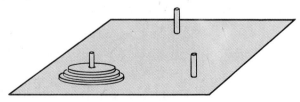

Problem Solving – Applications
Jobs for Teenagers

Example

DELICATESSEN PRICES

ITEM	PRICE PER POUND
CHEESE	3.80
TURKEY	3.84
ROAST BEEF	6.46
PASTRAMI	6.43
BOLOGNA	3.20

How much should Faith charge a customer for an order of 2.5 lb of pastrami and 1.4 lb of cheese?

First find the cost of each. Then find the total.

2.5 lb pastrami
```
   $6.43    ← pastrami $6.43/lb
 ×  2.5    ← 2.5 lb
   3215
   1286
  $16.075  ← cost
```

1.4 lb cheese
```
   $3.80
 ×  1.4
   1520
    380
  $5.320
```

Round 16.0 **7** 5 one cent higher to $16.08.

```
 $16.08    Add to find the total.
 + 5.32
 $21.40
```

So, the total bill is $21.40.

In business, costs are usually rounded up to the next higher cent unless there are 0's after the cents place.

$7.3241 $12.6492 $14.7800 ← only 0's after 8

rounds to $7.33   rounds to $12.65   stays at $14.78

PROBLEM SOLVING

Use the chart for the price per pound.

DELICATESSEN PRICES	
Item	Price per Pound
Cheese	$3.80
Turkey	$3.84
Roast beef	$6.46
Pastrami	$6.43
Bologna	$3.20

1. José ordered 2.5 lb of bologna and 1.7 lb of turkey. Find the cost of the order.

2. Maureen ordered 2.4 lb of roast beef and 1.5 lb of bologna. Find the cost of the order.

3. Miss Riggio called for 1.6 lb of roast beef and 2.7 lb of turkey. How much will her bill be?

4. Mr. Rothstein called for 0.6 lb of pastrami and 1.5 lb of roast beef. Find the cost of the order.

5. Janet is planning a party. She ordered 2.7 lb of cheese, 3.2 lb of turkey, and 0.6 lb of bologna. Find the cost of the order.

6. The Martinsons are planning a luncheon. They ordered 4.5 lb of cheese, 2.5 lb of pastrami, and 3.6 lb of turkey. How much did the order cost?

7. Ms. Becker is having a party. She ordered 1.4 lb of cheese and bologna and 1.7 lb of roast beef and turkey. How much will her bill be?

8. Julia bought 0.6 lb of turkey and 1.2 lb of cheese. Find the cost. How much change did she get from a $20 bill?

9. Bill bought 1.4 lb of roast beef and 0.4 lb of pastrami. Find the cost. How much change did he get from a $20 bill?

10. Nick bought 0.6 lb of cheese, 0.5 lb of roast beef, and 0.3 lb of pastrami. Find the total cost. How much change did he get from a $20 bill?

11. The Serkins ordered 1.5 lb of cheese, 0.6 lb of turkey, and 1.1 lb of bologna. Find the total cost. How much change did they get from a $20 bill?

12. Mrs. White uses 0.1 lb of roast beef on a sandwich. How many lb of roast beef should she buy for 3 sandwiches? How much will she pay?

13. Lars uses 0.08 lb of bologna and 0.08 lb of cheese on a sandwich. He bought enough bologna and cheese for 4 sandwiches. How much did he pay?

14. The Bryants bought 6 rolls at 15 cents each and 0.5 lb of turkey. They made 6 sandwiches. How much did each sandwich cost them?

15. Mr. Snow bought a pickle for $0.15, an iced tea for $0.50, 2 rolls for $0.15 each, and 0.09 lb of pastrami. How much did his order cost?

Dividing Decimals

◇ OBJECTIVES ◇
To divide decimals by decimals
To round quotients to the nearest tenth or hundredth

◇ RECALL ◇
$10(1.4) = 14 \quad 100(1.4) = 140 \quad 1{,}000(1.4) = 1{,}400$

$8\overline{)16}$ means $\frac{16}{8}$

Example 1

Rewrite $4.12\overline{)12.365}$ so that the divisor is a whole number.
$4.12\overline{)12.365}$ means $\frac{12.365}{4.12}$ and $12.365 \div 4.12$

Multiply the numerator and denominator by 100.
$100(12.365) = 1{,}236.5$
$100(4.12) = 412$

$\frac{12.365(100)}{4.12(100)}$

$\frac{1236.5}{412}$ or $412\overline{)1236.5}$
↑
whole number

The divisor 412 is a whole number.

So, $4.12\overline{)12.365}$ can be rewritten as $412\overline{)1{,}236.5}$.

Example 1 suggests how to move decimal points when dividing by decimals.
$4.12\overline{)12.365}$

$4.12\overline{)12.365}$

Move the decimal point two places to the right to get a whole number.

Also move this decimal point the same number of decimal places to the right.

Example 2

Make the divisor, 0.03, a whole number. Move each decimal point two places to the right. Divide. Line up the decimal point in the quotient with the decimal point in the dividend.

Divide. $2.796 \div 0.03$
$0.03\overline{)2.796}$
two places

$\begin{array}{r} 93.2 \\ 3\overline{)279.6} \end{array}$

So, $2.796 \div 0.03 = 93.2$

practice ▷ Divide.

1. $0.04\overline{)3.684}$
2. $0.02\overline{)68.462}$
3. $3.575 \div 0.005$

Example 3

Divide. Round to the nearest tenth. 0.41 ÷ 0.07

Make the divisor, 0.07, a whole number. Move each decimal point two places to the right.

0.07̖)0.41̖

To find the answer to the nearest tenth, carry the division to hundredths, or two decimal places.

```
     5.85
 7)41.00    ← Write two 0's to get two
   35            decimal places.
   ──
    60
    56
    ──
     40
     35
     ──
      5
```

5.8 5 ← 5 or more. Round to 5.9.

5.85 rounded to the nearest tenth is 5.9.

So, 0.41 ÷ 0.07 = 5.9 rounded to the nearest tenth.

Example 4

Divide. Round to the nearest hundredth. 0.03375 ÷ 0.048

Make the divisor, 0.048, a whole number. Move each decimal point three places to the right.

0.048̖)0.033̖75

To find the answer to the nearest hundredth, carry the division to thousandths, or three decimal places.

48)33.750 ← Write one 0 to get three decimal places.

48 is close to 50. 50)337

5)33 Try 7.

```
    0
48)15     Write 0 above 5.
```

48)150; 48 is close to 50. Think: 50)150 = 3.

```
     0.7              0.70
 48)33.750         48)33.750
    33 6              33 6
    ────              ────
       15               150  ← Bring down the 0.

     0.703
 48)33.750
    33 6
    ────
       150
       144
       ───
         6
```

0.70 3 ← less than 5 Round to 0.70.

0.703 rounded to the nearest hundredth is 0.70.

So, 0.03375 ÷ 0.048 = 0.70 rounded to the nearest hundredth.

---

practice ▷ Divide. Round to the nearest tenth and to the nearest hundredth.

4. 0.06)0.23

5. 0.01393 ÷ 0.023

6. 0.46)0.3827

◇ ORAL EXERCISES ◇

Where should the decimal point be placed?

1. 0.4)2.848 → 712
2. 0.04)2.848 → 712
3. 0.004)2.848 → 712
4. 0.06)0.126 → 21
5. 0.005)4.555 → 911
6. 0.002)0.0486 → 243
7. 0.05)0.0055 → 11
8. 0.21)0.0714 → 34

◇ EXERCISES ◇

Divide.

1. 0.02)2.468
2. 0.03)66.939
3. 0.004)7.848
4. 0.05)0.045
5. 0.002)8.42
6. 0.03)8.43
7. 0.03)0.3249
8. 0.06)0.186
9. 0.003)273.9
10. 0.06)246
11. 0.008)82.48
12. 0.002)84.62

Divide. Round to the nearest tenth.

13. 23 ÷ 0.07
14. 0.77 ÷ 0.03
15. 1.337 ÷ 0.5
16. 7.62 ÷ 2.4
17. 24.6 ÷ 7.1
18. 0.786 ÷ 0.34
19. 0.70 ÷ 0.32
20. 0.4965 ÷ 0.043
21. 0.893 ÷ 0.63

Divide. Round to the nearest hundredth.

22. 0.754 ÷ 2.3
23. 0.4935 ÷ 0.42
24. 0.497 ÷ 0.41
25. 0.7577 ÷ 6.8
★ 26. 0.7856 ÷ 0.0462
★ 27. 0.04682 ÷ 1.19

Solve these problems.

28. It takes Jim 9.2 minutes to run 2.5 kilometers. How many kilometers can he run in 1 minute? Round to the nearest tenth.

29. Jane traveled by car 199.7 kilometers in 4.5 hours. Find her average speed. Round to the nearest hundredth.

Challenge

A customer in Mr. Sole's shoe store bought a pair of shoes for $42 and paid with a $50 bill. Mr. Sole had no change. So he walked to Harry's Barber Shop to change the bill. Just after the customer left, Harry stormed into Mr. Sole's store calling the $50 bill counterfeit. Mr. Sole had to give him $50. Also Mr. Sole had given the crook $8 change. It looks like Mr. Sole lost $58 in all. Right or wrong?

DIVIDING DECIMALS

Scientific Notation and Large Numbers

◇ OBJECTIVES ◇
To multiply decimals by powers of ten
To write large numbers in scientific notation

◇ RECALL ◇
Powers of 10
$10^2 = 100$, $10^3 = 1{,}000$, $10^4 = 10{,}000$

Example 1

Multiply by 10: Move the decimal point 1 place right.
Multiply by 100: Move the decimal point 2 places right.

Multiply.
3.47×10
$3.47 \times 10 = 34.7$
1 zero   1 place

5.967×100
$5.967 \times 100 = 596.7$
2 zeros   2 places

So, $3.47 \times 10 = 34.7$ and $5.967 \times 100 = 596.7$.

---

practice ▷ Multiply.

1. 2.87×10
2. 4.385×100
3. $9.876 \times 1{,}000$

Example 2

$10^2 = 100$

Multiply.
5.876×10^2
5.876×10^2
$5.876 \times 100 = 587.6$

4.375×10^3
4.375×10^3
$4.375 \times 1{,}000 = 4{,}375$

$10^3 = 1{,}000$

So, $5.876 \times 10^2 = 587.6$ and $4.375 \times 10^3 = 4{,}375$.

---

practice ▷ Multiply.

4. 5.325×10^3
5. 9.6257×10^4
6. 10.4156×10^2

Example 3

7.36 multiplied by what power of 10 is 736?

What number should be substituted for n to make the sentence true?
$736 = 7.36 \times 10^n$
$736 = 7.36 \times 100$
$736 = 7.36 \times 10^2$

So, $n = 2$.

---

practice ▷ Find n to make the sentence true.

7. $547 = 5.47 \times 10^n$
8. $7{,}635.7 = 7.6357 \times 10^n$

Large numbers can be written as a product of a number between 1 and 10 and a power of 10. This is called *scientific notation*.

Example 4

To make 497 a number between 1 and 10, move the decimal point and multiply by a power of ten.

Write 497 and 5,000,000 in scientific notation.

$497 = 4.97 \times 10^n$
    2 places
$n = 2$
$497 = 4.97 \times 10^2$

So, $497 = 4.97 \times 10^2$ in scientific notation.

$5,000,000 = 5.000000 \times 10^n$
        6 places
$n = 6$
$5,000,000 = 5 \times 10^6$

So, $5,000,000 = 5 \times 10^6$ in scientific notation.

practice Write in scientific notation.

9. 234
10. 49,000
11. 30,000,000

◇ EXERCISES ◇

Multiply.

1. 3.65×100
2. 4.155×10
3. $9.8763 \times 1,000$
4. $8.1456 \times 10,000$
5. 1.3456×100
6. $12.432158 \times 100,000$
7. 4.872×10^2
8. 12.34125×10^3
9. 10.176235×10^4
10. 18.145×10
11. 5.449872×10^5
12. 27.92976×10^2

Write in scientific notation.

13. 487
14. 96,000
15. 3,000,000
16. 12,000,000
17. 582,000
18. 140,000,000
19. 750,000
20. 635,000,000
21. 8,050,000
22. 4,100,000
23. 2,000,000,000
24. 350,000,000,000

Write in scientific notation the combined populations of:

25. Philadelphia and Detroit.

26. Boston and Honolulu.

27. All four population areas.

CITY	POPULATION
Philadelphia	4,809,900
Detroit	4,434,300
Boston	3,918,400
Honolulu	691,200

SCIENTIFIC NOTATION AND LARGE NUMBERS

Decimal Equations

◆ OBJECTIVE ◆

To solve equations like
$1.62 + x = 4.39$, $0.05x = 7.5$,
$\frac{x}{4.6} = 9.71$, and $3x - 4.2 = 5.4$

◆ RECALL ◆

$(0.7)(10) = 7$

$(0.05)(100) = 5$

Example 1

Solve and check. $1.62 + x = 4.39$
$$1.62 + x = 4.39$$
Subtract 1.62 from each side.
$$1.62 - 1.62 + x = 4.39 - 1.62$$
$$x = 2.77$$

Substitute 2.77 for x.

Check: $\begin{array}{c|c} 1.62 + x = 4.39 \\ 1.62 + 2.77 & 4.39 \\ 4.39 = 4.39 \end{array}$

So, the solution is 2.77.

practice ▷ Solve.

1. $x + 1.8 = 4.7$
2. $y - 2.63 = 5.98$
3. $9.3 + z = 12.7$

Example 2

Solve. $0.05x = 7.5$
$$0.05x = 7.5$$
Divide each side by 0.05.
$$\frac{0.05x}{0.05} = \frac{7.5}{0.05}$$
$$x = 150$$

So, the solution is 150.

Example 3

Solve. $\frac{x}{4.6} = 9.71$
$$\frac{x}{4.6} = 9.71$$
Multiply each side by 4.6.
$$(4.6)\frac{x}{4.6} = (9.71)(4.6)$$
$$x = 44.666$$

So, the solution is 44.666.

practice ▷ Solve and check.

4. $4.1x = 11.48$
5. $\frac{y}{7.25} = 3.1$
6. $0.12x = 0.06$

Example 4

Solve. $3x - 4.2 = 5.4$

$3x - 4.2 = 5.4$

Add 4.2 to each side. $3x - 4.2 + 4.2 = 5.4 + 4.2$

$3x = 9.6$

Divide each side by 3. $\dfrac{3x}{3} = \dfrac{9.6}{3}$

$x = 3.2$

So, the solution is 3.2.

practice ▷ Solve.

7. $2x - 1.2 = 6.4$
8. $4y + 3.2 = 4.4$
9. $5x - 3.8 = 2.7$

◇ ORAL EXERCISES ◇

Tell what you would do to each side of the equation to solve it.

1. $x + 1.8 = 17.9$
2. $0.7y = 4.9$
3. $12.6 + y = 18.2$
4. $c - 4.2 = 16.3$
5. $\dfrac{x}{7.9} = 14.8$
6. $12.03z = 36.09$
7. $x - 3.8 = 14.9$
8. $\dfrac{x}{1.07} = 7.21$
9. $0.3y = 0.09$
10. $\dfrac{z}{8.1} = 5.9$
11. $r - 9.2 = 7.3$
12. $s + 0.09 = 0.26$

◇ EXERCISES ◇

Solve.

1. $x + 7.6 = 8.2$
2. $0.3 + y = 8.9$
3. $c + 0.5 = 9.1$
4. $r + 3.21 = 8.96$
5. $4.98 + z = 7.14$
6. $6.8 + b = 8.0$
7. $x + 3.5 = 12.8$
8. $7.02 + z = 12.51$
9. $x - 2.5 = 7.1$
10. $y - 4.07 = 6.31$
11. $z - 5.4 = 0.9$
12. $c - 3.02 = 8.15$
13. $r - 0.04 = 1.39$
14. $s - 2.8 = 4.13$
15. $x - 0.3 = 8.97$
16. $z - 0.06 = 0.08$
17. $0.3x = 0.06$
18. $0.8y = 0.24$
19. $0.9x = 0.54$
20. $0.7y = 0.63$
21. $0.06x = 0.03$
22. $0.04z = 1.6$
23. $4.2x = 11.34$
24. $2.3x = 11.04$
25. $\dfrac{x}{0.2} = 0.9$
26. $\dfrac{y}{1.9} = 6.7$
27. $\dfrac{z}{0.6} = 4.9$
28. $\dfrac{r}{2.8} = 0.4$
29. $\dfrac{a}{0.9} = 2.5$
30. $\dfrac{c}{4.3} = 2.85$
31. $2x + 1.9 = 3.7$
32. $4y - 7.2 = 5.2$
33. $6x - 4.2 = 7.2$
★ 34. $5.3 + 3z = 8.1$
★ 35. $\dfrac{a}{2.5} + 1.6 = 3.04$
★ 36. $\dfrac{r}{0.7} - 0.74 = 7.26$

Solve these problems.

37. In the formula $T = 1.12a$, find a if $T = \$196.00$.
38. In the formula $c = 3.14d$, find d if $c = 12.56$.

DECIMAL EQUATIONS

Problem Solving – Careers
Machinists

1. A rod for a machine must be 8.97 cm long. It is now 9.01 cm. How much must be filed down?

2. Three holes are to be drilled into a metal strip that is 13.6 cm long. Two of the holes are to be at either end of the strip, and the third hole is to be exactly in the middle. How far is the middle hole from the end?

3. A piece of sheet metal is 42.6 cm wide. It is to be cut into 3 pieces equal in width. How wide will each be?

4. On a blueprint, 1 dm represents 15 cm. The diameter of a gear on the blueprint measures 1.2 dm. Find the actual diameter of the gear.

5. A machinist earns $5.85 an hour. How much is earned by a machinist who works 48 hours in a week?

6. The length of a metal rod is 6.31 cm. It must be shortened to 6.02 cm. How much must be filed down?

7. A wheel has a diameter of 8.1 centimeters. A piece of wire is to be wrapped around the wheel. Find the length of the wire to the nearest tenth of a centimeter. Use $c = 3.14d$.

Non-Routine Problems

There is no simple method used to solve these problems. Use whatever methods you can. Do not expect to solve these easily.

1. Kathy wanted to find out the difference of the sum of all even numbers from 2 through 1,000 and all of the odd numbers from 1 through 999. To avoid the tedious work of finding the two sums, she arranged her work as follows:

    $$\begin{array}{r} 2 + 4 + 6 + \ldots + 996 + 998 + 1{,}000 \\ -1 + 3 + 5 + \ldots + 995 + 997 + 999 \\ \hline 1 + 1 + 1 + \ldots + 1 + 1 + 1 \end{array}$$

    Use the above approach and tell what the difference is.

2. Do you know that the number 96 can be divided into three parts in the ratio 1 to 3 to 8? Find these three numbers.

3. The sum of two numbers is 105. One-half of one number is equal to one-third of the other number. What are the two numbers?

4. A father gave some money to his three sons. To the first son he gave half the money plus one-half dollar. To the second son he gave half of what was left plus one-half dollar. To the third son he gave half of what was left plus one-half dollar. Having done that, he had no money left. How much money did the father give away in all?

5. The number of bacteria in a bottle triples each minute. The bottle is completely filled after 15 minutes. After how many minutes was the bottle one-third full?

6. Choose any three consecutive numbers. Find their product. Is the product divisible by 6? Now do the same with other triplets of consecutive numbers. Is the product in each case divisible by 6? Do you think it is always true? Can you find an exception?

COMPUTER ACTIVITIES

Recall that commission is a way many businesses encourage their employees to work harder in order to increase their incomes. The more you sell, the greater your earnings. The program below illustrates an application of the branching concept of computer programming taught in the computer section at the end of this text.

Example Joan is paid a regular salary of $145 per week plus a commission of $7.00 for each T.V. she sells over 6. Write a program to find her total pay. Then type and **RUN** the program to find her total pay for 10 sales; then for 3 sales.

```
                   Allows for two RUNS     10   FOR K = 1 TO 2
                                           20   INPUT "TYPE NUMBER SOLD ";N
  If more than 6 are sold, computer goes   30   IF N > 6 THEN 60
             to line 60 to compute total.  40   LET T = 145
                                           50   GOTO 70
                                           60   LET T = 145 + 7 * (N - 6)
                                           70   PRINT "TOTAL PAY IS "T
                                           80   NEXT K
                                           90   END

                                           ]RUN
                                           TYPE NUMBER SOLD 10
                                           TOTAL PAY IS 173
                                           TYPE NUMBER SOLD 3
   Total is $145 + 7 for each sold over 6  TOTAL PAY IS 145
```

See the computer section beginning on page 420 for more information.

Exercises

1. **RUN** the program above to find the total pay for 13 sales; for 5 sales.
2. Suppose the way Joan is paid is changed to a regular salary of $185 per week plus a commission of $10 for each sale she makes after 8 or more. Make the necessary changes in lines of the program. Then **RUN** the program to find the total pay for 6 sales; then for 8 sales; and for 12 sales.

★ 3. José is paid a basic salary of $220 plus a commission of $6 for each sale up to 5; $8 for each sale over 5 up to 9 sales; $9 for each sale when 10 or more. Write and **RUN** a program to find his total pay for 3 sales; then for 9 sales; and for 10 sales.

Chapter Review

Write a numeral for each word name. [165]

1. sixty-eight and ninety-four ten-thousandths
2. nine million, seven hundred fifty thousand and six tenths

Write a word name for each numeral. [165]

3. 700.96
4. 3,200,000.001

Change to decimals. [165]

5. $\frac{85}{100}$ 6. $\frac{9}{10}$ 7. $\frac{8}{100}$ 8. $\frac{8}{1{,}000}$ 9. $\frac{77}{100}$ 10. $\frac{77}{1{,}000}$

Round to the nearest tenth and to the nearest hundredth. [165]

11. 0.934
12. 9.758
13. 12.436
14. 10.952

Change to a decimal to the nearest tenth and to the nearest hundredth. [168]

15. $\frac{2}{5}$ 16. $\frac{1}{3}$ 17. $\frac{4}{11}$ 18. $\frac{3}{7}$ 19. $\frac{3}{11}$ 20. $\frac{5}{9}$

Add. [171]

21. $149.83 + 0.52 + 88.43$
22. $67.35 + 0.005 + 8.473$

Subtract. [171]

23. $38.43 - 22.78$
24. $17.14 - 8.384$

Multiply. [173]

25. $(71.2)(3.4)$
26. 41.6×0.036
27. $(2.45)(0.014)$
28. 218×0.31

Evaluate. [171, 173]

29. $x - 5.23$ if $x = 19.1$
30. $4.3y$ if $y = 7.01$

Divide. Round to the nearest tenth or nearest hundredth. [177]

31. $0.04\overline{)2.448}$
32. $0.24 \div 0.07$
33. $0.024\overline{)0.01384}$

34. Multiply. 3.6453×10^4 [180]
35. Write 532,000 in scientific notation. [180]
36. Solve. $x - 4.8 = 9.21$ [182]
37. Solve. $0.3x = 0.93$ [182]
38. Horace saves $4.25 each week. How much has he saved after 52 weeks? [173]
39. Orlando traveled 156 kilometers in 2.5 hours. Find his average speed. [177]

Chapter Test

Write a numeral for each word name.

1. five hundred twenty-eight and three thousandths
2. forty-two and three hundred one millionths

Write a word name for the numeral.

3. 1,462.0075

Change to decimals.

4. $\frac{3}{10}$ 5. $\frac{75}{100}$ 6. $\frac{5}{1,000}$

Round to the nearest tenth and to the nearest hundredth.

7. 0.732   8. 6.845

Change to decimals.

9. $\frac{2}{11}$ Round to the nearest tenth.   10. $\frac{5}{7}$ Round to the nearest hundredth.

Add.

11. 258.43 + 0.62 + 99.28   12. 72.36 + 0.004 + 9.416

Subtract.

13. 78.42 − 19.68   14. 19 − 3.14

Multiply.

15. (84.3)(6.2)   16. 249 × 0.012   17. 49.3567 × 10^3

Evaluate.

18. $y + 16.3$ if $y = 2.78$ 19. $2.03x$ if $x = 6.4$

Divide.

20. $0.04 \overline{)7.284}$ 21. 6.4862 ÷ 0.02

22. Divide. Round to the nearest tenth. 0.01372 ÷ 0.021

23. Write in scientific notation. 38,000,000

24. Solve. $x - 4.8 = 18.9$ 25. Solve. $\frac{x}{8.2} = 0.7$

26. Potatoes cost $1.18 per kilogram. What is the cost of 2.5 kilograms?

27. Norma traveled 279.3 kilometers in 3.8 hours. Find her average speed.

Measurement

8

The Columbia *space shuttle at lift-off weighs 4.5 milllion pounds.*

Centimeters and Millimeters

◆ **OBJECTIVES** ◆

To measure to the nearest centimeter

To measure to the nearest millimeter

◆ **RECALL** ◆

To measure the length of segment AB:
1. Place the end of a ruler at point A.
2. Read the number of units at point B.

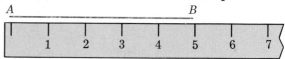

Segment AB measures 5 units.

The centimeter (cm) is a unit for measuring length.

1 cm

The ruler below is marked in centimeters.

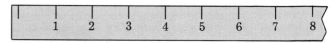

Example 1   Use a centimeter ruler to measure segment AB.

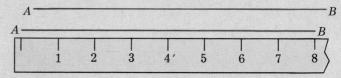

Read $\overline{AB}$ as segment AB.   So, $\overline{AB}$ measures 8 cm.

Example 2   Measure $\overline{CD}$ to the nearest centimeter.

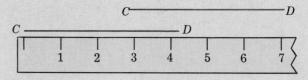

$\overline{CD}$ is between 4 cm and 5 cm but closer to 4 cm.
Measurements are approximate.   So, $\overline{CD}$ measures 4 cm to the nearest centimeter.

practice ▷   **Measure to the nearest centimeter.**

1. R————S   2. X————————Y

The millimeter (mm) is used to measure lengths with greater accuracy. There are 10 mm in 1 cm.

1 cm = 10 mm, or
1 mm = 0.1 cm

Example 3   Use a ruler marked in millimeters to measure $\overline{EF}$ to the nearest millimeter.

So, $\overline{EF}$ measures 68 mm to the nearest millimeter.

practice ▷ Measure to the nearest millimeter.

3. P ——————— Q   4. M ————————————— N

◇ EXERCISES ◇

Measure to the nearest centimeter.

1. A ———————————— B   2. C ———————————— D
3. E ————— F   4. G ———— H

Measure to the nearest millimeter.

5. K ————————————— L   6. M ———————————— N
7. P ——————— Q   8. R ————————————— S

Measure to the nearest centimeter.

9. The length of your right foot.
10. The length of your left index finger
11. Your height
12. The length of a chalkboard eraser

Measure to the nearest millimeter.

13. The length of your right thumbnail
14. The length of a paper clip
15. The length of your pencil
16. The width of your math book

CENTIMETERS AND MILLIMETERS

Units of Length

◇ **OBJECTIVES** ◇

To state the meaning of metric prefixes

To express one metric unit of length in terms of another

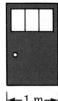
|←1 m→|

|←2 dm→|

A baseball bat is about 1 m long.

◇ **RECALL** ◇

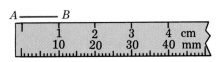

$\overline{AB}$ measures 1 cm.
$\overline{AB}$ measures 10 mm.

$$1 \text{ cm} = 10 \text{ mm, or}$$
$$1 \text{ mm} = 0.1 \text{ cm}$$

The decimeter (dm) is another unit for measuring length.

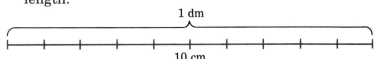

There are 10 cm in 1 dm.

$$1 \text{ dm} = 10 \text{ cm, or}$$
$$1 \text{ cm} = 0.1 \text{ dm}$$

Example 1

How many millimeters are there in a decimeter? A millimeter is what part of a decimeter?

$$1 \text{ dm} = 10 \text{ cm}$$
$$1 \text{ cm} = 10 \text{ mm}$$

So, 1 dm = 10 × (10 mm), or 100 mm.

$$1 \text{ cm} = 0.1 \text{ dm}$$
$$1 \text{ mm} = 0.1 \text{ cm}$$

So, 1 mm = 0.1 × (0.1 dm), or 0.01 dm.

The meter (m) is the basic unit for measuring length. There are 10 dm in 1 m.

$$1 \text{ m} = 10 \text{ dm, or}$$
$$1 \text{ dm} = 0.1 \text{ m}$$

Example 2

Make true sentences.
$1\text{ m} = \underline{\ ?\ }\text{ cm}\qquad 1\text{ cm} = \underline{\ ?\ }\text{ m}$
$1\text{ m} = 10\text{ dm}\qquad 1\text{ cm} = 0.1\text{ dm}$
$1\text{ dm} = 10\text{ cm}\qquad 1\text{ dm} = 0.1\text{ m}$

So, $1\text{ m} = 10 \times (10\text{ cm})$, or 100 cm, and
$1\text{ cm} = 0.1 \times (0.1\text{ m})$, or 0.01 m.

Example 3

Make true sentences.
$1\text{ m} = \underline{\ ?\ }\text{ mm}\qquad 1\text{ mm} = \underline{\ ?\ }\text{ m}$
$1\text{ m} = 10\text{ dm}\qquad 1\text{ mm} = 0.1\text{ cm}$
$1\text{ dm} = 10\text{ cm}\qquad 1\text{ cm} = 0.1\text{ dm}$
$1\text{ cm} = 10\text{ mm}\qquad 1\text{ dm} = 0.1\text{ m}$

So, $1\text{ m} = 10 \times 10 \times 10\text{ mm}$, or $1{,}000\text{ mm}$, and
$1\text{ mm} = 0.1 \times 0.1 \times 0.1\text{ m}$, or 0.001 m.

practice ▷ **Make true sentences.**

1. $1\text{ cm} - \underline{\ ?\ }\text{ mm}$
2. $1\text{ mm} - \underline{\ ?\ }\text{ dm}$

METRIC PREFIXES

kilo thousand	hecto hundred	deka ten	basic unit	deci tenth	centi hundredth	milli thousandth

The chart below shows the units for measuring length. The units that are used most often are shaded.

kilometer km 1,000 m	hectometer hm 100 m	dekameter dam 10 m	meter m	decimeter dm 0.1 m	centimeter cm 0.01 m	millimeter mm 0.001 m

Example 4

Use the chart.

Make true sentences.
$1\text{ hm} = \underline{\ ?\ }\text{ m}\qquad 1\text{ m} = \underline{\ ?\ }\text{ hm}$
$1\text{ hm} = 100\text{ m}\qquad 1\text{ m} = 0.1 \times 0.1\text{ hm}$, or 0.01 hm

So, $1\text{ hm} = 100\text{ m}$ and $1\text{ m} = 0.01\text{ hm}$.

practice ▷ **Make true sentences.**

3. $1\text{ dam} = \underline{\ ?\ }\text{ m}$
4. $1\text{ m} = \underline{\ ?\ }\text{ km}$

UNITS OF LENGTH

◇ EXERCISES ◇

What is the meaning of each prefix?

1. deka
2. centi
3. kilo
4. milli
5. hecto
6. deci

Make true sentences.

7. 1 m = __?__ cm
8. 1 m = __?__ dam
9. 1 km = __?__ m
10. 1 cm = __?__ mm
11. 1 m = __?__ dm
12. 1 mm = __?__ m
13. 1 hm = __?__ m
14. 1 m = __?__ mm
15. 1 cm = __?__ m
16. 1 dm = __?__ m
17. 1 mm = __?__ cm
18. 1 m = __?__ hm

Solve these problems.

19. The distance across Jim's thumbnail is 1 cm. How many mm is this?
20. Juanita walked 1 km to school. How many m is this?
22. Marie is 1 m tall. How many cm is this?
22. A housing complex is 1,000 m long. How many km is this?
23. A lightning bug is 10 mm long. How many cm is this?
24. Jan's baseball bat is 100 cm long. How many m is this?

Algebra Maintenance

Evaluate.

1. $3a + 4b$ if $a = 4$ and $b = \frac{3}{4}$
2. $3mr^2$ if $m = 5$ and $r = \frac{1}{5}$
3. $3.05x$ if $x = 5.6$
4. $\frac{1}{3}c + \frac{3}{4}d$ if $c = 6$ and $d = \frac{1}{2}$
5. $4(x + 3)$ if $x = 0.27$
6. $0.3(x - 0.2)$ if $x = 5.1$

Solve.

7. $2x + 5 = 13$
8. $3x - 7 = 26$
9. $\frac{2}{3}x - 5 = 1$

Solve these problems.

10. Rita's age of 14 is the same as her father's age divided by 3 and the quotient decreased by 2. What is her father's age?
11. 5 more than 6 times a number is 29. What is the number?

Changing Units of Length

◇ **OBJECTIVE** ◇

To change units of length within the metric system

◇ **RECALL** ◇

To multiply 5.63 by 100, move the decimal point two places to the right.

$$5.63 \times 100 = 563$$

To divide 28 by 1,000, move the decimal point three places to the left.

$$28 \div 1,000 = 0.028$$

Example 1

Use the chart of units of length.

Change 4.25 meters to centimeters.

km	hm	dam	m	dm	cm	mm
			4.25		?	

2 places to the right →

$1 \text{ m} = 100 \text{ cm}$
To multiply by 100, move the decimal point two places to the right.

To get from m to cm, move two places to the right.

4.25

So, 4.25 m = 425 cm.

Example 2

Change 5 kilometers to meters.

km	hm	dam	m	dm	cm	mm
5			?			

3 places to the right

$1 \text{ km} = 1,000 \text{ m}$
To multiply by 1,000, move the decimal point three places to the right.

To get from km to m, move three places to the right.

5.000

So, 5 km = 5,000 m.

---

practice ▷ Change as indicated.

1. 12.5 m to cm
2. 7 km to m

Example 3

Change 625 millimeters to meters.

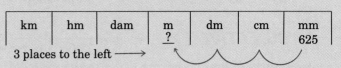

1 mm = 0.001 m
To multiply by 0.001, move the decimal point three places to the left.

To get from mm to m, move three places to the left.
0625.

So, 625 mm = 0.625 m.

practice ▷ Change as indicated.

3. 432 mm to m
4. 1,500 mm to m

Example 4

Change 4 meters to kilometers.

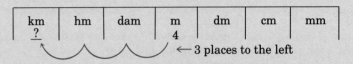

1 m = 0.001 km
To multiply by 0.001, move the decimal point three places to the left.

To get from m to km, move three places to the left.
0004.

So, 4 m = 0.004 km.

practice ▷ Change as indicated.

5. 529 m to km
6. 82 m to km

◇ ORAL EXERCISES ◇

To change as indicated, tell how the decimal point must be moved.

1. m to cm
2. km to m
3. m to mm
4. mm to cm
5. mm to m
6. cm to mm
7. m to km
8. cm to m
9. m to dm

◇ EXERCISES ◇

Change as indicated.

1. 8 m to cm
2. 3,000 mm to m
3. 6.5 km to m
4. 4 cm to mm
5. 7 m to mm
6. 800 m to km
7. 55 mm to cm
8. 63 cm to m
9. 420 mm to m
10. 7.5 m to cm
11. 8.1 cm to mm
12. 9.85 km to m
13. 43 m to km
14. 2.6 m to mm
15. 359 cm to m
16. 12 km to m
17. 2 mm to cm
18. 4 cm to m
19. 5 mm to m
20. 30 cm to mm
21. 6 m to km

Solve these problems.

22. Pat bought 2.5 meters of fabric. How many centimeters is this?

23. A caterpillar is 3.4 cm long. How many millimeters is this?

24. José walked 0.75 km to the store. How many meters is this?

25. Mary won the 100-meter dash. How many kilometers did she run?

★ 26. John jogged 800 m on Monday, 1,300 m on Tuesday, and 1,150 m on Wednesday. How many kilometers did he jog in the 3 days?

★ 27. Amy is making a bookcase. She needs 3 shelves each 850 cm long and 2 end pieces each 1,200 cm long. How many meters of wood are needed?

Work with a partner to measure both your height and the distance between your fingertips when your arms are outstretched. Use a tape measure, or a piece of string and a meter stick. Give measurements to the nearest centimeter. What do you notice about the measurements of your height and your arm span?

CHANGING UNITS OF LENGTH

Problem Solving – Applications
Jobs for Teenagers

Example

Read

Plan

Solve

Interpret

When taking a road trip, the Adams family fills up the gas tank each time they stop for gas. On the last stop, it took 18.6 gal to fill up the tank. They traveled 604.5 mi since the last fill-up. How many miles per gallon did they get?

To compute mileage, use the formula:

$$mpg = \frac{m}{g}, \text{ where}$$

mpg = miles per gallon (mi/gal)
m = miles traveled
g = gallons of gas used

$mpg = \frac{604.5}{18.6}$ Replace m by 604.5 and g by 18.6.

= 32.5 mi/gal.

So, the Adams got 32.5 mi/gal.

practice ▷ What is the mileage of a limousine that traveled 225.6 mi on 23.5 gal of gas? Use the formula given above to solve.

Solve these problems.

1. Pete's motorcycle traveled 304.5 mi on 5.8 gal of gas. What was the mileage?

2. On the last trip, your family got 29.5 mi/gal with your car. Getting the same mileage, how many gallons of gas will you need to travel 531 mi?

3. The Pavuks traveled 396.8 mi averaging 24.8 mi/gal. They paid $1.179/gal of gas. How much did the gas cost for this trip?

4. The Howells traveled 592.2 mi getting 28.2 mi/gal. They paid $26.25 for the gas. What was the price of the gas?

5. Ms. Lowery's car gets 18 mi/gal of gas. She paid $1.079/gal for gasoline. What is the cost of gas for a trip of 306 mi?

6. Mrs. Claire travels on the average 18 mi/day. She pays $1.099/gal for gas, and the car gets 28 mi/gal. What is the cost of gas for a 7-day week?

7. Chris works from 3:30 P.M. to 6:30 P.M. on Monday to Friday and from 8:00 A.M. to 4:00 P.M., with an hour off, on Saturday. She earns $3.25 per hour. How much does she earn in a week?

8. Chris wants to buy a stereo system with speakers that costs $119.95. How many weeks will she have to work before she can buy the system? How much money will she have left?

9. Soon, Chris will get a raise of $0.25/hr. How much more per week will she earn after the raise? How much per week will she earn then?

10. Leroy changes Mr. Haley's oil filter every 10,000 mi. His odometer read 28,397 when the filter was last changed. When should it be changed again?

11. Alan pumped 18 gal of gas into Anthony's car. Gas sells for $1.29/gal. How much did Anthony pay?

12. Ms. Gibson's car gets about 8 mi/gal of gas. Gas costs $1.29/gal. What is the cost for a trip of 192 k/mi?

13. Mrs. Rodriguez brought her car to the station for an oil change that cost $6.95, a new oil filter that cost $4.50, and a lubrication that cost $5.75. What was her total bill?

14. Jerry needed a new windshield-wiper blade and some windshield cleaning fluid. The blade cost $2.50 and the fluid cost $1.95. What was his change from a $10 bill?

15. The transmission in Jack's car has a leak. He must add a quart of transmission fluid every month. The fluid costs $2.50/qt. How much will Jack spend on transmission fluid in a year?

16. Eric needs to replace the water pump in his car. A new pump costs $35 plus 5% sales tax. The labor will cost $15. How much money will Eric need to replace the pump?

PROBLEM SOLVING

Area

◇ OBJECTIVES ◇

To change units of area within the metric system

To compute areas of rectangles and squares

◇ RECALL ◇

Area is measured in square units. The *square centimeter* (cm²) is a unit of area.

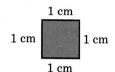

1 square centimeter
1 cm²

Example 1

Formula for the area of rectangle

To change from mm² to cm², divide by 100.

A rectangle measures 42 mm by 16 mm. Find the area in cm².

$A = l \cdot w$
42 mm · 16 mm
672 mm²
672 mm² = ___?___ cm²
6̰7̰2̰ To divide by 100, move the decimal point two places to the left.

So, the area is 6.72 cm².

Example 2

Formula for the area of a square

To change from m² to cm², multiply by 10,000.

A square measures 5.8 m on each side. Find the area in cm².

$A = s^2$
$(5.8 \text{ m})^2$
33.64 m²
33.64 m² = ___?___ cm
33.6400̰ To multiply by 10,000, move the decimal point four places to the right.

So, the area is 336,400 cm².

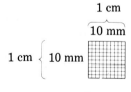

To change from cm² to mm², *multiply* by 100.
To change from mm² to cm², *divide* by 100.

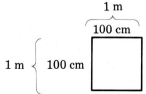

To change from m² to cm², *multiply* by 10,000.
To change from cm² to m², *divide* by 10,000.

practice Make true sentences.

1. $288 \text{ mm}^2 = \underline{} \text{ cm}^2$
2. $36 \text{ cm}^2 = \underline{} \text{ mm}^2$
3. $5.9 \text{ m}^2 = \underline{} \text{ cm}^2$
4. $15{,}890 \text{ cm}^2 = \underline{} \text{ m}^2$

Larger metric units of area are the are (a) and the hectare (ha). $1 \text{ a} = 100 \text{ m}^2$ $1 \text{ ha} = 10{,}000 \text{ m}^2$

Example 3

A rectangular lot measure 160 m by 240 m. Find the area in ha.

$A = l \cdot w$

160 m · 240 m

38,400 m²

To change from m² to ha, divide by 10,000.

$38{,}400 \text{ m}^2 = \underline{} \text{ ha}$

38400. To divide by 10,000, move the decimal point four places to the left.

So, the area is 3.84 ha.

practice Make true sentences.

5. $460 \text{ m}^2 = \underline{} \text{ ha}$
6. $21{,}900 \text{ m}^2 = \underline{} \text{ ha}$

◇ EXERCISES ◇

Make true sentences.

1. $3 \text{ cm}^2 = \underline{} \text{ mm}^2$
2. $8 \text{ m}^2 = \underline{} \text{ cm}^2$
3. $6 \text{ ha} = \underline{} \text{ m}^2$
4. $70{,}000 \text{ cm}^2 = \underline{} \text{ m}^2$
5. $60{,}000 \text{ m}^2 = \underline{} \text{ ha}$
6. $400 \text{ mm}^2 = \underline{} \text{ cm}^2$
7. $6.1 \text{ m}^2 = \underline{} \text{ cm}^2$
8. $4.2 \text{ cm}^2 = \underline{} \text{ mm}^2$
9. $2 \text{ a} = \underline{} \text{ m}^2$
10. $7{,}600 \text{ m}^2 = \underline{} \text{ ha}$
11. $93{,}000 \text{ cm}^2 = \underline{} \text{ m}^2$
12. $39 \text{ m}^2 = \underline{} \text{ cm}^2$
13. $26 \text{ cm}^2 = \underline{} \text{ mm}^2$
14. $7{,}640 \text{ mm}^2 = \underline{} \text{ cm}^2$
15. $83 \text{ ha} = \underline{} \text{ m}^2$
16. $7.9 \text{ ha} = \underline{} \text{ m}^2$
17. $4{,}500 \text{ cm}^2 = \underline{} \text{ m}^2$
18. $380 \text{ mm}^2 = \underline{} \text{ cm}^2$
★ 19. $420 \text{ m}^2 = \underline{} \text{ a}$
★ 20. $34{,}000 \text{ m}^2 = \underline{} \text{ a}$
★ 21. $67 \text{ a} = \underline{} \text{ m}^2$
★ 22. $375 \text{ a} = \underline{} \text{ ha}$
★ 23. $42 \text{ km}^2 = \underline{} \text{ ha}$
★ 24. $7{,}100 \text{ a} = \underline{} \text{ km}^2$

Solve these problems.

25. A rectangle measures 4 cm by 8 cm. Find the area in mm².
26. A square measures 360 cm on each side. Find the area in m².
27. A square lot measures 160 m on each side. Find the area in ha.
28. A rectangle measures 50 mm by 110 mm. Find the area in cm².
29. A rectangle measures 4.2 m by 7.3 m. Find the area in cm².
30. A rectangular lot measures 630 m by 370 m. Find the area in ha.
31. A square deck measures 4 m on each side. Find the area in cm².
32. A rectangular countertop measures 230 cm by 58 cm. Find the area in m².
33. A rectangular park measures 170 m by 80 m. Find the area in ha.
34. A business card measures 9 cm by 6.2 cm. Find the area in mm².
★ 35. A square lot is 2.3 km on each side. Find the area in ha.
★ 36. A state park contains 5,300 ha. Express its area in km².

Capacity

◇ OBJECTIVES ◇

To choose the best unit for measuring the capacity of an object

To change between milliliters and liters

cm³ means *cubic centimeter*.

◇ RECALL ◇

200 mm = __?__ m

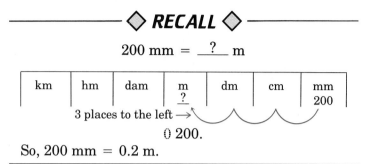

So, 200 mm = 0.2 m.

To measure the capacity of an object, we find out how much it holds.

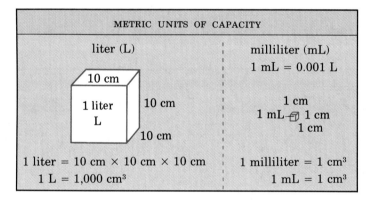

Example 1

Which unit, L or mL, is better for measuring the capacity of each?

gas tank of a car   dose of cough syrup

A gas tank is greater than 1 L, so use L. A dose of cough syrup is less than 1 L, so use mL.

L is better for measuring the capacity of a tank of gas, and mL is better for measuring the capacity of a dose of cough syrup.

---

practice ▷ Which unit, L or mL, is better for measuring the capacity of each?

1. a glass
2. a fishbowl

CHAPTER EIGHT

The chart below shows most of the units for measuring capacity. The units that are used most often are shaded.

kiloliter kL 1,000 L	hectoliter hL 100 L	dekaliter daL 10 L	liter L	deciliter dL 0.1 L	centiliter cL 0.01 L	milliliter mL 0.001 L

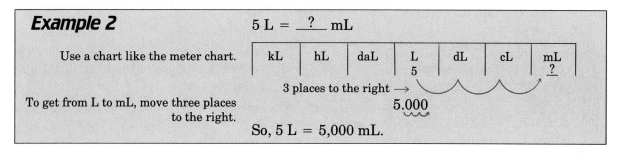

Example 2

Use a chart like the meter chart.

To get from L to mL, move three places to the right.

$5 \text{ L} = \underline{} \text{ mL}$

So, 5 L = 5,000 mL.

practice Make true sentences.

3. $8.2 \text{ L} = \underline{} \text{ mL}$

4. $30 \text{ L} = \underline{} \text{ mL}$

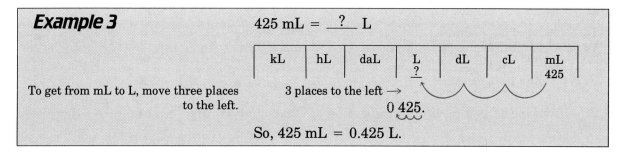

Example 3

$425 \text{ mL} = \underline{} \text{ L}$

To get from mL to L, move three places to the left.

So, 425 mL = 0.425 L.

practice Make true sentences.

5. $620 \text{ mL} = \underline{} \text{ L}$

6. $3,800 \text{ mL} = \underline{} \text{ L}$

Example 4

A punch recipe calls for $1\frac{1}{2}$ liters of juice. How many milliliters is this?

To get from L to mL, move three places to the right.

$1\frac{1}{2} \text{ L} = 1.5 \text{ L} = 1.500 \text{ mL}$

So, $1\frac{1}{2}$ liters is 1,500 milliliters.

practice Make true sentences.

7. $0.6 \text{ L} = \underline{} \text{ mL}$

8. $2.3 \text{ L} = \underline{} \text{ mL}$

◇ ORAL EXERCISES ◇

Make true sentences.

1. 1 mL = __?__ L
2. 1,000 mL = __?__ L
3. 1 L = __?__ mL
4. 0.001 L = __?__ mL

◇ EXERCISES ◇

Which unit, L or mL, is better for measuring the capacity of each?

1. a cup of hot chocolate
2. bottled water for a drinking fountain
3. a bathtub
4. a baby's bottle
5. a spoon
6. a water tank
7. a glass of milk
8. a bottle of cough syrup

Make true sentences.

9. 8,000 mL = __?__ L
10. 6 L = __?__ mL
11. 1,500 mL = __?__ L
12. 12 L = __?__ mL
13. 900 mL = __?__ L
14. 8.2 L = __?__ mL
15. 365 mL = __?__ L
16. 4.25 L = __?__ mL
17. 10.4 L = __?__ mL
★ 18. 3,600 L = __?__ kL
★ 19. 2.1 kL = __?__ L
★ 20. 3 kL = __?__ mL

Solve these problems.

21. A recipe calls for $\frac{1}{2}$ liter of milk. How many milliliters is this?

22. Judy drank $\frac{1}{4}$ liter of juice. How many milliliters is this?

Calculator

Find the capacity of the rectangular solid in liters. Round to the nearest tenth.

7.1 cm
9.8 cm
16.3 cm

Formula $V = l \cdot w \cdot h$

PRESS   16.3 ⊗ 9.8 ⊗ 7.1 ⊘ 1,000 ⊜
DISPLAY   1.134154

So, the capacity is 1.1 liters.

To change from cm^3 to L, divide by 1,000.

Find the capacity of the rectangular solid in liters.

1. 24 cm by 11 cm by 7 cm
2. 18.1 cm by 10.5 cm by 8.4 cm

Weight

◇ OBJECTIVES ◇

To choose the best unit for weighing an object
To change between milligrams, grams, kilograms, and tons

◇ RECALL ◇

1 mL — 1 cm³

1 cm³ = 1 mL

1 L, 10 cm × 10 cm × 10 cm

1,000 cm³ = 1 L

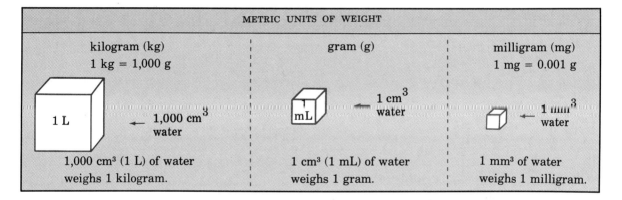

METRIC UNITS OF WEIGHT		
kilogram (kg) 1 kg = 1,000 g	gram (g)	milligram (mg) 1 mg = 0.001 g
1 L ← 1,000 cm³ water	mL ← 1 cm³ water	← 1 mm³ water
1,000 cm³ (1 L) of water weighs 1 kilogram.	1 cm³ (1 mL) of water weighs 1 gram.	1 mm³ of water weighs 1 milligram.

Example 1

Which unit, kg, g, or mg, is best for weighing each object?

   a person   a mosquito   an apple   a cat

A person weighs more than 1 kg, so use kg. A mosquito weighs more than 1 mg, but less than 1 g, so use mg.

So, kg is the best unit for weighing a person and a cat, mg is the best unit for weighing a mosquito, and g is the best unit for weighing an apple.

practice ▷

Which unit, kg, g, or mg, is best for weighing each object?

1. a mouse
2. 50 grains of sand

The chart below shows most of the units for measuring weight. The units used most often are shaded.

kilogram kg 1,000 g	hectogram hg 100 g	dekagram dag 10 g	gram g	decigram dg 0.1 g	centigram cg 0.01 g	milligram mg 0.001 g

WEIGHT

Example 2

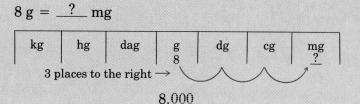

Use a chart like the meter chart.

To get from g to mg, move three places to the right.

So, 8 g = 8,000 mg.

practice ▷ Make true sentences.

3. 5.6 g = __?__ mg
4. 3 kg = __?__ g

Example 3

375 g = __?__ kg

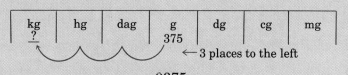

To get from g to kg, move three places to the left.

So, 375 g = 0.375 kg.

practice ▷ Make true sentences.

5. 80 g = __?__ kg
6. 400 mg = __?__ g

The metric ton (t) is used for weighing very heavy things, such as trucks or concrete.
  1 metric ton = 1,000 kilograms
  1 t = 1,000 kg, or 1 kg = 0.001 t

Example 4

To multiply by 1,000, move the decimal point three places to the right. To multiply by 0.001, move the decimal point three places to the left.

Make true sentences.

3.2 t = __?__ kg	750 kg = __?__ t
1 t = 1,000 kg	1 kg = 0.001 t
3.2 t = 3.2 × 1,000 kg	750 kg = 750 × 0.001 t
3.200, or 3,200	0750, or 0.75

So, 3.2 t = 3,200 kg and 750 kg = 0.75 t.

practice ▷ Make true sentences.

7. 4.5 kg = __?__ t
8. 63 t = __?__ kg

◇ ORAL EXERCISES ◇

Make true sentences.

1. 1 g = __?__ kg
2. 1 t = __?__ kg
3. 1 mg = __?__ g
4. 1 kg = __?__ g
5. 1 kg = __?__ t
6. 1 g = __?__ mg

◇ EXERCISES ◇

Which unit, mg, g, kg, or t, is best for weighing each object?

1. a bicycle
2. a fly
3. an elephant
4. a baby
5. a quarter
6. a magazine
7. an aspirin
8. a boxcar
9. a bus
10. a dog
11. a pinch of salt
12. a potato

Make true sentences.

13. 7.5 g = __?__ mg
14. 300 mg = __?__ g
15. 8.2 kg = __?__ g
16. 6.4 t = __?__ kg
17. 8,700 g = __?__ kg
18. 4,600 kg = __?__ t
19. 12 kg = __?__ g
20. 7,200 mg = __?__ g
21. 5,500 kg = __?__ t
22. 4.9 t = __?__ kg
23. 30 g = __?__ kg
24. 16 g = __?__ mg
★ 25. 6 t = __?__ g
★ 26. 12,000 mg = __?__ kg
★ 27. 1,000,000 g = __?__ t

Solve these problems.

28. A pork roast weighs 2.1 kg. How many grams is this?
29. A car weighs 888 kg. How many tons is this?
★ 30. Bill must take 4 mg of a sinus pill 3 times a day. How many grams will he take in a week?
★ 31. Meredith takes an iron tablet every day. Each tablet contains 63 mg of iron. How many grams of iron does she take in a year?

The smallest unit in our system of weight is the grain. A grain is equal to 0.0648 gram.

1. An aspirin tablet weighs 5 grains. How many milligrams is this?
2. How many grams do 50 aspirin tablets weigh?
3. How many milligrams do 100 aspirin tablets weigh?
4. How many kilograms do 1,000 aspirin tablets weigh?

WEIGHT

Problem Solving – Careers
Stonemasons

1. A marble floor measures 12 meters by 18 meters. How many square meters is the floor?

2. Mrs. Spencer is a stonemason. She earns $8.75 an hour. She worked 7 hours one day. How much did she earn for the day?

3. Mr. Alvarez works as a stonemason. He builds stone exteriors of buildings. A building requires 1,200 granite stones. Each stone weighs 96 kg. How many kg of granite are needed for the building?

4. Ms. Hayes is a cost estimator for a stonemasonry contractor. One job requires 430 stones to be cut. A stonemason needs about $\frac{1}{2}$ hour to cut each stone. How many hours should she estimate will be needed for the job?

5. Ms. Tsu cuts marble for marble floors. A floor of a theater lobby measures 50 m by 28 m. How many square meters of marble are needed to cover the floor?

6. To repair a church Mr. Black must replace 5 damaged stones in the exterior. He needs 3 hours to replace each stone. He is paid $8.60 per hour. How much will he earn?

7. Ms. Newman is a stonemason's apprentice. When she completes her apprenticeship, her wages will increase by $3.00 per hour. How much more will she earn in a 40-hour week?

Temperature: Celsius and Fahrenheit

To change between degrees Celsius and degrees Fahrenheit

Evaluate for y if $x = 6$.

$y = \frac{4}{3}x$

$y = \frac{4}{3} \cdot 6 = 8$

The table shows a comparison between Celsius and Fahrenheit readings.

C	0°	9.9°	15.4°	20.9°	26.4°	37°	48.2°	100°
F	32°	50°	60°	70°	80°	98.6°	120°	212°

Water freezes — Normal body temperature — Water boils

Example 1

Find an approximate relationship of the Fahrenheit (F) scale to the Celsius (C) scale.

C	Twice C	Add 30°	F
0°	0°	30°	32° ← exact
9.9°	19.8°	49.8°	50°
15.4°	30.8°	60.8°	60°
20.9°	41.8°	71.8°	70°
26.4°	52.8°	82.8°	80°

So, F is approximately twice C plus 30°.

Celsius to Fahrenheit formula

$$F = \frac{9}{5}C + 32°$$

Example 2

Change C = 15° and C = 20° to Fahrenheit. Use the Celsius to Fahrenheit formula.

Substitute the Celsius degrees for C in the formula.

C = 15°
$F = \frac{9}{5} \cdot 15 + 32$
$= 27 + 32$
$= 59°$

C = 20°
$F = \frac{9}{5} \cdot 20 + 32$
$= 36 + 32$
$= 68°$

So, F = 59° if C = 15°, and F = 68° if C = 20°.

TEMPERATURE: CELSIUS AND FAHRENHEIT

practice ▷ **Change Celsius to Fahrenheit.**

1. C = 10°   2. C = 25°   3. C = 18°

Fahrenheit to Celsius formula
$$C = \frac{5}{9}(F - 32°)$$

Example 3   Change F = 86° and F = 40° to Celsius. Use the Fahrenheit to Celsius formula.

$$F = 86°$$

Substitute the Fahrenheit degrees for F in the formula.
$$C = \frac{5}{9}(86 - 32)$$
$$= \frac{5}{9} \cdot 54$$
$$= 30°$$

$$F = 40°$$
$$C = \frac{5}{9}(40 - 32)$$
$$= \frac{5}{9} \cdot 8$$

$4.\overline{4}$ means 4.444 . . . $= \frac{40}{9}$ or $4.\overline{4}°$

So, C = 30° if F = 86°, and C = $4.\overline{4}°$ if F = 40°.

---

practice ▷ **Change Fahrenheit to Celsius.**

4. F = 80°   5. F = 60°   6. F = 50°

◇ EXERCISES ◇

Change to Fahrenheit.

1. 30°C   2. 12°C   3. 0°C   4. 24°C
5. 8°C   6. 35°C   7. 16°C   8. 32°C
9. 5°C   10. 27°C   11. 34°C   12. 29°C

Change to Celsius.

13. 32°F   14. 37°F   15. 68°F   16. 49°F
17. 72°F   18. 41°F   19. 59°F   20. 96°F
21. 54°F   22. 60°F   23. 78°F   24. 84°F
★ 25. 18°F ★ 26. 0°F   ★ 27. 10°F ★ 28. 24°F

COMPUTER ACTIVITIES

There are three commonly used methods for putting data into a computer.

INPUT Allows you to change the data each time you **RUN** the program.

READ/DATA The data is typed in as part of the program. The program can be **RUN** for only that *data*.

SEQUENTIAL INPUT The data consists of numbers in *sequence*.

Example 1 The formula for changing from Celsius to Fahrenheit is $F = \frac{9}{5}C + 32$. Write a program to find F for any 6 different values of C. Decide which of the three methods described above is most appropriate. Use **INPUT** since you wish to be able to change the data with each **RUN**.

```
Allows for 6 values.      10  FOR K = 1 TO 6
                          20  INPUT "TYPE IN A VALUE FOR C ";C
        Formula in BASIC. 30  LET F = 9 / 5 * C + 32
                          40  PRINT "FAHRENHEIT IS "F
                          50  NEXT K
```

Example 2 Choose the best method to write a program to find F for the following values of C: 25, 30, 35, 40, 45, 50. The data are in *sequence* 25, 30, 35, 40, 45, 50, each increased by 5.
Use **SEQUENTIAL INPUT**.

```
    Data are stepped up by 5.   10  FOR C = 25 TO 50 STEP 5
  No INPUT statement is needed. 20  LET F = 9 / 5 * C + 32
   The variable C in line 20 is 30  PRINT "FAHRENHEIT IS "F
   the same as C in line 10.    40  NEXT C
```

See the Computer Section beginning on page 420 for more information.

Exercises

1. Write a program to find F for the following values of C: 10, 8, 26, 45, 67, 90. Why would the **READ/DATA** format be the best choice? **RUN** the program.

Write and **RUN** a program to find F for the following values of C:
2. Any 7 different values of C   3. 14, 16, 18, 20, 22

COMPUTER ACTIVITIES

Chapter Review

Measure to the nearest centimeter. [190]

1. _____   2. _____

Measure to the nearest millimeter. [190]

3. _____   4. _____

Make true sentences. [192]

5. 1 cm = __?__ mm   6. 1 dm = __?__ m   7. 1 km = __?__ m

Change as indicated. [195]

8. 32 mm to cm   9. 4.65 km to m   10. 3 cm to m

Make true sentences. [200]

11. 9.2 m^2 = __?__ cm^2 12. 460 mm^2 = __?__ cm^2 13. $80,000 \text{ m}^2$ = __?__ ha

Solve this problem. [200]

14. A square measures 7.2 cm on each side. Find the area in mm^2.

Which unit, L or mL, is better for measuring the capacity of each? [202]

15. a soup bowl   16. the kitchen sink

Make true sentences. [202]

17. 4 L = __?__ mL   18. 23 mL = __?__ L   19. 5.62 L = __?__ mL

Solve these problems.

20. The wingspan of a moth is 42 mm. How many cm is this? [195]

21. Ramon drank $\frac{1}{2}$ liter of lemonade. How many milliliters is this? [202]

Which unit, mg, g, kg, or t, is best for weighing each object? [205]

22. a truck   23. a fox   24. a glass of water

Make true sentences. [205]

25. 8.4 g = __?__ mg   26. 4,700 g = __?__ kg   27. 25 kg = __?__ t

Solve these problems. [209]

28. Change 28°C to Fahrenheit.   29. Change 83°F to Celsius.

Chapter Test

Measure to the nearest centimeter.

1. _____

Measure to the nearest millimeter.

2. _____

Make true sentences.

3. 1 m = __?__ cm
4. 1 km = __?__ cm
5. 1 mm = __?__ m

Change as indicated.

6. 2.4 m to cm
7. 6 mm to m
8. 8.42 km to m

Make true sentences.

9. 73 cm^2 = __?__ mm^2
10. 37,500 cm^2 = __?__ m^2
11. 91 ha = __?__ m^2

Solve this problem.

12. A rectangle measures 70 mm by 420 mm. Find its area in cm^2.

Which unit, L or mL, is better for measuring the capacity of each?

13. a bucket
14. a soup spoon

Make true sentences.

15. 6.8 L = __?__ mL
16. 420 mL = __?__ L

Solve these problems.

17. Jane rode 0.85 km on her bike. How many m is this?
18. A recipe for pancakes calls for $\frac{1}{4}$ liter of milk. How many milliliters is this?

Which unit, mg, g, kg, or t, is best for weighing each object?

19. an orange
20. a baby
21. a flea

Make true sentences.

22. 4.3 kg = __?__ g
23. 28 mg = __?__ g

Solve these problems.

24. Change 21°C to Fahrenheit.
25. Change 71°F to Celsius.

Cumulative Review

Find the least common multiple (LCM).

1. 4; 14
2. 7; 9
3. 2; 5; 25
4. 4; 6; 10

Add or subtract. Simplify if possible.

5. $\frac{1}{2} + \frac{2}{7}$
6. $3\frac{3}{4} + 7\frac{5}{6}$
7. $\frac{4}{7} - \frac{1}{3}$
8. $5\frac{1}{4} - 2\frac{7}{10}$

Evaluate.

9. $\frac{2}{3}x + \frac{4}{5}y$ if $x = \frac{1}{2}$ and $y = \frac{5}{2}$
10. $1.06x$ if $x = 2.5$

11. Make a line graph to show Karen's scores on 6 mathematics tests: Test 1, 92; Test 2, 71; Test 3, 89; Test 4, 99; Test 5, 78; Test 6, 90.

Perform the indicated operations.

12. $4.78 + 2.006 + 10.4$
13. $57.04 - 12.678$
14. $(54.8)(7.4)$
15. $0.07 \overline{) 2.492}$

16. Write using scientific notation. 47,000,000

17. Solve. $0.05n = 1.345$

Make true sentences.

18. 3 m = __?__ cm
19. 6 mm = __?__ cm
20. 6.8 kg = __?__ g
21. 45 mg = __?__ g
22. 3.5 L = __?__ mL
23. 680 mL = __?__ L
24. Change 24°C to Fahrenheit.
25. Change 56°F to Celsius.

Solve these problems.

26. Carla worked $5\frac{2}{5}$ hours on Thursday and $6\frac{1}{4}$ hours on Friday. How many hours did she work on both days?

27. Fernando set himself a goal of working $45\frac{1}{4}$ hours during the 5-day week. He worked $38\frac{3}{4}$ hours during the 4 days. How many more hours does he have to work to meet his goal?

28. Lyle traveled 259.9 km in 4.6 hours. What was his average speed?

29. The length of a rectangle is 29 cm. Its perimeter is 97 cm. Find the width.

Percents

9

Chicago's famous skyline is dominated by some of the world's tallest buildings.

Introduction to Percent

◇ OBJECTIVES ◇
To change percents to decimals
To change decimals to percents
To change fractions to percents
To solve problems using percents

◇ RECALL ◇
$0.62 = \frac{62}{100}$ Change $\frac{2}{5}$ to a decimal.

Divide. $5\overline{)2.0}$ with quotient 0.4

42% means 42 out of 100 or 0.42.

Example 1

59% means 59 out of 100.

Change 59% to a decimal.
Drop the % symbol.
$0.59 \leftarrow$ Move the decimal point two places to the left.
So, 59% = 0.59.

practice ▷ Change to decimals.

1. 73%   2. 62%   3. 35%   4. 56%   5. 85%

Example 2

Drop the % symbol. Move the decimal point two places to the left.

Change to decimals. 38.3%, $45\frac{1}{2}$%, 175%

| 38.3% | $45\frac{1}{2}$% | 175% |
| 0.38,3 | $0.45\frac{1}{2}$ | 1.75 |

So, 38.3% = 0.383, $45\frac{1}{2}$% = $0.45\frac{1}{2}$, and 175% = 1.75.

practice ▷ Change to decimals.

6. 42.6%   7. $67\frac{1}{3}$%   8. 225%   9. 200%   10. 71.3%

Example 3

Drop the % symbol. Move the decimal point two places to the left.

Change to decimals. 6%, 0.6%, 6.2%

| 6% | 0.6% | 6.2% |
| 0.06 | 0.00,6 | 0.06,2 |

So, 6% = 0.06, 0.6% = 0.006, and 6.2% = 0.062.

216   CHAPTER NINE

practice ▷ **Change to decimals.**

11. 4% **12.** 0.4% **13.** 4.3% **14.** 5% **15.** 0.7%

Example 4

Change to percents. 0.03, 0.625, $0.33\frac{1}{3}$, 4

| 0.03 | 0.625 | $0.33\frac{1}{3}$ | 4 |

To change a decimal to a percent, move the decimal point two places to the right. Then write the % symbol.

| 003. | 62.5 | $33\frac{1}{3}$ | 400. |
| 3% | 62.5% | $33\frac{1}{3}$% | 400% |

So, 0.03 = 3%, 0.625 = 62.5%, $0.33\frac{1}{3} = 33\frac{1}{3}$%, and 4 = 400%.

---

practice ▷ **Change to percents.**

16. 0.07 **17.** 0.46 **18.** 0.667 **19.** $0.35\frac{1}{2}$ **20.** 8 **21.** 0.4

Example 5

Change $\frac{3}{5}$ to a percent. Change $\frac{2}{3}$ to a percent.

Change each to a decimal. Carry the division to hundredths.

$\frac{3}{5}$: $5\overline{)3.}$ | $\frac{2}{3}$: $3\overline{)2.}$

Write two 0's for hundredths.

$$5\overline{)3.00} \quad\quad 3\overline{)2.00}$$
$$0.60 \quad\quad\quad 0.66 = 0.66\frac{2}{3}$$
$$\underline{3\,0} \quad\quad\quad\quad \underline{1\,8}$$
$$0 \quad\quad\quad\quad 20$$
$$\underline{0} \quad\quad\quad\quad \underline{18}$$
$$0 \quad\quad\quad\quad 2$$

To change a decimal to a percent, move the decimal point two places to the right. Then write the % symbol.

$\frac{3}{5} = 0.60$ | $\frac{2}{3} = 0.66\frac{2}{3}$

60% | $66\frac{2}{3}$%

So, $\frac{3}{5} = 60$% and $\frac{2}{3} = 66\frac{2}{3}$%.

---

practice ▷ **Change to percents.**

22. $\frac{1}{2}$ **23.** $\frac{3}{4}$ **24.** $\frac{1}{3}$ **25.** $\frac{2}{7}$ **26.** $\frac{2}{9}$

◇ EXERCISES ◇

Change to decimals.

1. 48%
2. 59%
3. 79%
4. 66%
5. 88%
6. 43%
7. 45.3%
8. 37.9%
9. $52\frac{1}{4}$%
10. $68\frac{1}{3}$%
11. 180%
12. 400%
13. 5%
14. 2%
15. 0.1%
16. 0.3%
17. 7.9%
18. 6.3%

Change to percents.

19. 0.77
20. 0.63
21. 0.48
22. 0.66
23. 0.38
24. 0.51
25. 0.02
26. 0.08
27. 0.05
28. 0.449
29. 0.912
30. 0.551
31. $0.45\frac{1}{3}$
32. $0.66\frac{1}{2}$
33. $0.22\frac{2}{9}$
34. 0.046
35. 0.073
36. 0.018
37. 0.1
38. 0.5
39. 0.7
40. 6
41. 1
42. 2
43. $\frac{4}{5}$
44. $\frac{3}{4}$
45. $\frac{2}{5}$
46. $\frac{4}{9}$
47. $\frac{3}{7}$
48. $\frac{5}{6}$

Solve these problems.

49. Jake got 32 out of 100 correct on a test. What percent did he get correct?

50. Mary got 7 out of 10 correct on a quiz. What percent did she get correct?

★ 51. Find $\frac{1}{4} + \frac{1}{2}$. Change the answer to a percent.

★ 52. Find $\frac{1}{3} + \frac{1}{6}$. Change the answer to a percent.

Algebra Maintenance

Solve. Check.

1. $a + 3 = 21$
2. $x - 3 = 6$
3. $12 = 4z$
4. $\frac{t}{5} = 7$
5. $2x + 9 = 13$
6. $13 = 5s - 7$
7. $0.5m = 2.5$
8. $0.04k = 1.6$
9. $9 = 13 - \frac{1}{2}y$

Evaluate for $l = 13$ and $w = 9$.

10. $2l + 3w$
11. $7(l + w)$

Solve.

12. One-tenth of the cost of a book increased by $2 is $3.90. What is the cost of the book?

13. If you multiply the cost of a football ticket by 5 and add $2.50, you get $50. What is the cost of the ticket?

Percent of a Number

◇ **OBJECTIVE** ◇

To find a percent of a number

◇ **RECALL** ◇

Change 46.3% to a decimal.
46.3%
0.46̲3   Drop the % symbol.
         Move the decimal point
         two places to the left.
So, 46.3% = 0.463.

Example 1

26% of 246 is what number?

26% of 246 is what number?
         ↓   ↓
       0.26 × 246 = n

Write 26% as a decimal.
Of means *multiply*.
Multiply.

```
   246
 ×0.26
  1476
  492
 63.96
```

246 ← 0 decimal places
×0.26 ← 2 decimal places
63.96 ← 2 decimal places

63.96 = n
So, 26% of 246 is 63.96.

practice    1.   35% of 146 is what number?   2.   8% of 216 is what number?

Example 2

3.5% of 118.43 is what number?

3.5% of 118.43 is what number?
         ↓   ↓
       0.035 × 118.43 = n

Write 3.5% as a decimal.
Of means *multiply*.
Multiply.

```
  118.43
 ×0.035
   59215
  35529
 4.14505
```

118.43 ← 2 decimal places
×0.035 ← 3 decimal places
4.14505 ← 5 decimal places

4.14505 = n
So, 3.5% of 118.43 is 4.14505.

PERCENT OF A NUMBER

practice ▷ **3.** 3.4% of 48.24 is what number? **4.** 21.5% of 217 is what number?

Example 3

$6\frac{1}{4}$% of 175 is what number?

$6\frac{1}{4}$% of 175 is what number?

$\frac{1}{4} = 0.25;\ 6\frac{1}{4}\% = 6.25\%$ 6.25% of 175 = n

0.06 25 × 175 = n

Multiply.
$$\begin{array}{r} 175 \leftarrow \text{0 decimal places} \\ \times 0.0625 \leftarrow \text{4 decimal places} \\ \hline 875 \\ 350 \\ 1050 \\ \hline 10.9375 \leftarrow \text{4 decimal places} \end{array}$$

10.9375 = n

So, $6\frac{1}{4}$% of 175 is 10.9375.

---

practice ▷ **5.** $3\frac{3}{4}$% of 185 is what number? **6.** $5\frac{1}{2}$% of 19.7 is what number?

Example 4

A high school basketball team played 40 games. The team won 35% of the games they played. How many games did the team win?

The team won 35% of 40 games.
35% of 40 is what number?
0.35 × 40 = n

Multiply.
It is easier to multiply by 40.

$$\begin{array}{r} 40 \\ \times 0.35 \\ \hline \end{array}$$ or $$\begin{array}{r} 0.35 \leftarrow \text{2 decimal places} \\ \times 40 \leftarrow \text{0 decimal places} \\ \hline 14.00 \leftarrow \text{2 decimal places} \end{array}$$

14 = n
So, the team won 14 games.

---

practice ▷ **7.** A team played 40 games. They won 25% of them. How many games did they win? **8.** A test had 20 questions. Chris got 85% correct. How many were correct?

◇ ORAL EXERCISES ◇

To find each percent, by what decimal would you multiply?

1. 43% of 186
2. 32% of 45
3. 26% of 78
4. 49% of 16
5. 4% of 124
6. 6% of 89
7. 7% of 96
8. 3% of 88
9. 2.6% of 98
10. 3.1% of 59
11. 2.25% of 84
12. 1.3% of 95
13. 5.3% of 88
14. 4.5% of 96
15. 6.2% of 112
16. 8.1% of 19

◇ EXERCISES ◇

Compute.

1. 27% of 135 is what number?
2. 43% of 192 is what number?
3. 54% of 87 is what number?
4. 24% of 164 is what number?
5. 32% of 24.4 is what number?
6. 26% of 32.8 is what number?
7. 43% of 24.7 is what number?
8. 28% of 19.3 is what number?
9. 4.3% of 185 is what number?
10. 2.8% of 19.6 is what number?
11. 1.5% of 14.7 is what number?
12. 4.7% of 79 is what number?
13. $4\frac{1}{2}$% of 145 is what number?
14. $3\frac{1}{4}$% of 243 is what number?
15. $4\frac{3}{4}$% of 91 is what number?
16. $6\frac{1}{2}$% of 78 is what number?
★ 17. $5\frac{1}{8}$% of 196.12 is what number?
★ 18. $\frac{1}{16}$% of 19 is what number?

Solve these problems.

19. A basketball team played 45 games. They won 60% of them. How many games did they win?
20. A test had 50 items. Ronnie got 70% correct. How many were correct?
21. A town has a population of 60,000. If 15% of the population is unemployed, how many people are unemployed?
22. The senior class consists of 260 students. If 90% of the students will graduate, how many students will graduate?

Calculator

Compute each percent to the nearest penny.

1. 65% of $189.45
2. 45.3% of $2,478.32
3. 49.5% of $246.85
4. 62.3% of $11.48

PERCENT OF A NUMBER

Finding Percents

◇ OBJECTIVE ◇
To find the percent one number is of another

◇ RECALL ◇
Solve. $n \cdot 6 = 24$

$n = 24 \div 6$, or 4 $n = \frac{24}{6}$, or 4

Example 1   What % of 80 is 20?

What % of 80 is 20?

Let n = the percent.
Of means *multiply*. $n \cdot 80 = 20$

Carry the division to hundredths. $n = 20 \div 80$, or $80 \overline{)20.00}$

$$\begin{array}{r} 0.25 \\ 80\overline{)20.00} \\ \underline{16\ 0} \\ 4\ 00 \\ \underline{4\ 00} \\ 0 \end{array}$$

Write the percent for 0.25. $0.25 = 25\%$
So, 25% of 80 is 20.

Example 2   4 is what % of 7?

4 is what % of 7?

Let n = the percent.
Of means *multiply*. $4 = n \cdot 7$

Carry the division to hundredths. $4 \div 7 = n$, or

$7\overline{)4.00}$ = 0.57 or $0.57\frac{1}{7}$

Write the percent for $0.57\frac{1}{7}$. $0.57\frac{1}{7} = 57\frac{1}{7}\%$

So, 4 is $57\frac{1}{7}\%$ of 7.

practice   1. What % of 40 is 20?   2. 5 is what % of 6?

Example 3

The Celtics played 20 basketball games. They lost one game. What % of the games played did they lose?

Think: Loss is what % of games played?

$$1 = n \cdot 20$$

$$1 \div 20 = n, \text{ or } 20\overline{)1.00}$$

$$\begin{array}{r} 0.05 \\ 20\overline{)1.00} \\ \underline{1\ 00} \\ 0 \end{array}$$

Carry the division to hundredths.

$$\begin{array}{r} 0.0 \\ 20\overline{)1.00} \end{array} \qquad \begin{array}{r} 0.05 \\ 20\overline{)1.00} \\ \underline{1\ 00} \\ 0 \end{array}$$

$0.05 = 5\%$

So, the Celtics lost 5% of the games played.

practice ▷ 3. A basketball team played 25 games. They lost 2 games. What % of the games played did they lose?

◇ EXERCISES ◇

Compute.

1. What % of 80 is 40?
2. What % of 10 is 8?
3. What % of 40 is 30?
4. What % of 70 is 14?
5. What % of 20 is 18?
6. What % of 30 is 12?
7. 5 is what % of 7?
8. 5 is what % of 9?
9. 40 is what % of 60?
10. 20 is what % of 90?
11. 15 is what % of 21?
12. 21 is what % of 32?
13. What % of 62 is 45?
14. What % of 52 is 32?
★ 15. What % of 3.2 is 1.6?
★ 16. What % of $24.75 is $4.95?

Solve these problems.

17. The Blue Devils played 15 games. They lost 3 games. What % of the games played did they lose?
18. John's salary was $50. He got a raise of $2. What % of his salary was his raise?
19. Jane was at bat 80 times. She scored 6 hits. What % of the time did she score hits?
20. A bookkeeping test had 30 questions. Bill got 24 right. What % of the questions did he get right?
21. Concha's salary is $140 a week. She saves $28 a week. What % of her salary does she save?
22. Wendy weighed 60 kilograms. She lost 5 kilograms. What % of her weight did she lose?
23. Kazuo had $36. He spent $12 for a wallet. What % of his money did he spend?
24. In a class of 30 students, 7 students were absent. What % of the class was absent?

Finding the Number

◇ OBJECTIVE ◇

To find a number given a percent of the number

◇ RECALL ◇

Round to the nearest whole number.

73.2	73.5	73.8
↑	↑	↑
less than 5	5	more than 5
rounds to 73	74	74

Example 1 15 is 30% of what number?

15 is 30% of what number?

Let n be the number. $15 = 30\%$ of n
30% of n means $0.30 \cdot n$. $15 = 0.30 \cdot n$
Divide both sides by 0.30.

$15 \div 0.30$ means $0.30\overline{)15.}$ $15 \div 0.30 = n$ $0.30\overline{)15.00} \longrightarrow 30\overline{)1500}$
Move the decimal point two places to the right.

$$\begin{array}{r} 50 \\ 30\overline{)1500} \\ \underline{150} \\ 0 \\ \underline{0} \\ 0 \end{array}$$

$50 = n$
So, 15 is 30% of 50.

---

practice ▷ 1. 18 is 20% of what number? 2. 14 is 70% of what number?

Example 2 60% of what number is 40? Round the answer to the nearest whole number.

Let n be the number.
60% of n means $0.60 \cdot n$. 60% of what number is 40?
Divide both sides by 0.60. $0.60 \cdot n = 40$

$40 \div 0.60$ means $0.60\overline{)40.}$ $n = 40 \div 0.60$ $0.60\overline{)40.00} \longrightarrow 60\overline{)4000.0}$
Carry the division to tenths.

$$\begin{array}{r} 66.6 \\ 60\overline{)4000.0} \\ \underline{360} \\ 400 \\ \underline{360} \\ 400 \\ \underline{360} \\ 40 \end{array}$$

66.6
↑
more than 5
66.6 rounds to 67. $67 = n$
So, 40 is 60% of 67.

practice ▷ Round the answer to the nearest whole number.

   3. 70% of what number is 40?   4. 7 is 3% of what number?

Example 3

Mr. Roth sold a TV and made a profit of $70. The profit was 20% of the cost. Find the cost.

$70 profit is 20% of cost.

Let c = cost. $70 = 20\%$ of c
20% of c means $0.20 \cdot c$. $70 = 0.20 \cdot c$

Divide both sides by 0.20. $70 \div 0.20 = c$ $0.20\overline{)70.00} \rightarrow 20\overline{)7000}$
$70 \div 0.20$ means $0.20\overline{)70}$.

$$\begin{array}{r} 350 \\ 20\overline{)7000} \\ \underline{60} \\ 100 \\ \underline{100} \\ 0 \\ \underline{0} \\ 0 \end{array}$$

So, the cost of the TV was $350.

---

practice ▷ 5. Ms. Giglio sold a stereo and made a profit of $70. The profit was 25% of the cost. Find the cost.

◇ EXERCISES ◇

Compute.

1. 16 is 40% of what number?
2. 45 is 30% of what number?
3. 18 is 90% of what number?
4. 14 is 50% of what number?
5. 6 is 8% of what number?
6. 32 is 40% of what number?

Compute. Round each answer to the nearest whole number.

7. 70% of what number is 50?
8. 60% of what number is 8?
9. 62% of what number is 30?
10. 41% of what number is 24?
11. 19 is 31% of what number?
12. 14 is 45% of what number?
★ 13. 4.2 is 12.5% of what number?
★ 14. 2.3 is 30.5% of what number?

Solve these problems.

15. Joe sold a car and made a profit of $850. The profit was 10% of the cost. What was the cost?

16. Mona got an $8.00 raise. The raise was 5% of her salary. What was her salary?

17. Luis won 18 chess games. The games won were 40% of the games played. How many games did he play?

18. Ms. Stein saves $40 a week. The amount she saves is 20% of her salary. What is her salary?

Commission

◇ OBJECTIVES ◇
To find commissions
To find total wages based on commission plus salary

◇ RECALL ◇
7% as a decimal is 0.07.

It is common practice to give salespeople extra money, called *commission*, on the amount of sales. The commission is computed using the formula $c = r \times s$.

Example 1

John is paid a 7% commission on his camera sales. This week he sold $275.45 worth of cameras. Find his commission.

Let c = the commission.
c = 7% of $275.45
c = 0.07 × $275.45

7% of means 0.07 times.

$275.45 ← sales
× 0.07 ← rate of commission
$19.2815 ← commission

Round down to the lower cent.

c = $19.2815 or $19.28

So, his commission is $19.28.

practice ▷ 1. Henrietta earns 6% commission on sales of $185.43. Find her commission.

Example 2

Felicia is paid $125.00 a week plus 9% commission on sales. Find her total weekly earnings if her sales are $245.

Find her commission.
9% of $245 means 0.09 times $245.

c = 9% of $245
$245 ← sales
× 0.09 ← rate of commission
$22.05 ← commission

Now find her total weekly earnings.

$125.00 ← weekly salary
+ 22.05 ← commission
$147.05 ← total weekly earnings

So, her total weekly earnings are $147.05.

practice ▷ **2.** Bill is paid $215.00 a week plus 7% commission on sales. Find his total earnings if his sales are $1,275.

Example 3

Mona earns $115.45 a week plus 17% commission on sales. Find her total weekly earnings if she sold 3 TV's at $145.46 each, 2 freezers at $189.00 each, and 5 stereos at $139.85 each.

First, find the total sales.

Multiply to find the total sales for each item.

$145.46 $189 $139.85
×   3   × 2   ×   5
$436.38 $378 $699.25

Find the total sales. Add. Keep the decimal points in line with each other.

$ 436.38
  378.00
+ 699.25
$1,513.63 ← total sales

Second, find her commission.

Find 17% of $1,513.63.
Multiply 0.17 × $1,513.63.

$ 1,513.63 ← total sales
×   0.17 ← rate of commission
  1059541
  151363
$257.3171 ← commission

Round down to the lower cent.

$257.3171 rounds down to $257.31.

Third, find her total earnings.
$ 257.31 ← commission
+ 115.45 ← regularly weekly salary
$ 372.76
So, her total weekly earnings are $372.76.

---

practice ▷ **3.** Mr. Martinez is paid $165.65 a week plus 9% commission on sales. This week he sold 3 armchairs at $215 each, 4 end tables at $64.49 each, and 3 loveseats at $279.95 each. Find his total weekly earnings.

◇ EXERCISES ◇

Solve these problems.

1. Gary earns 7% commission on sales of $149.93. Find his commission.

2. Marietta earns 9% commission on sales of $275.68. Find her commission.

3. Mr. Jenkins is paid a 15% commission on furniture sales. This week his sales were $546.53. Find his commission.

4. Ms. Olson is paid a 6% commission on each sewing machine sold. A sale was $189.63. Find the commission.

5. Laura is paid a 3% commission on reorders of tune-up kits. Find her commission on a reorder of $1,545.

6. Elliot is paid a 9% commission on each carpet sold. A sale was $415.49. Find the commission.

7. Wanda is paid $135 a week plus 8% commission on sales. Find her total earnings if her sales are $345.

8. José is paid $245 a week plus 9% commission on sales. Find his total earnings if his sales are $645.

9. Leona is paid $185 a week plus 7% commission on sales. Find her total earnings if her sales are $415.

10. Frank makes $275 a week plus 5% commission on sales. Find his total earnings if his sales are $775.

11. Mr. Langella earns $355 a week plus 15% commission on sales. Find his total earnings if his sales are $418.45.

★ 12. Mrs. Whitney earns $265 a week plus $8\frac{1}{2}$% commission on sales. Find her total earnings if her sales are $1,385.64.

13. Juan earns $179.95 a week plus 19% commission on sales. Find his total earnings if he sold 3 stereos at $114.95 each, 2 dishwashers at $314.65 each, and 4 TV's at $118.45 each.

14. Mrs. Hirsch earns $265.45 a week plus 16% commission on sales. Find her total earnings if she sold 3 chairs at $245 each, 4 end tables at $43.55 each, and 3 loveseats at $265.95 each.

Reading in Math

Match the expression in the left-hand column with the corresponding expression in the right-hand column.

____ 53%
____ 53% of 100
____ 5.3 is what percent of 100?
____ 530 is what percent of 100?
____ $\frac{53}{1,000}$

a. 53
b. 530
c. 0.053
d. 0.53
e. 5.3

Discounts

◇ OBJECTIVE ◇

To find the cost of an item if a discount is given

◇ RECALL ◇

$\frac{1}{3}$ of 6 means

$\frac{1}{3} \times \frac{6}{1}$

$\frac{6}{3}$, or 2

Shortcut

$\frac{1}{3}$ of 6

$$3\overline{)6}^{\,2}$$

$\frac{1}{3}$ of 6 is 2.

Many stores try to increase sales by advertising a *discount,* or a cut, in the regular price.

Example 1

Find the cost, after the discount is deducted, of a $29.95 radio at a 45% discount.

cost = regular price − discount

Let c = cost. $c = \$29.95 - (45\% \times \$29.95)$

45% means 0.45.

$\begin{array}{r} \$29.95 \leftarrow \text{regular price} \\ \times 0.45 \leftarrow \text{rate of discount} \\ \hline 14975 \\ 11980 \end{array}$

Always round in favor of the storekeeper. $\$13.4775 = \$13.47 \leftarrow$ discount

$c = \$29.95 - \$13.47$

Subtract. $\begin{array}{r} \$29.95 \leftarrow \text{regular price} \\ -13.47 \leftarrow \text{discount} \\ \hline \$16.48 \leftarrow \text{cost after the discount is deducted} \end{array}$

So, the cost after the discount is deducted is $16.48.

---

practice ▷ **Find the cost after the discount is deducted.**

1. $17.95 pocket radio
   15% discount

2. $129.95 coat
   20% discount

Example 2

Find the cost of the following purchases at a 35% discount: suit, $79.95; jacket, $46.75; shirt, $17.85.

First, find the total cost. Add.
Then, find 35% of $144.55 or (0.35)($144.55).

```
 $ 79.95         $ 144.55   ← total cost
   46.75          × 0.35    ← rate of discount
 + 17.85          7 2275
 $144.55          43 365
                 $50.5925   ← discount
```

$50.5925 rounds down to $50.59.

The discount is $50.59.

Subtract the discount from the total cost.

```
 $144.55   ← total cost
 - 50.59   ← discount
 $ 93.96
```

So, the cost is $93.96.

practice ▷ 3. Find the cost of the following purchases at a 15% discount: $4.97, record; $8.95, book; $37.65, hair dryer.

Some discounts are advertised as fractions.

Example 3

A drugstore advertised "$\frac{1}{3}$ off" on all prices. Find the cost of a $4.94 bottle of cologne.

Find $\frac{1}{3}$ of $4.94.

$\frac{1}{3}$ × $4.94 means $\frac{\$4.94}{3}$.

Divide $4.94 by 3.

```
    $1.64
 3)$4.94
    3
    19
    18
    14
    12
     2
```

Round in favor of the store. The discount rounds down to $1.64.

Subtract.
```
 $4.94   ← original marked price
 -1.64   ← discount
 $3.30
```
So, the bottle of cologne cost $3.30.

practice ▷ 4. A music store advertised a "$\frac{1}{3}$ off" sale. Find the cost of a $6.95 record.

◇ ORAL EXERCISES ◇

Find the cost after the discount is deducted.

1. $0.50 pen
   $0.10 discount
2. $5.00 record
   $1.00 discount
3. $250 TV
   $100 discount
4. $1,200 used car
   $300 discount
5. $75 bike
   $20 discount
6. $135 suit
   $45 discount

◇ EXERCISES ◇

Find the cost after the discount is deducted.

1. $19.75 shirt
   15% discount
2. $5.95 book
   45% discount
3. $49.95 tire
   35% discount
4. $175.45 dishwasher
   12% discount
5. $245.95 stereo
   35% discount
6. $345.49 sofa
   45% discount
7. $79.95 suit
   25% discount
8. $749.95 used car
   33% discount
9. $0.68 bottle of juice
   10% discount
★ 10. $10,275 car
   16.5% discount
★ 11. $4,995.95 boat
   14.3% discount

12. Find the total cost of the following purchases at a 25% discount: $15.45, shirt; $7.50, tie; $34.65, pair of shoes.

13. Find the total cost of the following purchases at a 45% discount: $7.45, perfume; $3.95, razor; $49.68, hair dryer.

14. Find the total cost of the following items at a 35% discount: $39.75, dress; $16.49, blouse; $25.75, sweater.

15. Find the total cost of the following items at a 15% discount: $49.68, mixer; $19.63, coffee maker; $8.43, frying pan.

16. Find the cost of a $7.19 record at a "$\frac{1}{3}$ off" sale.

17. Find the cost of a $37.95 sports jacket at a "$\frac{1}{4}$ off" sale.

How many pennies can be placed next to each other with their edges touching for a distance of 1 kilometer? Give your answer to the nearest thousand.

DISCOUNTS

Problem Solving – Applications
Jobs for Teenagers

Example

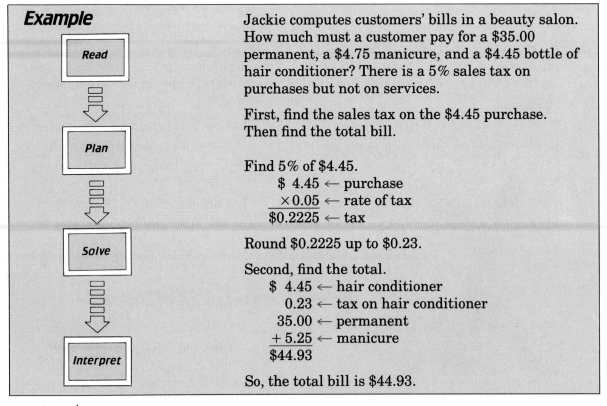

Jackie computes customers' bills in a beauty salon. How much must a customer pay for a $35.00 permanent, a $4.75 manicure, and a $4.45 bottle of hair conditioner? There is a 5% sales tax on purchases but not on services.

First, find the sales tax on the $4.45 purchase. Then find the total bill.

Find 5% of $4.45.

$\quad\;\;$ $ 4.45 ← purchase
$\quad\;\;$ ×0.05 ← rate of tax
$\quad\;\;$ $0.2225 ← tax

Round $0.2225 up to $0.23.

Second, find the total.
$\quad\;\;$ $ 4.45 ← hair conditioner
$\quad\;\;\;\;\;\;$ 0.23 ← tax on hair conditioner
$\quad\;\;\;\;$ 35.00 ← permanent
$\quad\;$ + 5.25 ← manicure
$\quad\;\;$ $44.93

So, the total bill is $44.93.

practice ▷ Find the total cost of a $28.95 permanent, a $5.75 manicure, and a $6.75 bottle of skin conditioner. There is a 5% sales tax on purchases but not on services.

Solve these problems.

1. Find the total cost for a $49.95 permanent, a $15.75 manicure, and a $6.95 bottle of hair conditioner. There is a 5% sales tax on purchases only.

2. Find the total cost for a $19.95 haircut, a $15.25 manicure, and a $5.95 bottle of skin lotion. There is a 6% sales tax on purchases only.

3. Find the total cost for a $25.50 cut and blow dry, a $16.25 manicure, and a $3.75 bottle of hand lotion. There is a 5% sales tax on purchases only.

4. Find the total cost for a $53.50 permanent, a $10.25 facial massage, and a $6.25 bottle of hair conditioner. There is a 6% sales tax on purchases only.

5. Find the total cost of a $45.95 permanent and a $15.15 manicure. Find the change from four $20 bills.

6. Find the total cost of a $52.50 permanent and a $10.50 manicure. Find the change from seven $10 bills.

7. Find the total cost for a $44.75 permanent and a $6.25 bottle of lotion. There is a 5% sales tax on purchases only. Find the change from three $20 bills.

8. Find the total for this bill and the change from a $100 bill.

CURLY'S	
Permanent	$54.25
Manicure	15.25
Lotion	5.30
Tax	.25
Total	

9. Janet works part time for a hairdresser. She is paid $5.15 per hour. Find her earnings for a 12-hour week.

10. Henry works part time for a beauty salon. He is paid $4.25 per hour. He worked 13 hours last week. How much did he earn?

11. Vincent's advertises a 20% discount on all permanents. Jane works at the register. How much should she charge a customer for a regularly priced $48 permanent?

12. This week, Cut 'n Curl is featuring a 15% discount on haircuts. Ivan is at the register this week. How much should he bill a customer for a haircut that regularly costs $15.00?

13. As a grand opening special, Hair Fair is featuring a 25% discount. Debbie is at the register. How much should she charge a customer for a cut and blow dry regularly priced at $25.60 and a frosting regularly-priced at $30.25?

14. On Wednesdays at Cerise's, all manicures are given at $\frac{1}{10}$ off. The regular price of a manicure is $15.00. How much is it on Wednesday?

15. On Tuesdays, senior citizens get a 15% discount at the Clip 'n Snip. The cost of a wash, cut, and blow dry is regularly $20.00. How much does a senior citizen pay on Tuesday? How much change will the senior citizen get from a $20 bill?

PROBLEM SOLVING

Problem Solving – Careers
Sales Representatives

1. Mr. Skool is a yearbook sales representative. He figures that the cost per yearbook will be $4.75. What will be the total bill for a senior class of 275 students?

2. Julius is a traveling sales rep. He is allowed to claim $0.15 per kilometer for car expenses. Find his claim for a sales trip of 450 kilometers.

3. Maria sells encyclopedias. This month, the company is offering a 20% discount. How much should she charge a customer for a set that usually costs $485?

4. Goldie sells cosmetics. A customer bought a bottle of perfume for $24.75. The sales tax was 5%. The customer paid with three $10 bills. How much change should Goldie give?

5. Bill was told that there are 800 potential new car buyers in his sales district. He hopes to capture 25% of all sales. How many new cars does he expect to sell?

6. Mr. Martinson attended a sales convention. His expenses were $46 a night for 3 nights at a hotel, $81.72 for meals, and $65.49 for car expenses. Find his total expenses.

Non-Routine Problems

There is no simple method used to solve these problems. Use whatever methods you can. Do not expect to solve these easily.

1. Ms. Sobel traveled for 2 h at the rate of 50 mi/h. She made her return trip at the rate of 40 mi/h. What was the average speed for the entire trip? (*Hint:* It is *not* 45 mi/h.)

2. A car travels 1 mile uphill at 20 mi/h. How fast should it travel 1 mile back downhill in order to end up with a 40 mi/h average speed for the entire 2-mile stretch?

3. The Dearing family starts out on their vacation trip. They average 30 mi/h. The Osborne family starts out one hour later, and they average 40 mi/h. If they follow the same route, after how many hours of travel time are the Osbornes going to catch up with the Dearings?

4. Two cars 400 mi apart start traveling toward each other. One car averages 45 mi/h and the other 35 mi/h. After how many hours will the two cars meet?

5. A train covered a distance of 200 mi in 5 h. How fast must it travel another 200 mi in order to finish an entire trip with an average speed of 50 mi/h?

6. Two trains started out traveling toward each other. They passed each other after 2 h of travel at a point 90 mi from the starting point of the first train. The second train's speed is 10 mi/h faster than the first train's speed. How many miles apart were the starting points of the two trains?

7. You can bicycle at a speed of 15 mi/h. Your friend can bicycle at 12 mi/h. You are going to bicycle the distance of 3 mi to town, both of you starting from the same point. How much later should you start out bicycling in order for both of you to arrive there at the same time?

NON-ROUTINE PROBLEMS

COMPUTER ACTIVITIES

Mona deposits $1,000 in a bank that pays interest at a rate of 15% compounded annually (each year). This means that not only will the $1,000 draw interest, but the interest itself will, also. The following formula is used to compute the amount that the deposit will be worth at the end of any number of years:

$$A = D(1 + R)^Y \leftarrow \text{number of Years}$$

Amount — Deposit — Rate as a decimal

Example Type in the following program for finding A in the compound interest formula $A = D(1 + R)^Y$. Then **RUN** the program to find the Amount a Deposit of $1,000 will be worth after 3 Years at a Rate of 15% (.15). First write the formula in BASIC.

$A = D \cdot (1 + R)^Y$ becomes $A = D*(1 + R) \wedge Y$.

```
10  INPUT "HOW MUCH IS DEPOSITED? ";D
20  INPUT "WHAT IS THE RATE AS A DECIMAL? ";R
30  INPUT "FOR HOW MANY YEARS? ";Y
40  LET A = D * (1 + R) ^ Y
50  PRINT "THE AMOUNT OF $"D" IS NOW WORTH $"A
60  END

]RUN
HOW MUCH IS DEPOSITED? 1000
WHAT IS THE RATE AS A DECIMAL? .15
FOR HOW MANY YEARS? 3
THE AMOUNT OF $1000 IS NOW WORTH $1520.875
```

You type 1000. No comma or $.
You type .15.
You type 3.
Round this to $1,520.88

See the Computer Section beginning on page 420 for more information.

Exercises

1. **RUN** the program to find A if $4,000 is deposited at a rate of 12% for 5 years.
2. **RUN** the program to find A if $10,000 is deposited at a rate of 13% for 20 years.
3. Suppose the deposit in Exercise 2 above is doubled. Is the amount doubled for the same rate and years? Confirm by **RUN**ning the program.
4. Suppose the rate in Exercise 2 above is doubled to 26%. Is the amount doubled for the same deposit and years? Confirm your conclusion by **RUN**ning the program.

Chapter Review

Change to decimals. [216]

1. 74%   2. 43.7%   3. $36\frac{1}{3}$%   4. 500%   5. 8%   6. 0.8%

Change to percents. [216]

7. 0.04   8. 0.733   9. $0.46\frac{1}{3}$ 10. 8   11. $\frac{2}{5}$ 12. $\frac{4}{7}$

Compute. [219]

13. 45% of 156 is what number?
14. 3.4% of 52.23 is what number?
15. $7\frac{1}{4}$% of 185 is what number?

Compute. [222]

16. What % of 50 is 40?
17. 4 is what percent of 9?

Compute. [224]

18. 12 is 60% of what number?
19. 40 is 5% of what number?

Round each answer to the nearest whole number. [224]

20. 3% of what number is 11?
21. 20 is 70% of what number?

Solve these problems.

22. [219] The Razors played 50 games. They won 30% of the games. How many games did they win?

23. [222] The Waves played 15 games. They lost 4 games. What % of the games played did they lose?

24. [224] Maria got a $6 raise. The raise was 4% of her salary. What was her salary?

25. [226] Henry earns 7% commission on all sales. This week his sales were $165.43. Find his commission.

26. [226] Jane earns $345 a week plus 7% commission on sales. Find her total earnings if her sales are $765.

27. [229] Find the cost, after the discount is taken, of a $29.95 radio at a 25% discount.

28. [229] Find the cost of the following purchases at a 25% discount: $14.95, shirt; $8.75, tie; $44.95, boots.

29. [229] Find the cost of a $19.95 electric razor at a "$\frac{1}{4}$ off" sale.

Chapter Test

Change to decimals.

1. 29%   2. 62.3%   3. $47\frac{1}{3}\%$ 4. 800%   5. 0.9%

Change to percents.

6. 0.09   7. $0.49\frac{1}{3}$ 8. 5   9. $\frac{3}{5}$ 10. $\frac{4}{9}$

Compute.

11. 27% of 519 is what number?

12. 4.2% of 46.83 is what number?

13. $8\frac{1}{4}\%$ of 495 is what number?

14. What % of 50 is 30?

15. 5 is what % of 9?

16. 18 is 60% of what number?

17. 8% of what number is 21? Round the answer to the nearest whole number.

Solve these problems.

18. A test had 40 questions. Martina got 60% of them right. How many did she get right?

19. The Jingles played 12 games. They lost 3 games. What % of the games played did they lose?

20. Juan got a $15 raise. The raise was 5% of his salary. What was his salary?

21. Michele earns 7% commission on all sales. Last week her sales were $814.37. Find her commission.

22. Adam earns $385 a week plus 8% commission on sales. Find his total earnings if his sales are $735.

23. Find the cost, after the discount is taken, of a $49.95 broiler at a 35% discount.

24. Find the cost of the following purchases at a 15% discount: $19.95, slacks; $39.95, jacket; $14.35, shirt.

25. Find the cost of a $17.95 hair dryer at a "$\frac{1}{2}$ off" sale.

Statistics

10

Ohio's Great Serpent Mound, built by prehistoric Indians as a burial mound, is $\frac{1}{4}$ mile long.

Range, Mean, Median, Mode

To find the range, mean (average), median, and mode for a set of data

Scores: 8, 9, 7, 6, 10, 8, 9
Arrange in order: 10, 9, 9, 8, 8, 7, 6
Add: 10 + 9 + 9 + 8 + 8 + 7 + 6 = 57

The *range* is the difference between the highest and lowest numbers.

Example 1

Find the range. Daily high temperatures: 24°, 25°, 20°, 18°, 21°, 23°, and 22°.

Arrange the numbers in order. 25°   24°   23°   22°   21°   20°   18°

To find the range, subtract the lowest number from the highest number.

25° − 18° = 7°

So, the range is 7°.

Example 2

Ivan scored 92, 97, 88, 93, and 100 on his mathematics tests. What was his average?

Add the scores.
Find out how many scores there are.

```
Add.     92        Divide.      94
         97                  5)470
         88                     45
There are 5 scores.  93                     20
Divide the sum by 5.+100                    20
                    ---                      0
                    470
```

94 is called the *mean*.   So, Ivan's average was 94.

The *mean* is the average. To find the mean, add the numbers and divide by the number of terms.

practice ▷ Find the range. Find the mean.

1. 91, 95, 96, 84   2. 48, 50, 46, 45, 44   3. 32, 30, 35, 31, 27

Example 3

What was Ivan's middle score?

Arrange the scores in order.   100   97   93   92   88
                                2 above   middle score   2 below

93 is called the *median*.   So, 93 is the middle score.

The *median* is the middle number. To find the median, arrange the numbers in order and choose the middle one.

Example 4

When there are 2 middle scores, find their average.

$$\begin{array}{r} 7.5 \\ 2\overline{)15.0} \\ \underline{14} \\ 10 \\ \underline{10} \\ 0 \end{array}$$

Joe scored 7, 9, 9, 8, 6, and 5 in a dart game. What is his median score?

9 9	8 7	6 5
2 above	Middle scores	2 below

Average the middle scores: $\frac{8+7}{2} = \frac{15}{2}$

So, the median score is 7.5.

practice ▷ Find the median.

4. 8, 7, 9, 6, 4, 8, 5   5. 4, 9, 8, 5, 10, 11   6. 99, 90, 88, 97, 86

Example 5

Group like scores. Find the score that occurs most often.

8 is called the *mode*.

Maxine scored 9, 8, 6, 7, 8, 10, 8, 9, and 10 on her English quizzes. What score did she achieve most often?

10 10	9 9	8 8 8	7 6
twice	twice	3 times	once

So, 8 is the score achieved most often.

The *mode* is the number that occurs most often. To find the mode, list each number as many times as it occurs and choose the one that occurs most frequently.

Example 6

Group like scores.

The high temperatures for a week were 31°, 29°, 30°, 31°, 30°, 28°, and 27°. The low temperatures were 17°, 19°, 18°, 20° 16°, 21°, and 15°. Find the mode(s) if they exist.

31° 31°	30° 30°	29° 28° 27°
twice	twice	once

So, the modes for the high temperatures are 30° and 31°. There is no mode for low temperatures.

practice ▷ Find the mode(s) if they exist.

7. 8, 9, 9, 8, 7, 9   8. 3, 4, 6, 5, 2, 7   9. 10, 10, 9, 8, 7, 9, 9, 10

RANGE, MEAN, MEDIAN, MODE

◇ **ORAL EXERCISES** ◇

Find the range, mean (average), median, and mode(s).

1. 5, 3, 3, 3, 1
2. 5, 4, 3, 2, 1
3. 10, 8, 6, 6, 3, 3

◇ **EXERCISES** ◇

Find the range. Find the mean.
1. 87, 84, 93, 92
2. 10, 12, 8, 14, 16
3. 18, 20, 22, 16, 18, 14
4. 99, 87, 99, 100, 95
5. 14, 12, 10, 14, 15
6. 16, 19, 18, 22, 19, 20
7. 26, 33, 32, 41, 58
8. 89, 78, 73, 72, 83, 91
9. 68, 72, 69, 76, 70, 77

Find the median.
10. 97, 86, 93, 84, 94
11. 89, 91, 78, 47, 96
12. 68, 71, 78, 48, 63
13. 8, 14, 17, 16, 19, 22
14. 84, 78, 63, 94, 96, 85
15. 14, 16, 12, 10, 9, 18
16. 3, 7, 14, 6, 5
17. 2, 4, 6, 3, 5
18. 4, 8, 13, 14, 87, 46

Find the mode(s) if they exist.
19. 96, 94, 96, 93, 100, 97
20. 98, 87, 86, 98, 87
21. 9, 8, 8, 9, 8, 7, 6
22. 24, 26, 23, 19, 18
23. 10, 12, 10, 9, 8, 11
24. 97, 96, 96, 97, 98, 100
25. 86, 84, 100, 84, 83
26. 46, 48, 53, 47, 49
27. 97, 96, 100, 98, 99

Find the range, mean, median, and mode(s).
28. 29, 36, 45, 28, 36, 54
29. 9, 8, 9, 8, 9, 7, 9
30. 48, 47, 52, 56, 52, 58
31. 94, 91, 88, 91, 94, 96
32. 94, 90, 90, 87, 87
33. 8, 8, 9, 9, 8, 8, 7, 6, 9
34. 8, 10, 10, 9, 9, 8, 10
35. 28, 30, 30, 27, 26
36. 99, 98, 86, 100, 98, 100

★ 37. A basketball team averaged 76 points for 8 games. The scores for 7 games were 82, 78, 72, 76, 81, 80, and 73. What was the team's score in the eighth game?

★ 38. A worker averaged $102 per week for 10 weeks. The earnings for 9 of the weeks were $98, $105, $100, $95, $115, $101, $99, $100, and $106. What was the pay for the tenth week?

$$9(9) = 81$$
$$99(99) = 9801$$
$$999(999) = 998001$$
$$9999(9999) = 99980001$$

Look for a pattern; then use it to find these products.

1. 99999(99999)
2. 999999(999999)

Mean of Grouped Data

◇ OBJECTIVES ◇
To make a frequency table
To find a mean of grouped data

◇ RECALL ◇
Find the mean.   10, 9, 8, 9, 7, 5
$10 + 9 + 8 + 9 + 7 + 5 = 48$
$48 \div 6 = 8$
So, the mean is 8.

Example 1

Make a frequency table.
Scores for 18 Holes of Golf

3 5 5 4 4 7 5 6 3
3 4 4 5 4 6 3 7 6

Label the columns.
Tally.
Count the tallies to get the frequency.

Score (s)	Tally	Frequency (f)	Sum (f · s)
3	////	4	12
4	/////	5	20
5	////	4	20
6	///	3	18
7	//	2	14
Total		18	84

Sum = frequency × score.

Example 2

Make a frequency table. Then find the mean.
Class Test Scores

95 85 100 95 75 90 100 75 100 90
90 90   85 90 85 95   90 70   95 70
100 95   80 75 80 100   75 95   90 95

Label the columns.
Tally.
Count the tallies to get the frequency.
Sum = frequency × score.
The mean is the sum divided by the frequency.

```
    88.3
30)2,650.0
    2 40
      250
      240
       100
        90
        10
```

Score (s)	Tally	Frequency (f)	Sum (f · s)
100	/////	5	500
95	///// //	7	665
90	///// //	7	630
85	///	3	255
80	//	2	160
75	////	4	300
70	//	2	140
Total		30	2,650

So, the mean = $\frac{2,650}{30}$, or 88.3.

practice ▷ 1. Make a frequency table, and then find the mean.

7 8 9 10 10 8 8 9 8 8 9 10 9 10 9 9 7 8
10 7

Sometimes it is desirable to group the data in intervals.

Example 3

Group the scores in intervals of 5. Make a frequency table. Find the mean.

Scores

48	24	26	30	34	45	28	13	21	21
6	22	31	44	36	37	12	10	27	34
35	41	48	7	14	29	31	36	42	43
9	15	23	26	29	32	36	43	10	18
20	43	30	29	27	21	27	33	28	27

Label the columns.

Choose a convenient interval by which to group the scores. For these scores, use 5.
The interval midpoint is the middle number of each interval.
Use the interval midpoint for s.
Sum = frequency × interval midpoint.

Score(s)	Tally	Frequency(f)	Interval(s) Midpoint	Sum($f \cdot s$)
45–49	///	3	47	141
40–44	///// /	6	42	252
35–39	/////	5	37	185
30–34	///// ///	8	32	256
25–29	///// ///// /	11	27	297
20–24	///// //	7	22	154
15–19	//	2	17	34
10–14	/////	5	12	60
5–9	///	3	7	21
0–4		0	2	0
Total		50		1,400

Mean = $\frac{\text{sum}}{\text{total frequency}} = \frac{1{,}400}{50}$, or 28.

So, the mean is 28.

practice ▷ 2. Group the scores. Choose a convenient interval. Make a frequency table. Find the mean.

22 22 22 21 21 20 19 19 19 19 18 18
17 16 16 16 16 16 15 14 14 13 13 13
12 12 11 11 11 11 10   9   8   8   8   8
 8   7   7   7   6   6   6   5   4   4   3   2

◇ EXERCISES ◇

1. Copy and complete the frequency table. Then find the mean.

Scores (s)	Tally	Frequency (f)	Sum (f · s)
90	///		
80	##//		
70	##// //		
60	////		
50	/		
Total			

Use these test scores for Exercises 2 and 3.

```
95   90   85   90   95   100  80   85
75   85   90   100  70   100  85   90
70   75   95   75   75   95   70   100
```

2. Make a frequency table.

3. Find the mean.

Use these scores for Exercises 4 and 5.

```
44   43   43   43   42   42   41   41   41   41   40   40
40   39   39   38   37   37   36   36   36   35   34   34
34   34   34   33   33   32   32   31   31   31   31   30
30   30   29   29   28   28   27   27   27   26   26
```

4. Make a frequency table. Group the scores in intervals of three.

5. Find the mean.

Calculator

Use a calculator to find the sums and the mean.

Score (s)	Frequency (f)	Sum (f · s)
100	60	
95	75	
90	64	
85	103	
Total		

MEAN OF GROUPED DATA

Analyzing Data

◇ OBJECTIVES ◇

To construct a bar graph
To calculate mean
To collect data

◇ RECALL ◇

The mean is found by adding the numbers and dividing the sum by the number of terms.

$$\underbrace{70 + 80 + 90 + 100 + 110 + 120}_{6 \text{ terms}} = 570$$

$\frac{570}{6} = 95 \longleftarrow$ average

Example

The Wildcats football team defeated all of their opponents. Scores were: 24–8, 42–6, 26–18, 35–10, 44–30, and 21–18. Construct a bar graph to compare the scores. Find the average score for the Wildcats and for their opponents.

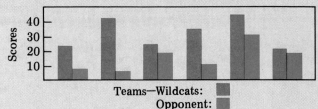

SCORES OF FOOTBALL GAMES

To find the mean, add the scores.

Wildcats: 24 + 42 + 26 + 35 + 44 + 21 = 192
Opponent: 8 + 6 + 18 + 10 + 30 + 18 = 90

Divide the sum by the number of scores, which is 6.

Wildcats   Opponent
$\frac{192}{6} = 32$ $\frac{90}{6} = 15$

So, the Wildcats averaged 32 points per game, and their opponents averaged 15 points per game.

◇ EXERCISES ◇

Construct a bar graph to compare scores. Calculate the averages.

1. Bears vs opponents. Scores:
   18–6, 37–32, 14–21, 12–8

2. Panthers vs opponents. Scores:
   6–8, 12–12, 24–6, 42–21

3. Collect data from 5 team games in your school. Construct bar graphs and calculate the averages.

Problem Solving – Applications
Jobs for Teenagers

1. Each scarf in a school store is to be sold at a discount of 10%. The marked price of each scarf is $9.00. The sales tax is 6%. What is the total sale price?

2. The marked price of a box of pencils is $1.20. The pencils are sold at a discount of 20%. The sales tax is 4%. Find the total sale price.

3. School sweaters are sold at a discount of 15%. The marked price on a school sweater is $12. The sales tax is 7%. What is the total sale price?

4. The marked price on a pair of sneakers in the school store is $24. There is a 25% discount on sneakers. The sales tax is 5%. What is the total sale price?

5. To cover the cost of operations, a school store marks up the selling price at 20% over the cost. A pen costs $4. What is the selling price?

6. A banner costs the school store $2.25. The store marks up the selling price at 10% over the cost. What is the selling price of the banner?

7. A spiral notebook costs the school store $1.10. The store marks up the selling price at 30% over the cost. Find the selling price.

8. A school store marks up the selling price at 10% over the cost. A gym suit costs the school store $8. What do they sell the gym suit for?

9. A package of construction paper costs the school store $1.50. The store marks up the selling price at 20% over the cost. What is the selling price of the construction paper?

10. A ruler costs the school store $1.30. The store marks up the selling price at 20% over the cost. The store gives a 10% discount on rulers. The sales tax is 4%. Find the total sale price.

PROBLEM SOLVING

Simple Events

◆ OBJECTIVES ◆

To find the probability of a simple event

To determine the probability of an event that is certain to occur and one that cannot happen

◆ RECALL ◆

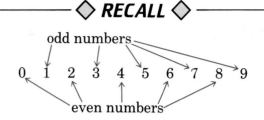

Example 1

Each part of the circle is the same size.
List all possible outcomes.
Find the number of favorable outcomes—the number of times red can occur.

What is the chance that the spinner will stop on red?

There are four possible outcomes: blue, red, green, red.
There are two favorable outcomes; red occurs twice.

Write a fraction. $\dfrac{\text{Number of favorable outcomes}}{\text{Total number of all possible outcomes}} = \dfrac{2}{4}$

2 out of 4 is the same as 1 out of 2, or $\dfrac{1}{2}$.

So, the chance that the spinner will stop on red is $\dfrac{2}{4}$, or 2 out of 4.

---

practice ▷

1. What is the chance that the spinner will stop on green?

2. What is the chance that the spinner will stop on blue?

$$\text{Probability of an event} = \dfrac{\text{Number of favorable outcomes}}{\text{Total number of all possible outcomes}}$$

Example 2

Each part of the circle is the same size.
Assume that the spinner will not stop on a line.

What is the probability that the spinner will stop on 1? on 2? on an even number?

There are 8 possible outcomes.
The number 1 occurs twice.
The number 2 occurs 3 times.
An even number occurs 4 times.

$P(1)$ means the probability that the spinner will stop on 1.

So, $P(1) = \dfrac{2}{8}$, $P(2) = \dfrac{3}{8}$, $P(\text{even number}) = \dfrac{4}{8}$.

The probability of an event that is certain to occur is 1.
The probability of an event that cannot occur is 0.

Example 3

$P(\text{event}) = \dfrac{\text{Favorable outcomes}}{\text{All possible outcomes}}$

Find. $P(\text{Red})$
Red occurs twice.

So, $P(\text{Red}) = \dfrac{2}{2}$, or 1.

Find. $P(\text{Blue})$
Blue does not occur.

So, $P(\text{Blue}) = \dfrac{0}{2}$, or 0.

Example 4

$P(1) = \dfrac{1}{2}$, or 1 out of 2.

About $\dfrac{1}{2}$ of the 50 spins should stop on 1.

$\dfrac{1}{2} \cdot 50 = 25$

Spin the arrow 50 times. About how many times should it stop on 1?
There are 8 possible outcomes.
The number 1 occurs 4 times.
$P(1) = \dfrac{4}{8}$, or $\dfrac{1}{2}$

So, for 50 spins, the arrow should stop on 1 about $\dfrac{1}{2} \cdot 50$, or 25 times.

practice ▷ Use the spinner in Example 4. Spin the arrow 80 times.

3. About how many times should it stop on 1?

4. About how many times should it stop on 2?

◇ ORAL EXERCISES ◇

1. How many possible outcomes?
2. How many blue outcomes?
3. How many green outcomes?
4. How many red outcomes?
5. What is the chance that the spinner will stop on green?
6. What is the chance that the spinner will stop on red?

◇ EXERCISES ◇

Find the probability for each event.

1. blue   2. green   3. red   4. white

Compute.

5. $P(1)$ 6. $P(2)$ 7. $P(3)$
8. $P(\text{even number})$ 9. $P(\text{odd number})$

10. Spin the arrow 100 times. About how many times should it stop on 1?

11. Spin the arrow 75 times. About how many times should it stop on 2?

SIMPLE EVENTS

Independent and Dependent Events

◇ OBJECTIVES ◇
To determine the probability of independent events
To determine the probability of dependent events

◇ RECALL ◇

$$P(\text{event}) = \frac{\text{number of favorable outcomes}}{\text{Total number of all possible outcomes}}$$

Find the probability that the spinner will stop on an odd number.

$P(\text{odd number}) = \frac{4}{8}$ or $\frac{1}{2}$

Tree diagrams can be used to determine all possible outcomes of independent events.

Example 1

Toss one coin, and roll one die. What is the probability of getting a head on the coin and a 4 on the die? Use tree diagrams to find all possible outcomes.

Toss a coin.
Roll a die.
All possible outcomes.

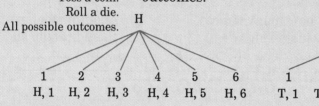

There is 1 favorable outcome, H, 4.
There are 12 possible outcomes.

$P(H, 4)$ means $P(H \text{ and } 4)$.

$$P(\text{event}) = \frac{\text{Number of favorable outcomes}}{\text{Total number of all possible outcomes}}$$

$P(H, 4) = \frac{1}{12}$

So, the probability of getting a head on the coin and a 4 on the die is $\frac{1}{12}$.

practice ▷ Toss a coin and roll a die. Use a tree diagram to find the probability.

1. $P(H, 6)$
2. $P(T, 1)$
3. $P(H, 3)$

Tossing a coin and rolling a die are *independent events*. One event does not depend on the other. If A and B are independent events,
$P(A \text{ and } B) = P(A) \cdot P(B)$

Example 2

A white die and a red die are rolled. What is the probability of rolling a 6 on the red die and a 2 on the white die?

The events are independent. $P(6 \text{ on red}) = \frac{1}{6}$ $P(2 \text{ on white}) = \frac{1}{6}$

$P(A \text{ and } B) = P(A) \cdot P(B)$ $P(6 \text{ on red and } 2 \text{ on white}) = \frac{1}{6} \cdot \frac{1}{6} \text{ or } \frac{1}{36}$

So, the probability of rolling a 6 on the red die and a 2 on the white die is $\frac{1}{36}$.

practice 4. What is the probability of tossing a tail on a coin and rolling a 5 on a die? Use the formula $P(A \text{ and } B) = P(A) \cdot P(B)$.

Example 3

4 green marbles and 6 red marbles are in a bag. What is the probability of picking a green marble? If you do not put it back, what is the probability of picking a green marble on the second draw?

There are 4 green marbles. $P(\text{Green Marble}) = \frac{4}{10} \text{ or } \frac{2}{5}$ First draw
10 marbles altogether.

Now, there are 3 green marbles, $P(\text{Second Green Marble}) = \frac{3}{9} \text{ or } \frac{1}{3}$ Second draw
9 marbles altogether.

So, the probability of picking a green marble on the second draw if the first is not replaced is $\frac{1}{3}$.

Example 3 is an example of *dependent events*. The second event depends on the first event.

◇ EXERCISES ◇

One box contains color cards: a white, a green, and a yellow. Another box contains number cards: 1, 2, 3, and 4. Pick one card from each.
1. Draw a tree diagram.   2. Find $P(W, 4)$.   3. Find $P(G, 3)$.   4. Find $P(Y, 1)$.

Look at the spinners on the right. Use the formula to find each probability.

5. $P(1 \text{ and red})$ 6. $P(3 \text{ and white})$
7. $P(2 \text{ and blue})$ 8. $P(4 \text{ and red})$

A bag contains 9 white marbles and 6 blue marbles. A white marble is drawn and it is not put back. On the second draw find the probability

9. of picking a white marble.   10. of picking a blue marble.

INDEPENDENT AND DEPENDENT EVENTS

Counting Principle

◇ OBJECTIVE ◇

To apply the counting principle in finding the number of possible arrangements

◇ RECALL ◇

List all possible two-digit numbers using the digits 3, 5, or 7.

33	35	37
53	55	57
73	75	77

There are nine possible numbers.

Example

There are 4 candidates for president: Toni, Cal, Rory, and Juanita.
There are 3 candidates for vice-president: Mary, Pat, and Ed.
How many different ways can the two offices be filled?

One way: Make a tree diagram to find all possible arrangements.

All possible arrangements:

Toni	Toni	Toni
Mary	Pat	Ed
Rory	Rory	Rory
Mary	Pat	Ed
Cal	Cal	Cal
Mary	Pat	Ed
Juanita	Juanita	Juanita
Mary	Pat	Ed

Toni ⟨ Mary, Pat, Ed ⟩ 3 ways   Rory ⟨ Mary, Pat, Ed ⟩ 3 ways

Cal ⟨ Mary, Pat, Ed ⟩ 3 ways   Juanita ⟨ Mary, Pat, Ed ⟩ 3 ways

So, the offices can be filled in 12 different ways.

Another way: There are 4 choices for president and 3 choices for vice-president.
So, there are 4 · 3, or 12, ways that the offices can be filled.

counting principle
One event can happen in m ways. Another event can happen in n ways. The total number of ways both events can happen is $m \cdot n$ ways.

---

practice

1. There are 5 candidates for president and 4 candidates for vice-president. How many different ways can the offices be filled?

2. Lee has 4 watches and 4 watchbands. How many different watch arrangements can be made?

◇ EXERCISES ◇

1. There are 3 candidates for president of the student council and 4 candidates for vice-president. How many different ways can the two offices be filled?

2. In a restaurant, there are 5 choices of soup and 3 choices of salad. How many different soup and salad orders are there?

3. There are 2 positions and 5 candidates. How many different ways can the positions be filled?

4. A closet contains 8 blouses and 3 skirts. How many different outfits are possible?

5. A member of a baseball team has 3 jerseys and 3 pairs of pants. How many different outfits may she wear?

★ 6. At a restaurant, there are 4 main courses, 5 beverage choices, and 6 dessert choices. How many different dinners can be formed?

Find the number of different license plates that can be formed.

★ 7. The license plate must use 6 spaces. Letters and numerals can be used in any order and can be repeated. Assume that 0 cannot be used but the letter O can be used.

★ 8. The license plate must have 3 numerals followed by 3 letters. Letters and numerals can be repeated. Assume that 0 cannot be used but the letter O can be used.

Algebra Maintenance

1. Find the area: $A = l \cdot w$.

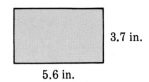

2. Find the perimeter: $P = 2(l + w)$.

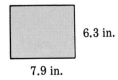

3. Find the area: $A = \dfrac{b \cdot h}{2}$

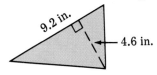

4. Find the perimeter: $P = a + b + c$.

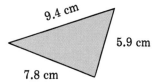

Solve and check.

5. $\dfrac{3}{4}x = 12$

6. $0.02y = 3.2$

7. $r + 3\dfrac{1}{2} = 7\dfrac{1}{4}$

COUNTING PRINCIPLE

Problem Solving – Careers
Manufacturing Inspectors

1. George Rey, an inspector at the Keystone Shirt Manufacturing Company, checks about 425 shirts in one day. About 12% of the shirts have minor defects. How many shirts have minor defects?

2. A shirt with no defects sells for $24. A shirt with a minor defect is discounted 20%. How much does a shirt with a minor defect sell for?

3. What is the total amount of money made on 250 shirts with minor defects?

4. A shirt with a major defect is discounted 60%. What is the total amount of money made on 125 shirts with major defects?

5. In one day, Agnes inspected 425 shirts and found that 34 of them had major defects. What percent of the shirts had major defects?

6. There are 5 inspectors at the Sunshine Records and Tapes Company. Each inspector checks about 275 cassettes a day. How many cassettes are inspected in a 5-day week?

7. In a batch of 650 records, 6% were defective. How many records were defective?

8. In a batch of 275 cassettes, 22 were defective. What percent of the batch of cassettes was defective?

Mathematics Aptitude Test

This test will prepare you for taking standardized tests.
Choose the best answer for each question.

1. The class average of 22 students was 82. If the two highest and two lowest scores were eliminated, the average of the remaining scores was 80. What was the average of the eliminated scores?
   a. 68　　　　　b. 72　　　　　c. 88　　　　　d. 91

2. How many positive whole numbers are factors of 48?
   a. 5　　　　　b. 7　　　　　c. 10　　　　　d. 12

3. If $A = \frac{3}{9} + \frac{3}{9} + \frac{3}{9} + \ldots$ and $B = \frac{4}{9} + \frac{4}{9} + \frac{4}{9} \ldots$, then which of the following is true?
   a. $A < B$　　　b. $B < A$　　　c. $A = B$　　　d. cannot be determined

4. Twenty minutes per pound is needed to cook a 4 lb 12 oz roast. The roast was placed in the oven at 6:00 P.M. What time should the roast be removed?
   a. 7:05 P.M.　　b. 7:30 P.M.　　c. 7:35 P.M.　　d. 7:36 P.M.

5. $\frac{5(0.2) + 4(0.25) - 2(0.5)}{8(0.125)} = ?$
   a. $\frac{1}{10}$　　　　b. $\frac{7}{8}$　　　　c. 1　　　　d. 2

6. A salesperson has a weekly salary of $300. She takes a 10% reduction in pay. She then receives a 10% increase in pay. What is the correct salary?
   a. $300　　　　b. $297　　　　c. $299.97　　　　d. $330

7. The cost of m ounces of nuts at n cents a pound is?
   a. $\frac{m(n)}{16}$　　　b. $\frac{16m}{n}$　　　c. $\frac{16n}{m}$　　　d. $16m(n)$

8. A student does $\frac{1}{4}$ of his homework before lunch. After lunch he completes $\frac{2}{3}$ of the remainder of his homework and then goes to a football game. What part of his homework must he still complete?
   a. $\frac{1}{6}$　　　b. $\frac{1}{2}$　　　c. $\frac{3}{4}$　　　d. $\frac{11}{12}$

9. Swimming 10 laps in a swimming pool is equivalent to swimming 2 km. How many laps are equivalent to 3.2 km?
   a. 6.25　　　b. 6.40　　　c. 16　　　d. 32

COMPUTER ACTIVITIES

If a die is tossed, any 1 of 6 different numbers can come up at *random*. The probability of getting a 4 is 1 out of 6, or $P(4) = \frac{1}{6}$. So, if you were to toss the die 240 times, a 4 could be expected $\frac{1}{6} \cdot 240 = 40$ times. This is a mathematical probability. Physically tossing a die 240 times will not necessarily produce exactly 40 4's.

The program below *simulates* (shows a model of) the tossing of a die 240 times. The number of times a 4 turns up is counted. It should be close to 40.

Exercise Write a program to produce 240 random numbers from 1 through 6. Use a counter to count the number of times a 4 occurs.

C is the counter. Start with C = 0.	10	LET C = 0
	20	FOR K = 1 TO 240
	30	LET R = INT (6 * RND (1) + 1)
Print toss number and the result.	40	PRINT "TOSS "K,R" COMES UP."
If a 4 occurs, count it. Increase the number of 4's by 1 each time.	50	IF R = 4 THEN C = C + 1
After 240 tosses, **END**.	60	NEXT K
Print the number of 4's after 240 tosses.	70	PRINT "4 OCCURS "C" TIMES."
	80	END

See the Computer Section beginning on page 420 for more information.

Exercises

1. Type and **RUN** the program above 5 times. Are the results close to the predicted 40 occurrences of 4's out of 240 tosses?
2. If a die is tossed 480 times, what is $P(4)$?
3. Modify the program above to allow for 480 tosses. **RUN** the program 5 times. Does increasing the number of tosses result in a closer degree of accuracy to the predicted $P(4)$?
★ 4. Write a program to simulate the tossing of a coin. Allow for 200 tosses of the coin. Allow for counting the number of times the coin comes up heads. **RUN** the program 5 times.

Chapter Review

Find the range. Find the mean. [240]

1. 12, 9, 13, 14, 17
2. 93, 97, 86, 76

Find the median. [240]

3. 78, 96, 70, 63, 58
4. 14, 18, 20, 12, 16, 23

Find the mode(s). [240]

5. 79, 63, 80, 79, 80, 79
6. 10, 9, 9, 8, 10, 7, 9, 8, 7, 8

Use these test scores for Exercises 7 and 8. [243]

```
100  95  100   85   90   95  100   95   80   90
 80  90   90  100   90   85   90  100   85   95
```

7. Make a frequency table.
8. Find the mean.

Use these scores for Exercises 9 and 10. [243]

```
99  99  98  97  97  97  96  95  95  95  95  95  94  93  93
93  92  91  91  91  91  90  90  89  89  89  88  87  87  87
86  86  86  85  85  85  85  85  84  83  83  82  81  80  80
78  78  77  75  74  74  74  73  72  72  72  71  71  71  70
```

9. Make a frequency table. Group the scores in intervals of 5.
10. Find the mean.
11. [246] Construct a bar graph to compare the scores. Tigers vs opponents: 58–46, 73–56, 67–72, 46–44. Calculate the averages.

Answer these questions about the spinner. [248]

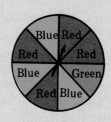

12. How many possible outcomes?
13. How many red outcomes?
14. blue outcomes?
15. P(Red)?
16. P(Blue)?
17. P(Green)?
18. Spin the arrow 100 times. About how many times should it stop on red?
19. P(Red and Blue) [250]
20. P(Green and Blue)

Solve these problems. [252]

21. A bag contains 10 green marbles and 6 red marbles. A red marble is drawn and not put back. Find the probability of picking another red marble.
22. A closet contains 9 shirts and 3 pairs of pants. How many different outfits are possible?

Chapter Test

Find the range. Find the mean.

1. 12, 9, 10, 17, 22
2. 83, 93, 86, 90

Find the median.

3. 26, 17, 83, 22, 71
4. 12, 16, 14, 18, 22, 10

Find the mode(s).

5. 8, 7, 8, 6, 7, 5, 7, 8
6. 27, 96, 27, 96, 96, 25

Use these quiz scores for Exercises 7 and 8.

```
10   7   6   8   9   10   10   8   6   7
 8   9   8   8   7   10    9   8   6   7
```

7. Make a frequency table.
8. Find the mean.

Use these scores for Exercises 9 and 10.

```
25  25  25  24  24  23  22  22  21  21  21  21  19
19  18  17  17  17  17  16  15  15  15  15  14  14
13  13  12  11  11  10  10  10   9   8   8   7
```

9. Make a frequency table. Group the scores in intervals of 3.
10. Find the mean.

11. Construct a bar graph to compare the scores. Owls vs opponents: 60–50, 48–42, 52–56, 70–68. Calculate the averages.

Answer these questions about the spinner.

12. How many possible outcomes?
13. P(1)? 14. P(2)? 15. P(3)?
16. Spin the arrow 500 times. About how many times should it stop on 1?
17. P(1 and 4) 18. P(2 and 3)

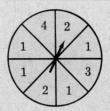

Solve these problems.

19. A bag contains 4 white marbles and 3 blue marbles. A white marble is drawn and not put back. Find the probability of picking another white marble.
20. A closet contains 12 blouses and 5 skirts. How many different outfits are possible?

Ratio and Proportion

11

Ten million bricks were used to build New York's famous Empire State Building.

Ratios

◇ OBJECTIVES ◇
To find ratios
To determine the fraction of a job completed
To write ratios in two forms

◇ RECALL ◇
$\frac{3}{5} = 3 \div 5$

A comparison of 3 parts to 5 parts

Example 1

Write the ratio in two ways. Find the number of shaded parts to the total number of parts.

All parts are the same size.

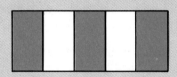

Find the number of shaded parts and the total number of parts. Compare the number of shaded parts to the total number of parts.

3 of 5 parts are shaded means

$\frac{3}{5}$ or $3:5$

Read: *three fifths* or *the ratio 3 to 5*

So, the ratio of the number of shaded parts to the total number of parts is $\frac{3}{5}$, or $3:5$.

A *ratio* is a comparison of two numbers by division. A fraction is a ratio.

Example 2

A soccer team won 5 games and lost 4 games.

Find the ratio: $\frac{\text{wins}}{\text{total games}}$; $\frac{\text{losses}}{\text{wins}}$.

won 5, lost 4, total 9

wins	to	total		losses	to	wins
↓	↓	↓		↓	↓	↓
5	:	9		4	:	5

So, the ratio of $\frac{\text{wins}}{\text{total games}}$ is $\frac{5}{9}$ or $5:9$; the ratio of $\frac{\text{losses}}{\text{wins}}$ is $\frac{4}{5}$ or $4:5$.

---

practice ▷ A volleyball team won 13 games and lost 3 games. Write each ratio in two ways.

1. wins to losses
2. losses to wins
3. wins to total

Example 3

Jennifer can cut the lawn in 5 hours. What part can she cut in 2 hours?

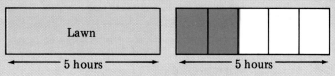

Write the ratio of the number of hours worked to the total number of hours.

$$\frac{\text{Number of hours worked}}{\text{Number of hours to do the job}} = \frac{2}{5}$$

So, Jennifer can cut $\frac{2}{5}$ of the lawn in 2 hours.

Example 4

George can wash the windows in 5 h. What part can he wash in 1 h? in 3 h? in n h?

Write the ratio:
$\frac{\text{Number of hours worked}}{\text{Number of hours to do the job}}$

1 h worked: $\frac{1}{5}$ of job   3 h worked: $\frac{3}{5}$ of job

n h worked: $\frac{n}{5}$ of job

So, George can wash $\frac{1}{5}$ of the windows in 1 hour, $\frac{3}{5}$ in 3 hours, and $\frac{n}{5}$ in n hours.

practice ▷ Lucille can clean the house in 7 hours. What part can she clean in the given number of hours?

4. 1   **5.** 3   **6.** 4   **7.** 7   **8.** n

Ratios can also be used to compare different types of rates, such as miles per hour or quantities such as number of items per cost.

Example 5

Write the ratio in two ways. Simplify when possible.

150 mi in 3 h

$\begin{array}{l}\text{miles} \to \\ \text{hours} \to\end{array} \frac{150}{3} = \frac{3 \cdot 50}{3 \cdot 1}$

$= \frac{50}{1}$ or 50:1

50 mi in 1 h

4 slices of pizza for $3.20

$\frac{4}{3.20}$ or $\frac{4}{320} \begin{array}{l}\leftarrow \text{slices} \\ \leftarrow \text{cents}\end{array}$

$\frac{\cancel{2} \cdot \cancel{2} \cdot 1}{\cancel{2} \cdot \cancel{2} \cdot 80} = \frac{1}{80}$ or 1:80

1 slice of pizza for 80 cents

practice ▷ Write each ratio in two ways. Simplify when possible.

9. 300 miles in 6 hours   **10.** 3 pounds of apples for $1.05

RATIOS

◇ ORAL EXERCISES ◇

A fruit basket contains 7 apples, 20 oranges, 5 bananas, and 3 peaches. Find each ratio.

1. apples to oranges
2. bananas to oranges
3. peaches to oranges
4. bananas to apples
5. apples to bananas
6. peaches to apples

◇ EXERCISES ◇

A hockey team won 12 games and lost 5 games. Find each ratio.

1. wins to losses
2. losses to wins
3. wins to total
4. total to losses
5. total to wins
6. losses to total

A farmer can milk the cows in 4 hours. What part of the milking can be completed in the given number of hours?

Alberta can paint a house in 13 days. What part of the painting can be completed in the given number of days?

7. 1   8. 2   9. 3   10. x 11. 4   12. 7   13. 10   14. n

Write each ratio in two ways. Simplify when possible.

15. 4 cans for 84 cents
16. 330 mi in 6 h
17. 16 km in 60 min
18. $66 for 2 shirts
19. 3 pounds for $6.45
20. 1,024 revolutions in 4 min
★ 21. 3.2 mi in $\frac{1}{3}$ h
★ 22. $2\frac{1}{2}$ cm per 2 m

Algebra Maintenance

Evaluate if $x = 5$, $y = 2\frac{3}{4}$, $a = 0.3$, and $b = 2.7$.

1. $2x + 3y$
2. $5y - 2x$
3. $2b - 4a$
4. $2(x - y)$
5. $\frac{x - y}{6}$
6. $\frac{b - 3a}{b}$

Solve. Check.

7. $x + 9 = 26$
8. $6 = 12 - \frac{2}{3}m$
9. $0.36k = 7.2$

Solve.

10. After spending twice as much as you intended to spend, you were left with $0.46 from a $20 bill. How much did you intend to spend?

11. Your friend gave you this puzzle. Start with a number, divide it by 2, multiply the quotient by 24. The result will be 6. What number should you start with?

Proportions

◇ **OBJECTIVES** ◇

To identify a proportion
To find the product of the means
  and of the extremes
To solve proportions

◇ **RECALL** ◇

Simplify. $\frac{12}{30}$

$\frac{\overset{1}{\cancel{2}} \cdot 2 \cdot \overset{1}{\cancel{3}}}{\underset{1}{\cancel{2}} \cdot \underset{1}{\cancel{3}} \cdot 5}$ Factor into primes.
         Divide out like factors.

$\frac{1 \cdot 2 \cdot 1}{1 \cdot 1 \cdot 5}$

$\frac{2}{5}$

Example 1   Are the ratios equal?

$\frac{18}{24}$ and $\frac{3}{4}$ | $\frac{2}{7}$ and $\frac{6}{20}$

Factor into primes.

$\frac{18}{24}$	$\frac{3}{4}$	$\frac{2}{7}$	$\frac{6}{20}$
$\frac{2 \cdot 3 \cdot 3}{2 \cdot 2 \cdot 2 \cdot 3}$	$\frac{3}{4}$	$\frac{2}{7}$	$\frac{2 \cdot 3}{2 \cdot 2 \cdot 5}$
$\frac{\cancel{2} \cdot \cancel{3} \cdot 3}{\cancel{2} \cdot 2 \cdot 2 \cdot \cancel{3}}$			$\frac{\cancel{2} \cdot 3}{\cancel{2} \cdot 2 \cdot 5}$
$\frac{1 \cdot 1 \cdot 3}{1 \cdot 2 \cdot 2 \cdot 1}$			$\frac{1 \cdot 3}{1 \cdot 2 \cdot 5}$
$\frac{3}{4}$			$\frac{3}{10}$

So, $\frac{18}{24} = \frac{3}{4}$, but $\frac{2}{7}$ does not equal $\frac{6}{20}$.

A *proportion* is an equation which states that two ratios are equal. $\frac{2}{3} = \frac{4}{6}$ or $2:3 = 4:6$ is a proportion.

Each is read: *two is to three as four is to six.*

practice ▷ Which equations are proportions? Read each proportion.

1. $\frac{2}{3} = \frac{12}{18}$ 2. $\frac{16}{20} = \frac{4}{5}$ 3. $\frac{1}{2} = \frac{7}{13}$ 4. $\frac{4}{3} = \frac{20}{15}$

Example 2

Identify the means and the extremes of the proportion $\frac{2}{3} = \frac{4}{6}$.

2:3 = 4:6
↑↑ ↑↑
means
extremes

$\frac{2}{3} = \frac{4}{6}$ means

$\frac{2}{3} = \frac{4}{6}$ extremes

So, the means are 3 and 4; the extremes are 2 and 6.

Example 3

For the proportion $\frac{2}{5} = \frac{8}{20}$, find the product of the means. Then find the product of the extremes.

Means: 5, 8
Extremes: 2, 20

$\frac{2}{5} = \frac{8}{20}$
$5 \cdot 8 = 40$

$\frac{2}{5} = \frac{8}{20}$
$2 \cdot 20 = 40$

So, the product of the means is 40; the product of the extremes is 40.

In a *true proportion,* the product of the means equals the product of the extremes.

practice ▷ **For each proportion, find the product of the means. Then find the product of the extremes.**

5. $\frac{1}{5} = \frac{4}{20}$ 6. $\frac{2}{3} = \frac{10}{15}$ 7. $\frac{18}{30} = \frac{3}{5}$ 8. $\frac{5}{2} = \frac{10}{4}$

Example 4

Solve the proportions.

$\frac{2}{3} = \frac{n}{6}$ $x:6 = 3:2$

Think: What value of n makes the ratios equal?
The product of the means equals the product of the extremes.
Divide 12 by 3.

$\frac{2}{3} = \frac{n}{6}$
$3n = 12$
$n = \frac{12}{3}$, or 4

$\frac{x}{6} = \frac{3}{2}$
$2x = 18$
$x = \frac{18}{2}$, or 9

So, the solution is 4.   So, the solution is 9.

practice ▷ **Solve each proportion.**

9. $\frac{2}{5} = \frac{n}{10}$ 10. $\frac{3}{7} = \frac{n}{21}$ 11. $\frac{2}{3} = \frac{6}{x}$ 12. $x:6 = 4:3$

◇ ORAL EXERCISES ◇

Read each proportion.

1. $\dfrac{2}{3} = \dfrac{10}{15}$ 2. $\dfrac{6}{5} = \dfrac{12}{10}$ 3. $\dfrac{1}{5} = \dfrac{6}{30}$ 4. $\dfrac{6}{15} = \dfrac{2}{5}$ 5. $\dfrac{2}{9} = \dfrac{6}{27}$

◇ EXERCISES ◇

Which equations are proportions?

1. $\dfrac{1}{2} = \dfrac{4}{8}$ 2. $\dfrac{9}{12} = \dfrac{3}{4}$ 3. $\dfrac{1}{3} = \dfrac{4}{11}$ 4. $\dfrac{5}{6} = \dfrac{15}{18}$ 5. $\dfrac{3}{7} = \dfrac{12}{28}$

For each proportion find the product of the means. Then find the product of the extremes.

6. $\dfrac{1}{4} = \dfrac{4}{16}$ 7. $\dfrac{5}{3} = \dfrac{15}{9}$ 8. $\dfrac{2}{5} = \dfrac{12}{30}$ 9. $\dfrac{3}{4} = \dfrac{18}{24}$ 10. $\dfrac{2}{9} = \dfrac{8}{36}$

11. $\dfrac{5}{10} = \dfrac{3}{6}$ 12. $\dfrac{4}{16} = \dfrac{5}{20}$ 13. $\dfrac{4}{12} = \dfrac{6}{18}$ 14. $\dfrac{2}{18} = \dfrac{3}{27}$ 15. $\dfrac{4}{20} = \dfrac{2}{10}$

Solve each proportion.

16. $\dfrac{2}{5} = \dfrac{n}{15}$ 17. $\dfrac{2}{7} = \dfrac{n}{28}$ 18. $\dfrac{3}{5} = \dfrac{x}{15}$ 19. $\dfrac{2}{3} = \dfrac{x}{12}$ 20. $\dfrac{4}{5} = \dfrac{x}{25}$

21. $\dfrac{1}{3} = \dfrac{5}{n}$ 22. $\dfrac{4}{5} = \dfrac{8}{n}$ 23. $\dfrac{3}{4} = \dfrac{15}{n}$ 24. $\dfrac{5}{6} = \dfrac{20}{x}$ 25. $\dfrac{1}{8} = \dfrac{3}{n}$

26. $\dfrac{n}{18} = \dfrac{4}{9}$ 27. $\dfrac{n}{24} = \dfrac{7}{8}$ 28. $\dfrac{x}{20} = \dfrac{2}{5}$ 29. $\dfrac{x}{15} = \dfrac{1}{3}$ 30. $\dfrac{x}{16} = \dfrac{3}{4}$

31. $x:12 = 5:6$ 32. $x:15 = 2:3$ 33. $x:21 = 3:7$

Calculator

Solve. $\dfrac{2}{1.78} = \dfrac{n}{49.31}$ Round to the nearest hundredth. A calculator can be used, as shown below, to solve the proportion.

$$\dfrac{2}{1.78} = \dfrac{n}{49.31}$$
$$1.78\,n = 2 \cdot 49.31$$
$$n = \dfrac{2 \cdot 49.31}{1.78}$$

Press 2 ⊗ 49.31 ⊘ 1.78 ⊜
Display 55.404494
Round to the nearest hundredth, 55.40.

Solve. Round to the nearest hundredth.

1. $\dfrac{3.2}{2.56} = \dfrac{n}{38.37}$ 2. $\dfrac{1.6}{5.8} = \dfrac{4.3}{n}$ 3. $\dfrac{n}{49.3} = \dfrac{6.2}{8.75}$

PROPORTIONS

Applying Proportions

 **OBJECTIVES**

To solve problems by using proportions
To determine if a table of data represents direct variation

 **RECALL**

A team won 7 games and lost 4 games. Find the ratio of losses to wins.

$$\frac{\text{losses}}{\text{wins}} = \frac{4}{7}$$

Example 1

Greta can buy a book for $6. Find the cost of 4 books. Then find the cost of 6 books.

1 book costs $6
4 · $6 = $24   4 books cost $24
6 · $6 = $36   6 books cost $36

Display this data in a table.

NUMBER OF BOOKS	COST
1	$6
4	$24
6	$36

Notice that the ratio of the number of books to the cost remains the same.

$$\frac{1}{6} = \frac{4}{24} = \frac{6}{36}$$

As the number of books increases, the cost also increases. This is an example of *direct variation*.

Example 2

From the table, determine if y varies directly as x.

x	y
3	9
2	6
1	2

Determine if the ratio $\frac{x}{y}$ remains the same.

$\frac{3}{9} = \frac{1}{3}$ $\frac{2}{6} = \frac{1}{3}$ $\frac{1}{2} \neq \frac{1}{3}$

So, y does not vary directly as x.

practice ▷ Determine if y varies directly as x.

1.
x	y
3	15
2	10
4	20

2.
x	y
7	14
4	8
2	4

3.
x	y
6	18
9	27
8	32

4.
x	y
2	8
3	12
5	25

Proportions can be used to solve problems involving direct variation.

Example 3

Lou can buy 2 book covers for 25¢. How many can he buy for 75¢?

Write a proportion. $\dfrac{2}{25} = \dfrac{n}{75}$ ← Let n be the number of covers he can buy for 75¢.

The product of the means = the product of the extremes. $25n = 2 \cdot 75$
$25n = 150$

Divide.
$6 \cdot 25 = 150$ $n = \dfrac{150}{25}$, or $25\overline{)150}^{\;6}$

So, Lou can buy 6 book covers for 75¢.

practice ▷ 5. Jorge can buy 2 pairs of trousers for $36. How much will 3 pairs cost?

Example 4

Clark received $15 for 6 hours of babysitting. How many hours did he sit if he earned $20?

Write a proportion.
Compare dollars to hours.
$6 \cdot 20 = 120$

pay →
hours → $\dfrac{15}{6} = \dfrac{20}{n}$ ← Let n be the number of hours he babysat.
$15n = 120$

Divide. $n = \dfrac{120}{15}$, or $15\overline{)120}^{\;8}$

So, Clark babysat for 8 hours.

practice ▷ 6. Anne got $6 every 3 days for feeding a neighbor's dog. How many days did she feed the dog if she earned $14?

APPLYING PROPORTIONS

◇ ORAL EXERCISES ◇

State each proportion.

1. 10 boxes for $0.70
   n boxes for $1.20

2. 17 cans for $3.90
   43 cans for $$n$

3. 8 hits for every 20 times at bat
   n hits for 465 times at bat

4. 3 shirts for $20
   5 shirts for $$n$

◇ EXERCISES ◇

Determine if y varies directly as x.

1.
x	y
1	3
8	24
2	8

2.
x	y
4	6
8	12
10	15

3.
x	y
4	5
12	15
2	5

4.
x	y
2	7
6	21
4	7

Solve each problem. Use a proportion.

5. Ben can buy 2 pairs of shoes for $60. How many pairs of shoes can he buy for $90?

6. Marge used 2 skeins of yarn to make a scarf. How many skeins of yarn will she need to make 3 scarves?

7. James was given $5 for 2 hours of babysitting. How many hours must he babysit in order to earn $7.50?

8. It takes 3 meters of cloth to make 2 skirts. How many meters of cloth are needed to make 3 skirts?

9. Walter types term papers for $1.25 per page. How many pages must he type in order to earn $13.75?

10. Arlene made $6.50 for 2 hours of cutting grass. How much did she earn for 3 hours of grass cutting?

11. Chipo saves $6 out of every $20 he earns. He earned $70 one month. How much did he save?

12. It cost Bernie $2.52 to bake 6 loaves of bread. How much did it cost him to bake 8 loaves?

13. Ms. Rich drove 200 km in 4 h. How long will it take her to drive 300 km?

14. Akiko drove 400 km in 6 h. How far can she drive in 9 h?

15. Cynthia bought 5 donuts for 75¢. How many donuts can she buy for $1.65?

16. A Chinese restaurant charges $2.70 for 6 egg rolls. How much should they charge for 7 egg rolls?

★ 17. A close-out sale features ties at $3.98 each, 6 for $19.99, and $44.99 a dozen. Which is the best buy?

★ 18. A supermarket sells apples at $1.56 a dozen and oranges at $1.44 a dozen. What is the total cost of 3 apples and 7 oranges?

Inverse Variation

◇ OBJECTIVE ◇
To determine if a table of data represents inverse variation

◇ RECALL ◇
Area of rectangle = length · width

$A = l \cdot w$
$36 = 36 \cdot 1$

Example 1

Find the area of each rectangle. Display this data in a table.

l	w	$A = l \cdot w$
36	1	36
18	2	36
12	3	36

Notice that the product of the length and width remains the same for all three rectangles. As the width increases, the length decreases. This is an example of *inverse variation*.

Example 2

From the table, determine if y varies inversely as x.

x	y
3	16
6	8
24	2

Determine if the product $x \cdot y$ remains the same.
$3 \cdot 16 = 48$
$6 \cdot 8 = 48$
$24 \cdot 2 = 48$

The product $y \cdot x$ is always 48. So, y varies inversely as x.

◇ EXERCISES ◇

Determine if y varies inversely as x.

1.
x	y
7	8
14	4
2	28

2.
x	y
4	25
10	10
30	3

3.
x	y
4	18
9	8
24	3

4.
x	y
7	12
21	4
6	12

INVERSE VARIATION

Circle Graphs

◇ OBJECTIVES ◇
To read circle graphs
To make circle graphs

◇ RECALL ◇
The sum of the measures of the central angles in a circle is 360°.

Example 1

The sum of the percents in a circle is 100%.
30% + 20% + 40% + 10% = 100%

The circle graph shows how a senior class spent $4,000. How much was spent for each event?

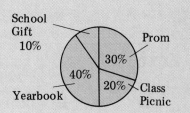

30% = 0.30; 20% = 0.20;
40% = 0.40; 10% = 0.10

Prom	Class Picnic	Yearbook	School Gift
$4,000	$4,000	$4,000	$4,000
×0.30	×0.20	×0.40	×0.10
$1,200.00	$800.00	$1,600.00	$400.00

So, $1,200 was spent on the prom, $800 on the class picnic, $1,600 on the yearbook, and $400 on the school gift.

Example 2

10% + 10% + 5% + 15.2% + 59.8% = 100%

One year, a family spent $1,500 on the use of energy. How much was spent for each use?

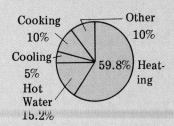

5% = 0.05; 10% = 0.10;
59.8% = 0.598; 15.2% = 0.152

$1,500
×0.598
‾‾‾‾‾
12 000
135 00
750 0
‾‾‾‾‾
$897.000

Cooling	Cooking	Other
$1,500	$1,500	$1,500
×0.05	×0.10	×0.10
$75.00	$150.00	$150.00

Heating	Hot Water
$1,500	$1,500
0.598	0.152
$897.00	$228.00

So, the family spent $75 on cooling, $150 on cooking, $150 on other, $897 on heating, and $228 on hot water.

practice ▷ 1. Total monthly budget: $1,200. How much budgeted for each expense?

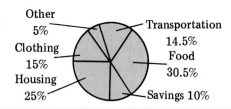

Example 3

Carlos earns about $60 per month. Make a circle graph to show how he spends his money.

$$\frac{1}{6} = 6\overline{)1.0000}^{\,0.1666}, \text{ or } 16.7\%$$

	Amount Spent	Fractional Part	Measure of Central Angle	Percent Spent
School supplies	$10	$\frac{10}{60}$, or $\frac{1}{6}$	$\frac{1}{6}(360°) = 60°$	16.7%
Clothing	$15	$\frac{15}{60}$, or $\frac{1}{4}$	$\frac{1}{4}(360°) = 90°$	25%
Miscellaneous	$5	$\frac{5}{60}$, or $\frac{1}{12}$	$\frac{1}{12}(360°) = 30°$	8.3%
Entertainment	$20	$\frac{20}{60}$, or $\frac{1}{3}$	$\frac{1}{3}(360°) = 120°$	33.3%
Savings	$10	$\frac{10}{60}$, or $\frac{1}{6}$	$\frac{1}{6}(360°) = 60°$	16.7%
Total	$60	1	360°	100%

From the table, find the measure of each central angle. Use a protractor and make each central angle in the circle. From the table, find the corresponding percent for each item. Label the graph.

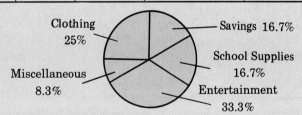

practice ▷ 2. Copy and complete the table. Then make a circle graph.

Crop	Hectares Planted	Fractional Part	Measure of Central Angle	Percent Planted
Soybean	50			
Wheat	25			
Corn	100			
Oats	25			
Total	200			

CIRCLE GRAPHS

◇ ORAL EXERCISES ◇

The circle graph shows an advertising budget. What percent was budgeted for each item?

1. TV
2. Radio
3. Direct mail
4. Newspaper
5. Other

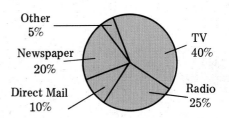

◇ EXERCISES ◇

How much was spent on each item?

1. Automobile expenses for a year: $3,000.

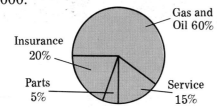

2. Summer living expenses: $4,400.

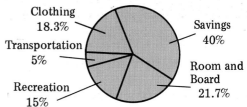

Make a circle graph.

3. Summer earnings: $300
   Mowing lawns, $150
   Washing cars, $30
   Babysitting, $90
   Walking dogs, $30

4. Audio equipment sales: $200,000
   TV's, $100,000
   CB's, $40,000
   Tape recorders, $25,000
   Radios, $25,000
   Other, $10,000

$\frac{1}{5} < \frac{1}{4}$ and $\frac{1}{6} < \frac{1}{5} \rightarrow \frac{1}{5} + \frac{1}{6} < \frac{1}{4} + \frac{1}{5}$

$\frac{1}{3} > \frac{1}{4}$ and $\frac{1}{2} > \frac{1}{3} \rightarrow \frac{1}{3} + \frac{1}{2} > \frac{1}{4} + \frac{1}{3}$

Replace __?__ with <, >, or = to make the sentence true. Explain why the sentence is true.

1. $\frac{1}{6} + \frac{1}{5} + \frac{1}{4} + \frac{1}{3}$ __?__ $\frac{1}{5} + \frac{1}{4} + \frac{1}{3} + \frac{1}{2}$

2. $\frac{1}{9} + \frac{1}{7} + \frac{1}{5} + \frac{1}{3}$ __?__ $\frac{1}{10} + \frac{1}{8} + \frac{1}{6} + \frac{1}{4}$

3. $\frac{1}{10} + \frac{1}{20} + \frac{1}{5} + \frac{1}{4}$ __?__ $0.10 + 0.05 + 0.20 + 0.25$

Problem Solving – Careers
Electric Sign Service

Hanssen's Electric Signs charges $25 for a service call and $18 per hour for services or labor, plus parts. Hanssen's charges 15¢/km after the first 15 km. There is no charge for the first 15 km.

1. It took a servicer 3 hours to repair a thermostat. The parts cost $24. What was the total bill?

2. A servicer traveled 105 kilometers. Find the travel charge.

3. Joanne Hanssen repaired an electric sign. She started at 9:20 A.M. and finished at 10:50 A.M. She traveled 58 kilometers, and the parts cost $18.25. What was the total bill?

4. Hanssen's sent two servicers to repair the Burger Time sign. They each worked 6 hours, the parts cost $23.45, and they traveled 126 kilometers. What was the total bill?

5. The odometer on the Hanssen's van read 29,977 when the servicer left Hanssen's and 30,019 when the servicer returned. What was the travel charge?

6. Two servicers installed a new sign. They each worked 9 hours. The odometer read 16,905 when they left Hanssen's and 16,978 when they returned. What was the total bill?

7. Monte Hanssen worked $4\frac{1}{2}$ hours repairing a sign. The odometer read 53,429 when he left and 53,492 when he returned. The parts cost $73. What was the total bill?

PROBLEM SOLVING

Scale Drawing

◆ OBJECTIVE ◆

To find distances by using a scale

◆ RECALL ◆

```
    30         16        2.5
  ×1.5      ×3.25       ×2.5
   15 0        80       1 25
   30         3 2        5 0
   45.0        48        6.25
              52.00
```

Example 1

The distance between two towns on a map is 2.9 cm. The scale on the map is 1 cm → 15 km. Find the distance between the towns.

Write a proportion. Compare cm to km.
The product of the means equals the product of the extremes.

$$\frac{1 \text{ cm}}{15 \text{ km}} = \frac{2.9 \text{ cm}}{n \text{ km}}$$

$n = 15(2.9)$
$n = 43.5$

$15 \times 2.9 = 4.35$

So, the distance between the towns is 43.5 km.

Example 2

Find the distance: from Fort Morris to Darby; from Darby to Westbury.

Use a centimeter ruler. Measure the distance between the towns in centimeters.

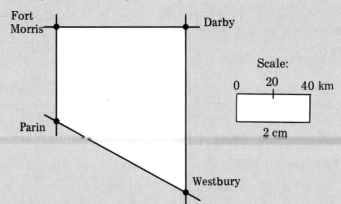

From the scale on the map,
2 cm → 40 km, or
1 cm → 20 km.

Between Fort Morris and Darby is about 3.5 cm.

Between Darby and Westbury is about 4.4 cm.

Multiply.
```
   20         20
  ×3.5       ×4.4
   10 0       8 0
   60         80
   70.0       88.0
```

$$\frac{1 \text{ cm}}{20 \text{ km}} = \frac{3.5 \text{ cm}}{n \text{ km}}$$

$1n = 20(3.5)$
$n = 70$

$$\frac{1 \text{ cm}}{20 \text{ km}} = \frac{4.4 \text{ cm}}{n \text{ km}}$$

$1n = 20(4.4)$
$n = 88$

So, it is about 70 km from Fort Morris to Darby and 88 km from Darby to Westbury.

practice ▷ **Find the distance. Use the map in Example 2.**

1. from Westbury to Parin
2. from Parin to Fort Morris

Example 3

Find the distance.
from A to B; from B to C

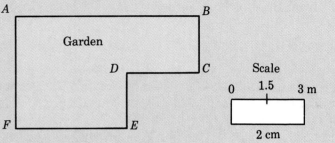

Between A and B is 5 cm.

$$\frac{1 \text{ cm}}{1.5 \text{ m}} = \frac{5 \text{ cm}}{n \text{ m}}$$

$1n = 1.5(5) = 7.5$

```
  1.5
 × 5
 ───
  7.5
```

Between B and C is 1.5 cm.

$$\frac{1 \text{ cm}}{1.5 \text{ m}} = \frac{1.5 \text{ cm}}{n \text{ m}}$$

$1n = 1.5(1.5) = 2.25$

```
   1.5
  ×1.5
  ───
   75
  15
  ────
  2.25
```

So, the distance between A and B is 7.5 m and the distance between B and C is 2.25 m.

practice ▷ **Find the distance. Use the drawing in Example 3.**

3. from D to E
4. from E to F
5. from F to A

◇ ORAL EXERCISES ◇

How far is it between towns with the given map distance?
Use the scale 1 cm → 30 km.

1. 2 cm
2. 3 cm
3. 4 cm
4. 5 cm
5. 10 cm

◇ EXERCISES ◇

How far is it between towns with the given map distance?
Use the scale 1 cm → 25 km.

1. 3 cm
2. 7 cm
3. 4.1 cm
4. 6.3 km
5. 10.5 cm

How far is it between towns with the given map distance?
Use the scale 2 cm → 60 km.

6. 6 cm
7. 9 cm
8. 2.7 cm
9. 5.4 cm
10. 11.1 cm

SCALE DRAWING

Find the distance.

11. from Bradford to Eaton
12. from Eaton to Cramer
13. from Cramer to Randall
14. from Randall to Sealy
15. from Sealy to Bradford
16. from Bradford to Cramer

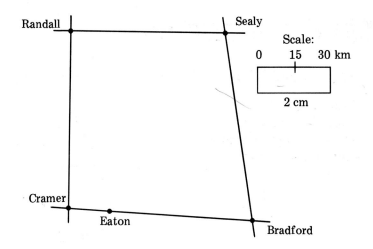

Find the distance from:

17. Landing to Filton
18. Landing to Battle
19. Battle to Kaldor
20. Kaldor to Grandville
21. Grandville to Filton
22. Landing to Grandville

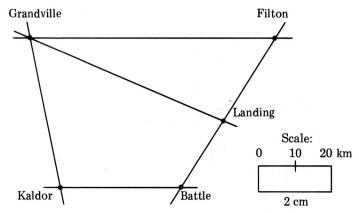

Find the distance.

23. from A to B
24. from B to C
25. from C to D
26. from D to E
27. from E to F
28. from F to A
29. Find the perimeter of the room.

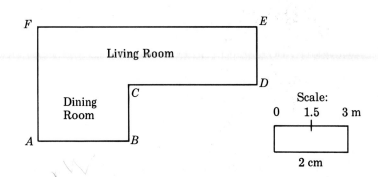

Make scale drawings.

	Item	Dimensions	Scale
★ 30.	soccer field	50 m by 100 m	1 cm → 10 m
★ 31.	table	1 m by 2 m	1 cm → 0.5 m
★ 32.	room	3.5 m by 4.5 m	1 cm → 1 m

Similar Triangles

◇ OBJECTIVES ◇

To find the lengths of the sides of similar triangles
To solve word problems using similar triangles

◇ RECALL ◇

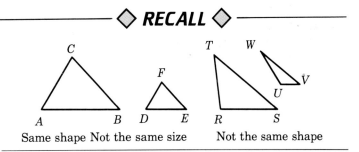

Same shape Not the same size   Not the same shape

Triangles *ABC* and *DEF* have the same shape. They are similar triangles.

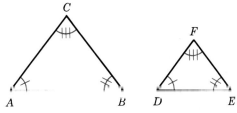

Corresponding sides:
$\overline{AB}$ and $\overline{DE}$,
$\overline{BC}$ and $\overline{EF}$,
$\overline{AC}$ and $\overline{DF}$
Corresponding angles:
$\angle A$ and $\angle D$,
$\angle B$ and $\angle E$,
$\angle C$ and $\angle F$

Read: Triangle *ABC* is similar to triangle *DEF*.
Write: $\triangle ABC \sim \triangle DEF$.

Example 1

$\angle R$ means angle R.
$m\angle R$ means measure of $\angle R$.
Corresponding angles:
$\angle R$ and $\angle U$,
$\angle S$ and $\angle W$,
$\angle T$ and $\angle W$
A right angle measures 90°.
$m\angle R = 90° = m\angle U$
$m\angle S = 37° = m\angle V$
$m\angle T = 53° = m\angle W$
Corresponding sides:
$\overline{RS}$ and $\overline{UV}$,
$\overline{ST}$ and $\overline{VW}$,
$\overline{TR}$ and $\overline{WU}$

$\triangle RST$ is similar to $\triangle UVW$. Show that the measures of the corresponding angles are equal. Show that the lengths of the corresponding sides have the same ratio.

$m\angle R = m\angle U$
$m\angle S = m\angle V$
$m\angle T = m\angle W$

$\dfrac{RS}{UV} = \dfrac{8}{4} = \dfrac{2}{1}$

$\dfrac{ST}{VW} = \dfrac{10}{5} = \dfrac{2}{1}$ same ratio

$\dfrac{TR}{WU} = \dfrac{6}{3} = \dfrac{2}{1}$

So, the measures of the corresponding angles are equal and the lengths of the corresponding sides have the same ratio.

SIMILAR TRIANGLES

If two triangles are *similar*, then the measures of the corresponding angles are equal and the lengths of the corresponding sides have the same ratio.

Example 2

$\triangle ABC \sim \triangle DEF$. Find x and y.

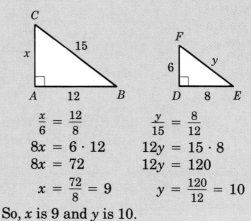

$\triangle ABC \sim \triangle DEF$.
The lengths of the corresponding sides have the same ratio.
Write and solve proportions.

The product of the means equals the product of the extremes.

Divide.

$\dfrac{x}{6} = \dfrac{12}{8}$ $\qquad$ $\dfrac{y}{15} = \dfrac{8}{12}$

$8x = 6 \cdot 12$ $\qquad$ $12y = 15 \cdot 8$

$8x = 72$ $\qquad$ $12y = 120$

$x = \dfrac{72}{8} = 9$ $\qquad$ $y = \dfrac{120}{12} = 10$

So, x is 9 and y is 10.

practice 1. $\triangle ABC \sim \triangle DEF$. Find x and y.

Example 3

A vertical meterstick casts a shadow 4 m long. At the same time, a pole casts a shadow 28 m long. How tall is the pole?

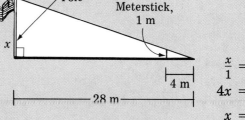

Two similar triangles are formed.
Let x = the height of the pole.
Write and solve a proportion.

$\dfrac{x}{1} = \dfrac{28}{4}$

$4x = 28 \cdot 1$

$x = \dfrac{28}{4}$, or 7

So, the pole is 7 m tall.

practice 2. A man 2 m tall casts a shadow 4 m long. At the same time, a pole casts a shadow 16 m long. How tall is the pole?

◇ ORAL EXERCISES ◇

△ ABC ~ △ DEF. Which side corresponds to the given side?

1. $\overline{AC}$ 2. $\overline{AB}$ 3. $\overline{BC}$
4. $\overline{DE}$ 5. $\overline{DF}$ 6. $\overline{EF}$

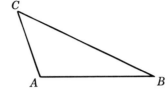

Which angle corresponds to the given angle?

7. ∠A 8. ∠C 9. ∠B
10. ∠E 11. ∠D 12. ∠F

◇ EXERCISES ◇

△ ABC ~ △ DEF. Find x and y.

1.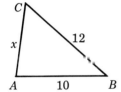

2.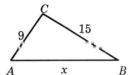

3.

4.

5.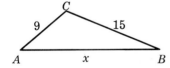

6.

7. A vertical 2-meter stick casts a shadow 10 m long. At the same time, a flagpole casts a shadow 15 m long. How tall is the flagpole?

8. A child 1 m tall casts a shadow 2 m long. At the same time, a telephone pole casts a shadow 20 m long. How tall is the telephone pole?

★ 9. A scout used similar triangles to find the width of a river. The scout constructed similar triangles and made measurements as shown. Find the width of the river.

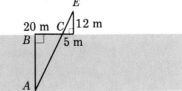

SIMILAR TRIANGLES

Trigonometric Ratios

◇ OBJECTIVES ◇

To find ratios of the sides of a right triangle

To compute tangent, sine, and cosine of an acute angle of a right triangle

◇ RECALL ◇

$\triangle ABC \sim \triangle DEF$

$\dfrac{a}{b} = \dfrac{d}{e}; \dfrac{a}{c} = \dfrac{d}{f}; \dfrac{b}{c} = \dfrac{e}{f}$

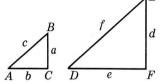

Trigonometry means *three-angle measure* or *triangle measure*. Only right triangles will be used. The side opposite the right angle in a right triangle is called the *hypotenuse*.

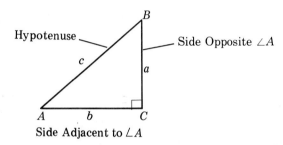

Example 1

What is the length of the side opposite $\angle R$? of the side adjacent to $\angle R$? of the hypotenuse?

r is the side *opposite* $\angle R$
t is the side *adjacent* to $\angle R$
s is the *hypotenuse*

So, the length of the side opposite $\angle R$ is 12, the length of the side adjacent to $\angle R$ is 5, and the length of the hypotenuse is 13.

practice ▷ Find the indicated lengths.

1. side opposite $\angle B$
2. side adjacent to $\angle B$
3. the hypotenuse

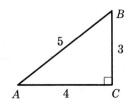

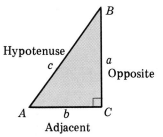

trigonometric ratios

In right △ABC, the ratios of the sides have special names:

tangent of ∠A = $\dfrac{\text{length of side opposite } \angle A}{\text{length of side adjacent to } \angle A}$

sine of ∠A = $\dfrac{\text{length of side opposite } \angle A}{\text{length of hypotenuse}}$

cosine of ∠A = $\dfrac{\text{length of side adjacent to } \angle A}{\text{length of hypotenuse}}$

tan stands for *tangent*.
sin stands for *sine*.
cos stands for *cosine*.

$\tan A = \dfrac{a}{b}$, $\sin A = \dfrac{a}{c}$, $\cos A = \dfrac{b}{c}$

Example 2

Find each ratio. tan A, sin A, and cos A

$\tan A = \dfrac{\text{opp. }(\angle A)}{\text{adj. (to }\angle A)}$ $\tan A = \dfrac{a}{b}$ or $\dfrac{5}{12}$

$\sin A = \dfrac{\text{opp. }(\angle A)}{\text{hyp.}}$ $\sin A = \dfrac{a}{c}$ or $\dfrac{5}{13}$

$\cos A = \dfrac{\text{adj. (to }\angle A)}{\text{hyp.}}$ $\cos A = \dfrac{b}{c}$ or $\dfrac{12}{13}$

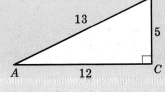

So, $\tan A = \dfrac{5}{12}$, $\sin A = \dfrac{5}{13}$, and $\cos A = \dfrac{12}{13}$.

practice ▷ Find each ratio. Use △ABC in Example 2.

4. tan B   5. sin B   6. cos B

Example 3

Find each ratio to three decimal places. tan 30°, sin 30°, and cos 30°

$\tan B = \dfrac{\text{opp. }(\angle B)}{\text{adj. (to }\angle B)}$ $\tan 30° = \dfrac{1}{1.732} = 0.577$

$\sin B = \dfrac{\text{opp. }(\angle B)}{\text{hyp.}}$ $\sin 30° = \dfrac{1}{2} = 0.500$

$\cos B = \dfrac{\text{adj. (to }\angle B)}{\text{hyp.}}$ $\cos 30° = \dfrac{1.732}{2} = 0.866$

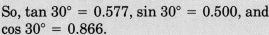

So, tan 30° = 0.577, sin 30° = 0.500, and cos 30° = 0.866.

practice ▷ Find each ratio to three decimal places. Use △ABC in Example 3.

7. tan 60°   8. sin 60°   9. cos 60°

TRIGONOMETRIC RATIOS

◇ ORAL EXERCISES ◇

1. What is the length of the side opposite ∠D?
2. What is the length of the side adjacent to ∠D?
3. What is the length of the side opposite ∠E?
4. What is the length of the side adjacent to ∠E?
5. What is the length of the hypotenuse?

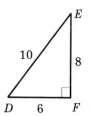

◇ EXERCISES ◇

Find each ratio. Leave answers in fractional form.

1. tan A
2. sin A
3. cos A
4. tan B
5. sin B
6. cos B

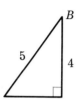

Find each ratio. Leave answers in fractional form.

7. tan D
8. sin D
9. cos D
10. tan E
11. sin E
12. cos E

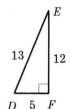

Find each ratio. Leave answers in fractional form.

13. tan R
14. sin R
15. cos R
16. tan S
17. sin S
18. cos S

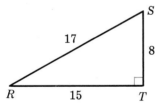

Find each ratio. Leave answers in fractional form.

19. tan U
20. sin U
21. cos U
22. tan V
23. sin V
24. cos V

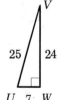

Find each ratio to three decimal places.

25. tan A
26. sin A
27. cos A
28. tan B
29. sin B
30. cos B

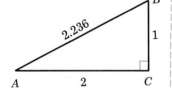

Find each ratio to three decimal places.

31. tan 45°
32. sin 45°
33. cos 45°

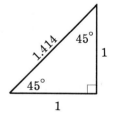

Show that each statement is true. Use the given figure.

★ 34. sin A = cos B
★ 35. sin B = cos A
★ 36. tan A = $\dfrac{1}{\tan B}$

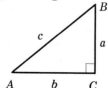

CHAPTER ELEVEN

Non-Routine Problems

There is no simple method used to solve these problems. Use whatever methods you can. Do not expect to solve these easily.

1. You have one of each of the following coins: 1¢, 5¢, 10¢, 25¢, and 50¢. How many different amounts of money can you obtain by forming all possible 2-coin combinations?

2. You have a choice of 2 different routes from Niceville to Warsaw and 3 different routes from Warsaw to Farmville. How many different routes can you take from Niceville to Farmville?

   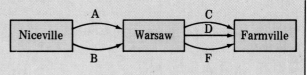

3. You have the five digits shown on the slips at the right.
   a. Form the smallest possible five-digit number using each of the digits once.
   b. Form the largest possible five-digit number using each of the digits once.
   c. How many different five-digit numbers can you form using each digit once?
   d. How many different four-digit numbers?
   e. Three-digit numbers?
   f. Two-digit numbers?

4. You have 5 different colors available with which to form 2-color flags. How many different flags can you form? (*Hint:* Number the colors 1, 2, 3, 4, 5; how many colors are available for the top portion? After that choice is made, how many colors are available for the lower portion?)

5. How many different automobile plates can be made using one letter followed by two digits? You can use the same digit twice. (*Hint:* How many choices do you have for the letter spot? After you choose the letter, how many choices do you have for the first digit? For the second digit?)

NON-ROUTINE PROBLEMS

COMPUTER ACTIVITIES

The program below determines whether an equation is a proportion. Writing such a program depends upon the ability of the programmer to state the general problem. Discovery of the generalization can be aided by first looking at several specific examples.

Example Determine if the equations are proportions.

$$\frac{3}{4} = \frac{6}{8} \qquad\qquad \frac{5}{10} = \frac{1}{2}$$

Check: $4 \cdot 6 = 3 \cdot 8$ Check: $10 \cdot 1 = 5 \cdot 2$
  $24 = 24$ $10 = 10$

Now in *general* $\frac{A}{B} = \frac{X}{Y}$ if $B \cdot X = A \cdot Y$

Write and **RUN** a program to determine if $\frac{5}{17} = \frac{10}{32}$ is a proportion.

```
                              10  INPUT "WHAT IS A? ";A
                              20  INPUT "WHAT IS B? ";B
                              30  INPUT "WHAT IS X? ";X
                              40  INPUT "WHAT IS Y? ";Y
         Computer checks if   50  IF B * X = A * Y THEN 80
product of means = product of extremes.  60  PRINT "THIS IS NOT A PROPORTION."
                              70  END
                              80  PRINT "THE EQUATION IS A PROPORTION."
```

See the Computer Section beginning on page 420 for more information.

Exercises

Use the program above to determine if each of the following is a proportion. In each case first make the determination on your own by computation.

1. $\frac{7}{14} = \frac{35}{70}$ 2. $\frac{14}{19} = \frac{98}{133}$ 3. $\frac{43}{11} = \frac{215}{55}$ 4. $\frac{17}{37} = \frac{68}{168}$

5. Modify the program of the lesson so that it is not necessary to type **RUN** each time. Allow for checking three equations.

6. Use the results of **Exercise 5** above to determine if the following are proportions.

$$\frac{37}{65} = \frac{105}{195}; \quad \frac{119}{207} = \frac{238}{415}; \quad \frac{39}{117} = \frac{156}{468}$$

284   CHAPTER ELEVEN

Chapter Review

A football team won 8 games and lost 3 games. Find each ratio. [260]

1. wins to losses
2. wins to total
3. losses to total

4. It takes 12 hours to plow a field. What part can be plowed in 5 hours? [260]

5. Write the ratio in two ways. Simplify. 8 pencils for 96 cents. [260]

Solve each proportion. [263]

6. $\dfrac{2}{3} = \dfrac{n}{9}$
7. $\dfrac{3}{4} = \dfrac{6}{x}$
8. $\dfrac{n}{10} = \dfrac{2}{5}$
9. $x:15 = 1:5$

Solve. [266]

10. A baker used 3 cups of flour to make 60 cookies. How many cups of flour are needed to make 90 cookies?

11. Mr. Flax drove 160 kilometers in 2 hours. How far can he drive in 6 hours?

12. The circle graph shows how a scout troop raised $3,000. How much did they earn from each event? [270]

13. Janet earned $2,000 by working part time. Make a circle graph to show how she spent her earnings. [270]

Clothing, $250
Savings, $1,000
Entertainment, $500
School supplies, $200
Charity, $50

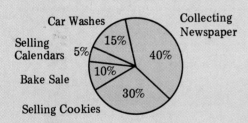

14. How far is it between Sandland and Auburn? [274]

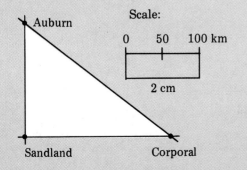

15. Find each ratio. Leave answers in fractional form. [280]
tan D
sin D
cos E

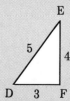

Chapter Test

A baseball team won 20 games and lost 7 games. Find each ratio.

1. losses to total
2. wins to losses

Solve.

3. It takes 5 hours to fill the pool. What part can be filled in 3 hours?
4. Write the ratio two ways. Simplify. 110 miles in 2 hours.

Solve each proportion.

5. $\frac{3}{7} = \frac{n}{21}$
6. $\frac{6}{x} = \frac{2}{5}$
7. $n:10 = 3:5$

Solve.

8. A dressmaker used 5 meters of silk to make 2 dresses. How many meters of silk are needed to make 8 dresses?

9. The circle graph shows how a band raised $4,000 for a trip. How much did they earn from each event?

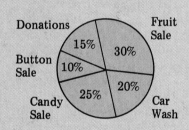

10. How far is it between Concord and Rochester?

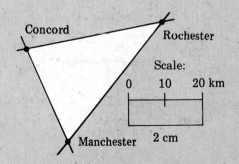

11. Find each ratio. Leave answers in fractional form.
    sin A
    cos B
    tan A

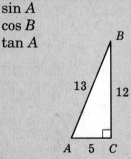

286   CHAPTER ELEVEN

Adding and Subtracting Integers

12

The main section of the Golden Gate Bridge spans 4,200 feet across San Francisco Bay.

Integers on a Number Line

◇ OBJECTIVES ◇

To tell what integer corresponds to a given point

To tell what integer comes just after or just before a given integer

To compare integers

To think of integers in terms of "trips" on a number line

◇ RECALL ◇

Whole numbers can be shown on a number line.

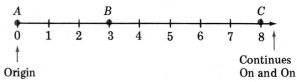

0 corresponds to point A, 3 corresponds to point B, and 8 corresponds to point C.

A number line can go on forever in both directions.

Integers on a number line

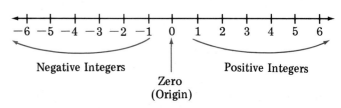

	Positive integers	1	2	3
		↑	↑	↑
Read.		positive 1	positive 2	positive 3

Zero is halfway between -1 and 1.   Zero is neither positive nor negative.

	Negative integers	-1	-2	-3
		↑	↑	↑
Read.		negative 1	negative 2	negative 3

Example 1

What integer corresponds to point A? to point M? to point R?

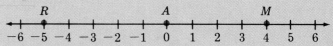

Read on the number line above.   So, 0 corresponds to point A, 4 corresponds to point M, and -5 corresponds to point R.

practice ▷ What integer corresponds to the point?

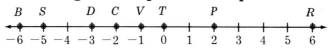

1. P 2. S 3. T 4. V 5. C 6. R 7. B 8. D

288   CHAPTER TWELVE

Example 2

On a number line, 4 is to the right of 3; 0 is to the right of −1; 1 is to the left of 2; and −5 is to the left of −4.

What integer comes just after 3? just after −1? just before 2? just before −4?

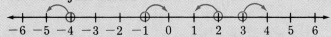

So, 4 comes just after 3, 0 comes just after −1, 1 comes just before 2, and −5 comes just before −4.

practice ▷ What integer comes just before the given integer? just after the given integer?

9. 4   10. 1   11. −3   12. −5   13. 0

Example 3

> means is *greater than*.
< means is *less than*.

4 comes after 3. 4 > 3
−2 comes before 1. −2 < 1

Read 4 > 3 as 4 *is greater than* 3.
Read 3 < 4 as 3 *is less than* 4.

Compare. Use > or <.

4 _?_ 3, 3 _?_ 4   0 _?_ −1, −1 _?_ 0
−2 _?_ 1, 1 _?_ −2   −4 _?_ −6, −6 _?_ −4

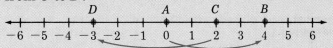

So, 4 > 3 and 3 < 4, 0 > −1 and −1 < 0, 2 < 1 and 1 > −2, and −4 > −6 and −6 < −4.

practice ▷ Compare. Use > or <.

14. 2 _?_ 3   15. −2 _?_ −3   16. 0 _?_ 1   17. −4 _?_ 0

Example 4

What integer corresponds to the "trip" from A to B? from C to D?

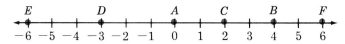

The trip from A to B is in the positive direction. The trip from A to B is 4. The trip from C to D is in the negative direction. The trip from C to D is −5. So, 4 corresponds to the trip from A to B and −5 corresponds to the trip from C to D.

practice ▷ What integer corresponds to the trip?

```
   E       D       A   C   B   F
───●───┼───●───┼───●───●───●───●───
  −6  −5  −4  −3  −2  −1   0   1   2   3   4   5   6
```

18. from A to F   19. from D to B   20. from C to D

INTEGERS ON A NUMBER LINE

◇ ORAL EXERCISES ◇

Read.

1. 4
2. −5
3. 3
4. −1
5. −2
6. 10
7. 8
8. −7

◇ EXERCISES ◇

What integer corresponds to the point?

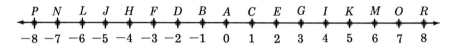

1. M
2. A
3. J
4. D
5. P
6. B
7. R
8. E
9. K
10. N
11. F
12. L

What integer comes just after the given integer?

13. 6
14. 0
15. −2
16. −8
17. −1
18. 7
19. −5
20. 1

What integer comes just before the given integer?

21. 7
22. 0
23. −2
24. 5
25. 4
26. −6
27. −1
28. 1

Compare. Use > or <.

29. 5 __?__ 6
30. −3 __?__ −2
31. 0 __?__ 1
32. −5 __?__ −1
33. 3 __?__ 10
34. 4 __?__ −4
35. 0 __?__ −5
36. −3 __?__ 2

What integer corresponds to the trip?

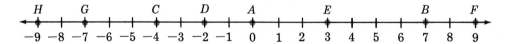

37. from A to B
38. from B to A
39. from A to C
40. from D to G
41. from E to F
42. from F to E
43. from H to F
44. from F to H
45. from B to G

Graph the numbers on a number line.

★ 46. The whole numbers less than 10.
★ 47. The odd integers between −10 and 10.
★ 48. The odd integers between −12 and −1.
★ 49. The even integers between −5 and 10.

Adding Integers

◇ OBJECTIVE ◇
To add integers

◇ RECALL ◇

$9 + 8 = 17 \qquad 7 + 6 = 13 \qquad 5 + 9 = 14$

$$\begin{array}{r}28\\+39\\\hline 67\end{array} \qquad \begin{array}{r}56\\+97\\\hline 153\end{array} \qquad \begin{array}{r}139\\+485\\\hline 624\end{array}$$

We can use trips on a number line to add integers.

Example 1

$2 + 6$ is read *positive two add positive six*.
A number line can be vertical.
A trip of 2 means move 2 units in the positive direction.
Start at 0.
The second trip starts where the first trip ended.

Add. $2 + 6$

$2 + 6$ means a trip of 2 followed by a trip of 6.

Think: Start at 0; move 2 in the positive direction, then 6 in the positive direction; finish at 8.

So, $2 + 6 = 8$.

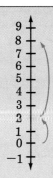

The sum of two positive integers is a positive integer.

Example 2

$-3 + (-4)$ is read *negative three add negative four*.
A trip of -3 means move 3 units in the negative direction.
Start at 0.
The second trip starts where the first trip ended.

Add. $-3 + (-4)$
$-3 + (-4)$ means a trip of -3 followed by a trip of -4.

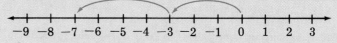

Think: Start at 0; move 3 in the negative direction, then 4 in the negative direction; finish at -7.
So, $-3 + (-4) = -7$.

The sum of two negative integers is a negative integer.

practice ▷ Add. Draw trips on a number line.

1. $3 + 5$
2. $4 + 3$
3. $-2 + (-1)$
4. $-4 + (-2)$

ADDING INTEGERS

Example 3

$-5 + 2$ is read *negative five add positive two*.
A trip of -5 means move 5 units in the negative direction.
A trip of 2 means move 2 units in the positive direction.
Start at 0.
The second trip starts where the first trip ended.

Add. $-5 + 2$
$-5 + 2$ means a trip of -5 followed by a trip of 2.

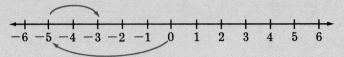

Think: Start at 0; move 5 units in the negative direction, then 2 units in the positive direction; finish at -3.
So, $-5 + 2 = -3$.

Example 4

$6 + (-3)$ is read *positive six add negative three*.
A trip of 6 means move 6 units in the positive direction.
A trip of -3 means move 3 units in the negative direction.
Start at 0.
The second trip starts where the first trip ended.

Add. $6 + (-3)$
$6 + (-3)$ means a trip of 6 followed by a trip of -3.

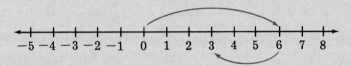

Think: Start at 0; move 6 units in the positive direction, then 3 units in the negative direction; finish at 3.
So, $6 + (-3) = 3$.

practice ▷ **Add. Draw trips on a number line.**

5. $-6 + 4$ 6. $-2 + 5$ 7. $8 + (-7)$ 8. $2 + (-6)$

Example 5

$-4 + 4$ is read *negative four add positive four*.
A trip of -4 means move 4 units in the negative direction.
A trip of 4 means move 4 units in the positive direction.
Start at 0.
The second trip starts where the first trip ended.

Add. $-4 + 4$
$-4 + 4$ means a trip of -4 followed by a trip of 4.

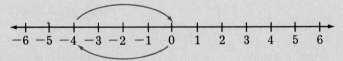

Think: Start at 0; move 4 units in the negative direction, then 4 units in the positive direction; finish at 0.
So, $-4 + 4 = 0$.

practice ▷ **Add. Draw trips on a number line.**

9. $2 + (-2)$ 10. $-1 + 1$ 11. $-3 + 3$ 12. $5 + (-5)$

◇ ORAL EXERCISES ◇

Read.

1. $3 + 9$
2. $6 + 8$
3. $9 + 4$
4. $-6 + (-2)$
5. $-7 + (-5)$
6. $3 + (-7)$
7. $1 + (-9)$
8. $-2 + 8$
9. $-7 + 3$
10. $8 + (-3)$
11. $-2 + 2$
12. $7 + (-7)$
13. $9 + 0$
14. $0 + (-2)$
15. $-9 + 0$

Add.

16. $2 + 3$
17. $3 + 7$
18. $4 + 9$
19. $7 + 8$
20. $9 + 9$
21. $-1 + (-2)$
22. $-2 + (-6)$
23. $-5 + (-7)$
24. $-9 + (-2)$
25. $-7 + (-8)$
26. $-2 + 3$
27. $-4 + 6$
28. $-7 + 2$
29. $-9 + 12$
30. $-8 + 1$
31. $5 + (-4)$
32. $7 + (-2)$
33. $9 + (-4)$
34. $2 + (-6)$
35. $3 + (-7)$
36. $7 + (-7)$
37. $-9 + 9$
38. $8 + (-8)$
39. $10 + (-10)$

◇ EXERCISES ◇

Add. Draw trips on a number line.

1. $4 + 5$
2. $6 + 8$
3. $9 + 0$
4. $8 + 5$
5. $7 + 7$
6. $8 + 11$
7. $13 + 4$
8. $14 + 6$
9. $5 + 20$
10. $8 + 19$
11. $-3 + (-1)$
12. $-2 + (-7)$
13. $-6 + (-2)$
14. $0 + (-4)$
15. $-5 + (-5)$
16. $-9 + (-4)$
17. $-7 + (-3)$
18. $-9 + (-9)$
19. $-6 + (-11)$
20. $-8 + (-12)$
21. $-4 + 2$
22. $-7 + 0$
23. $-5 + 9$
24. $-8 + 3$
25. $-1 + 6$
26. $-8 + 11$
27. $-10 + 3$
28. $-12 + 1$
29. $-5 + 14$
30. $-6 + 15$
31. $6 + (-2)$
32. $1 + (-8)$
33. $7 + (-6)$
34. $8 + (-9)$
35. $4 + (-7)$
36. $10 + (-3)$
37. $5 + (-12)$
38. $18 + (-9)$
39. $6 + (-19)$
40. $11 + (-15)$
41. $-6 + 6$
42. $11 + (-11)$
43. $-12 + 12$
44. $13 + (-13)$

Step 1 Write down the year of your birth.

Step 2 Add it to the year of some important event in your life.

Step 3 Add the age you will be this year.

Step 4 Add the number of years since the important event took place.

Step 5 Multiply the current year by 2.

Step 6 What do you discover? Try to explain why the answers in Steps 4 and 5 are the same.

Problem Solving – Applications
Jobs for Teenagers

Example

Percent can be used in describing a rate of increase or decrease.

NUMBER OF PAPERS SOLD		
	Lou	Maria
last week	28	96
this week	42	72

Find the amount of change.
Divide this amount by the original.
Write the percent.

$42 - 28 = 14$ | $96 - 72 = 24$
$n = 14 \div 28$ | $n = 24 \div 96$
$n = 0.50 = 50\%$ | $n = 0.25 = 25\%$

So, the number of papers sold by Lou increased by 50% while those sold by Maria decreased by 25%.

To find the % increase or decrease:
1. Find the amount of increase or decrease.
2. Find what % this amount is of the original.

Solve these problems.

1. Glenda delivered 45 papers yesterday. She delivered 54 papers today. Find the percent increase.

2. Joshua delivered 125 papers yesterday. He delivered 105 today. Find the percent decrease.

3. Lora's earnings increased from $23.50 to $29.61. Find the percent increase.

4. Paulo rode 12 km on his paper route yesterday. He rode only 8 km today. Find the percent decrease.

5. Sele earned $42 last week. He earned $48.30 this week. Find the percent increase.

6. Issac's earnings increased from $43.20 to $47.52. Find the percent increase.

Subtracting Integers

◇ OBJECTIVE ◇
To subtract integers

◇ RECALL ◇
Add. $3 + (-7)$

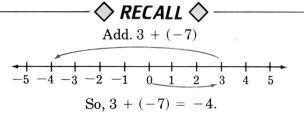

So, $3 + (-7) = -4$.

Example 1

$6 - 2$ is read *positive six subtract positive two*.

Show the trips on a number line.

The second trip starts where the first trip ended.

Subtract. $6 - 2$
Subtract a positive integer means move in the negative direction. $6 - 2$ means a trip of 6 followed by a trip of 2 in the negative direction.

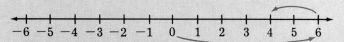

Think: Start at 0; move 6 units in the positive direction, then 2 units in the negative direction; finish at 4.
So, $6 - 2 = 4$.

Example 2

$3 - 7$ is read *positive three subtract positive seven*.

Show the trips on a number line.

Subtract. $3 - 7$
$3 - 7$ means a trip of 3 followed by a trip of 7 in the negative direction.

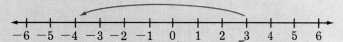

Think: Start at 0; move 3 units in the positive direction, then 7 units in the negative direction; finish at -4.
So, $3 - 7 = -4$.

Example 3

$-5 - 3$ is read *negative five subtract positive three*.

Show the trips on a number line.

Subtract. $-5 - 3$
$-5 - 3$ means a trip of -5 followed by a trip of 3 in the negative direction.

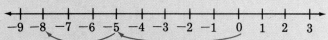

Think: Start at 0; move 5 units in the negative direction, then 3 units in the negative direction; finish at -8.
So, $-5 - 3 = -8$.

practice ▷ **Subtract. Draw trips on a number line.**

1. $5 - 1$ 2. $2 - 7$ 3. $-2 - 1$ 4. $-4 - 2$

When we subtracted a positive integer, we moved in the negative direction on a number line. We shall agree to move in the positive (opposite) direction on a number line when we subtract a negative integer.

Example 4

$-5 - (-3)$ is read *negative five subtract negative three.*
Think: To subtract a negative integer means to move in the positive direction.

Subtract. $-5 - (-3)$
$-5 - (-3)$ means a trip of -5 followed by a trip of 3 in the positive direction.

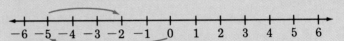

Think: Start at 0; move 5 units in the negative direction, then 3 units in the positive direction; finish at -2.
So, $-5 - (-3) = -2$.

Example 5

$2 - (-4)$ is read *positive two subtract negative four.*
Think: To subtract a negative integer means to move in the positive direction.

Subtract. $2 - (-4)$
$2 - (-4)$ means a trip of 2 followed by a trip of 4 in the positive direction.

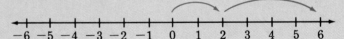

Think: Start at 0; move 2 units in the positive direction, then 4 units in the positive direction; finish at 6.
So, $2 - (-4) = 6$.

Example 6

Subtract. $0 - (-3)$.
$0 - (-3)$ means a trip of 0 followed by a trip of 3 in the positive direction.

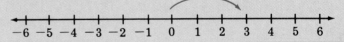

Think: Start at 0; finish at 3.
So, $0 - (-3) = 3$.

practice ▷ **Subtract. Draw trips on a number line.**

5. $-4 - (-2)$ 6. $3 - (-5)$ 7. $0 - (-5)$ 8. $0 - (-7)$

◇ ORAL EXERCISES ◇

Read.

1. $4 - 2$
2. $7 - 5$
3. $9 - 4$
4. $5 - 2$
5. $3 - 5$
6. $2 - 8$
7. $4 - 6$
8. $5 - 7$
9. $-2 - 3$
10. $-1 - 4$
11. $-3 - 7$
12. $-4 - 6$
13. $-4 - (-2)$
14. $-1 - (-1)$
15. $-2 - (-3)$
16. $-7 - (-2)$
17. $5 - (-1)$
18. $3 - (-3)$
19. $7 - (-2)$
20. $4 - (-6)$
21. $0 - (-2)$
22. $0 - 5$
23. $0 - (-4)$
24. $0 - 1$

◇ EXERCISES ◇

Subtract. Draw trips on a number line.

1. $6 - 1$
2. $5 - 3$
3. $4 - 2$
4. $6 - 4$
5. $2 - 4$
6. $3 - 4$
7. $2 - 6$
8. $5 - 8$
9. $-2 - 4$
10. $-1 - 1$
11. $-3 - 2$
12. $-5 - (-1)$
13. $-6 - (-2)$
14. $-4 - (-3)$
15. $-5 - (-1)$
16. $-2 - (-1)$
17. $-1 - (-7)$
18. $-4 - (-10)$
19. $-3 - (-6)$
20. $-5 - (-10)$
21. $2 - (-2)$
22. $4 - (-1)$
23. $7 - (-3)$
24. $8 - (-5)$
25. $3 - (-6)$
26. $1 - (-5)$
27. $4 - (-9)$
28. $2 - (-8)$
29. $5 - (-6)$
30. $0 - 3$
31. $0 - 8$
32. $0 - (-1)$
33. $0 - (-4)$
34. $0 - (-7)$
35. $5 - 0$
36. $-3 - 0$
37. $-4 - (-4)$
38. $6 - 6$
39. $-5 - 5$
40. $9 - (-6)$
41. $-8 - 2$
42. $-7 - (-6)$
43. $0 - (-8)$
44. $4 - (-3)$

Algebra Maintenance

Solve each proportion.

1. $\frac{4}{7} = \frac{k}{21}$
2. $\frac{12}{m} = \frac{3}{5}$
3. $x : 12 = 7 : 24$

Solve. Check.

4. $y - 2 = 12$
5. $6 = 62 - 4r$
6. $0.02g = 5.1$

7. What is the perimeter of a rectangle whose length is r and width is u?
8. What is the area of a circle whose radius is w cm long?

Problem Solving – Careers
Farm Equipment Mechanics

1. Ms. Contreras repairs an average of 15 tractors per week. How many tractors does she repair in a year?

2. Last year, Mr. Robinson repaired 625 diesel-powered tractors. About how many did he repair each week?

3. During August, Sato Repair received $3,000 for repairs of machinery. In September, Sato received $600 more than in August. In October, Sato received $800 less than in September. How much did Sato receive in October?

4. In May, Mr. Fitch spent 150% of what he spent in April to repair his crop dryer. In April he spent $30. How much did he spend in May?

5. In August, Ms. Royce spent 95% of what she spent in July to repair her combine. In July she spent $120. How much did she spend in August?

6. Ms. Bonk spends $80 per month on the maintenance of her machinery. How much does she spend in a year?

7. It takes George $1\frac{1}{2}$ hours to repair a corn picker. How many hours will it take him to repair 36 corn pickers?

8. Melinda can repair 3 hay balers in 2 hours. How many hay balers can she repair in 8 hours?

Opposites

◇ OBJECTIVES ◇

To identify the opposite of an integer

To find the missing integer given a sentence like $3 + \underline{\ ?\ } = 0$

To express real-life situations with integers

◇ RECALL ◇

Add. $-3 + 3$

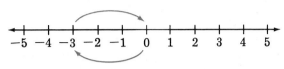

So, $-3 + 3 = 0$

Example 1

Find two integers that are each 4 units from 0.

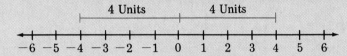

So, 4 and -4 are each 4 units from 0.

4 and -4 are opposites.   *Opposites* are the same distance from 0 on a number line. 0 is its own opposite.

Example 2

Give the opposite of 2 and of -5.

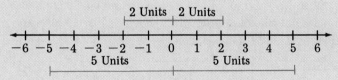

So, the opposite of 2 is -2 and the opposite of -5 is 5.

practice ▷ Give the opposite.

1. 3   2. -8 3. 7   4. -10 5. 21   6. 0

Example 3

Find the sum. $-5 + 5$

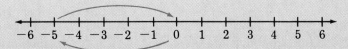

So, $-5 + 5 = 0$.

OPPOSITES

The sum of an integer and its opposite is 0.

practice ▷ Copy and complete.

7. $-9 + 9 = \underline{\ ?\ }$ 8. $\underline{\ ?\ } + (-25) = 0$ 9. $12 + \underline{\ ?\ } = 0$

Example 4

Express as integers: a gain of $10; a loss of $10
Gain of $10: 10
Loss of $10: -10
So, 10 shows a gain of $10 and -10 shows a loss of $10.

---

practice ▷ Express as an integer.

10. a gain of $25   11. a loss of 17 points   12. 33 m up

◇ ORAL EXERCISES ◇

What is the opposite of the given integer?

1. 6   2. -5 3. 1   4. -2 5. 0   6. -11

Express as an integer.

7. a gain of $5   8. a loss of $3   9. a gain of 20 points
10. 100 m above sea level   11. 5 km up   12. 64 m down

◇ EXERCISES ◇

What is the opposite of the given integer?

1. 15   2. -12 3. 25   4. 78   5. -56 6. -30
7. 100   8. -29 9. -32 10. 19   11. -20 12. -365

Copy and complete.

13. $-21 + 21 = \underline{\ ?\ }$ 14. $-60 + \underline{\ ?\ } = 0$ 15. $43 + \underline{\ ?\ } = 0$
16. $\underline{\ ?\ } + (-27) = 0$ 17. $\underline{\ ?\ } + 91 = 0$ 18. $82 + (-82) = \underline{\ ?\ }$

Express as an integer.

19. a gain of $950   20. a loss of 56 points   21. 1,200 m up
22. 3,000 m above sea level 23. 4,520 m below sea level   24. 38 km down
25. a loss of $75   26. a gain of 10 kg   27. 6 m up
28. a gain of 8 cm^3 29. a loss of 5 g   30. 742 m up

Subtracting by Adding

◇ OBJECTIVE ◇

To subtract an integer by adding its opposite

◇ RECALL ◇

−5 is the opposite of 5.
9 is the opposite of −9.
0 is the opposite of 0.
(0 is its own opposite.)

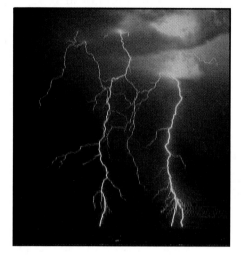

The number line on the left shows $7 - 4 = 3$.
The number line on the right shows $7 + (-4) = 3$.

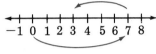 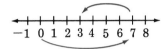

$7 - 4 = 3$ $7 + (-4) = 3$

This suggests that instead of subtracting, we can add.

$$7 - 4 = 7 + (-4)$$
          opposites

To subtract an integer, add its opposite.

Example 1

	Subtract. $3 - 9$
	$3 - 9$
−9 is the opposite of 9.	$3 + (-9)$
	-6
Check on a number line.	So, $3 - 9 = -6$

---

practice ▷ **Subtract by adding the opposite.**

1. $10 - 7$ 2. $12 - 3$ 3. $4 - 9$ 4. $9 - 15$

Example 2

	Subtract. $-8 - 7$
	$-8 - 7$
−7 is the opposite of 7.	$-8 + (-7)$
	-15
	So, $-8 - 7 = -15$.

---

practice ▷ **Subtract by adding the opposite.**

5. $-6 - 7$ 6. $-4 - 9$ 7. $-13 - 8$ 8. $-15 - 12$

Example 3

Subtract. $-12 - (-3)$

3 is the opposite of -3.
$-12 - (-3)$
$-12 + 3$
-9
So, $-12 - (-3) = -9$.

Example 4

Subtract. $-5 - (-14)$

14 is the opposite of -14.
$-5 - (-14)$
$-5 + 14$
9
So, $-5 - (-14) = 9$.

practice ▷ Subtract by adding the opposite.

9. $-9 - (-12)$ **10.** $-13 - (-4)$ **11.** $-10 - (-12)$ **12.** $-8 - (-13)$

Example 5

Subtract. $3 - (-9)$

9 is the opposite of -9.
$3 - (-9)$
$3 + 9$
12
So, $3 - (-9) = 12$.

practice ▷ Subtract by adding the opposite.

13. $8 - (-2)$ **14.** $10 - (-5)$ **15.** $4 - (-4)$ **16.** $2 - (-7)$

◇ EXERCISES ◇

Subtract by adding the opposite.

1. $12 - 5$	**2.** $16 - 9$	**3.** $15 - 7$	**4.** $18 - 9$
5. $4 - 10$	**6.** $7 - 27$	**7.** $3 - 24$	**8.** $12 - 26$
9. $-12 - 10$	**10.** $-30 - 15$	**11.** $-50 - 20$	**12.** $-42 - 31$
13. $-25 - 30$	**14.** $-46 - 70$	**15.** $-60 - 100$	**16.** $-20 - 130$
17. $-10 - (-5)$	**18.** $-32 - (-14)$	**19.** $-15 - (-12)$	**20.** $-50 - (-30)$
21. $-5 - (-20)$	**22.** $-10 - (-70)$	**23.** $-12 - (-26)$	**24.** $-31 - (-47)$
25. $4 - (-5)$	**26.** $11 - (-8)$	**27.** $14 - (-1)$	**28.** $7 - (-7)$
29. $21 - (-8)$	**30.** $38 - (-15)$	**31.** $42 - (-26)$	**32.** $19 - (-28)$
33. $29 - 0$	**34.** $0 - 37$	**35.** $-23 - 0$	**36.** $0 - (-80)$

★ **37.** Subtract around the track. What is the final result?

Start at 0. Subtract -5, and continue subtracting.

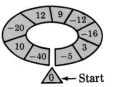

Using Integers

◇ **OBJECTIVE** ◇

To solve problems using integers

◇ **RECALL** ◇

4 + 5 = 9	5 + (−2) = 3
5 − 2 = 3	5 − 8 = −3
3 + 0 = 3	0 + (−5) = −5
5 + (−8) = −3	−2 + (−3) = −5
−2 − (−5) = 3	−2 − (−1) = −1
3 − 0 = 3	0 − (−2) = 2

Example 1

A football team gained 3 yards, lost 5 yards, and gained 12 yards. What was the net result?

Think of a gain as positive and a loss as negative. 3, −5, 12
Add the three numbers.
Positive 10

3 + (−5) + 12
−2 + 12
10

So, the net result was a gain of 10 yards.

practice ▷

1. A football team lost 6 yards, gained 18 yards, and lost 7 yards. What was the net result?

2. A football team gained 17 yards, lost 9 yards, and lost 8 yards. What was the net result?

Example 2

A mountain climber is located at 4,000 m above sea level. A submarine is located at 1,000 m below sea level. What is the difference between the levels?

Above sea level: positive
Below sea level: negative

Subtract to find the difference.

4,000 − (−1,000)
4,000 + 1,000
5,000

So, the difference between 4,000 m above sea level and 1,000 m below sea level is 5,000 m.

practice ▷

3. A mountain climber is located at 900 m above sea level. A submarine is located at 800 m below sea level. What is the difference between the levels?

4. A mountain climber is located at 1,200 m above sea level. A submarine is located at 1,200 m below sea level. What is the difference between the levels?

◇ EXERCISES ◇

1. $-4 + (-9) - 3$
2. $6 - 3 + (-5)$
3. $5 + 18 - (-15)$
4. $-17 - (-13) + 8$
5. $-3 - 7 + (-9)$
6. $4 + 0 - (-18)$
7. $-9 - (-4) + 15$
8. $10 + (-11) - 8$
9. $-13 - (-20) + (-7)$
10. $4 + 9 - (-12)$
11. $-11 - 7 - (-3)$
12. $1 - (-1) - 1$
13. $28 + (-8) - 17$
14. $-15 + (-9) - 12$
15. $22 + (-13) - 9$
16. $-9 + (-4) - (-13)$
17. $-30 + 19 - (-11)$
18. $23 - (-7) - (-32)$
19. $-2 + 21 - (-7)$
20. $-17 - (-9) + 56$

Solve these problems.

21. A football team gained 9 yards, lost 6 yards, and gained 11 yards. What was the net result?
22. A football team lost 15 yards, gained 9 yards, and lost 7 yards. What was the net result?
23. A mountain climber is located at 800 m above sea level. A submarine is located at 700 m below sea level. What is the difference between the levels?
24. A mountain climber is located at 3,000 m above sea level. A submarine is located at 1,500 m below sea level. What is the difference between the levels?
25. Brian started with 19 points. Then he won 9 points, lost 6 points, and lost 7 points. How many points does he now have?
26. Yoko started with 23 points. Then she won 12 points, lost 7 points, and won 2 points. How many points does she now have?
27. Jason had $400 in a checking account. He made deposits of $150 and $230. He has to write checks for $200, $250, and $300. Does he have enough money?
28. Muna had $250 in a checking account. She made deposits of $110 and $85. She has to write checks for $165, $140, and $145. Does she have enough money?
29. At noon, the temperature was 14°C. The temperature rose 4°C between noon and 1:00 P.M. and rose 5°C during the next hour. What was the temperature at 2:00 P.M.?
30. At 6:00 P.M., the temperature was 5°C. The temperature dropped 3°C between 6:00 P.M. and 7:00 P.M. and dropped 4°C during the next hour. What was the temperature at 8:00 P.M.?
★ 31. Five years ago, the population of a city was 120,000. During the last five years these changes took place: gained 5,000, lost 3,000, gained 2,000, gained 1,000, and lost 7,000. What is the population now?
★ 32. Four years ago, the population of a town was 8,650. During the last four years these changes took place: lost 150, gained 225, gained 175, and lost 100. What is the population now?
★ 33. Lucy's kite was flying at a height of 800 m. She lowered it 200 m, raised it 500 m, lowered it 300 m, and raised it 700 m. At what height is the kite flying now?
★ 34. Adam's kite was flying at a height of 650 m. He raised it 200 m, lowered it 350 m, raised it 125 m, and lowered it 175 m. At what height is the kite flying now?

Equations

◇ OBJECTIVE ◇
To solve equations like
$x - 5 = -2$, $x + 4 = -7$,
and $x + a = b$

◇ RECALL ◇
Solve.

$x + 9 = 12$	$x - 6 = 3$
$x + 9 - 9 = 12 - 9$	$x - 6 + 6 = 3 + 6$
$x = 3$	$x = 9$

Undo addition by subtraction.   Undo subtraction by addition.

Example 1

Undo addition. | Undo subtraction.
Subtract 5. | Add 8.

Solve and check.

$$x + 5 = -7 \qquad\qquad x - 8 = -15$$
$$x + 5 - 5 = -7 - 5 \quad x - 8 + 8 = -15 + 8$$
$$x = -12 \qquad\qquad x = -7$$

Check: $x + 5 = -7$ Check: $x - 8 = -15$
$\quad\quad -12 + 5 \mid -7 \qquad\qquad -7 - 8 \mid -15$
$\quad\quad\quad -7 \qquad\qquad\qquad\quad -15$
$\quad\quad -7 = -7 \qquad\qquad -15 = -15$
$\quad\quad\quad$ true $\qquad\qquad\qquad\quad$ true

So, -12 is the solution   So, -7 is the solution
of $x + 5 = -7$.   of $x - 8 = -15$.

practice ▷ Solve and check.

1. $x + 3 = -21$ 2. $x - 9 = -14$ 3. $x + 12 = 4$

Example 2

Solve. $x - (-4) = -3$
$\quad\quad x - (-4) = -3$
$\quad\quad\quad x + 4 = -3$

Undo addition by subtraction. $x + 4 - 4 = -3 - 4$
$\quad\quad\quad\quad x = -7$

Check on your own.   So, -7 is the solution of $x - (-4) = -3$.

practice ▷ Solve and check.

4. $x - (-7) = 10$ 5. $x + (-9) = -11$ 6. $x - (-5) = -3$

EQUATIONS

Summary

To solve an equation that shows addition, undo the addition by subtraction.
For example: $x + a = b$
$x = b - a$
So, $b - a$ solves the equation.

To solve an equation that shows subtraction, undo the subtraction by addition.
For example: $x - a = b$
$x = b + a$
So, $b + a$ solves the equation.

Opposites can also be used to solve equations involving integers.

Example 3
Add the opposite of -4 to each side of the equation.
$(-4) + 4 = 0$

Solve and check. $-4 + x = 11$
$-4 + x = 11$
$-4 + 4 + x = 11 + 4$
$x = 15$
Check: $\quad -4 + x = 11$
$\quad\quad -4 + 15 \mid 11$
$\quad\quad\quad\; 11$
$\quad\quad\quad\; 11 = 11 \quad$ true

So, 15 is the solution of $-4 + x = 11$.

practice   7. $-8 + x = 10$ 8. $-9 + x = -15$ 9. $3 + x = -5$

◇ EXERCISES ◇

Solve and check.

1. $x + 2 = -4$
2. $x + (-9) = -11$
3. $5 + x = -12$
4. $x + (-12) = -14$
5. $x + (-9) = -2$
6. $x + (-8) = -1$
7. $x + (-11) = -4$
8. $x + (-15) = -5$
9. $x + 4 = 3$
10. $x + 7 = -1$
11. $x + 5 = -9$
12. $x + 15 = 0$
13. $-5 + x = -2$
14. $-9 + x = -8$
15. $-10 + x = -1$
16. $-12 + x = 4$
17. $-9 + x = 4$
18. $-12 + x = 1$
19. $-20 + x = 10$
20. $-17 + x = 0$
21. $x - 5 = -2$
22. $x - 12 = -5$
23. $x - 13 = 15$
24. $x - 20 = 30$
25. $x - (-6) = 2$
26. $x - (-10) = -3$
27. $x - (-15) = 20$

Solve for x.

★ 28. $x + r = s$
★ 29. $x - p = q$
★ 30. $r = x + s$
★ 31. $-a + x = b$
★ 32. $c = x - d$
★ 33. $p = q + x$

COMPUTER ACTIVITIES

The **BASIC** language has a very useful device for adding or accumulating numbers. The **ACCUMULATOR** makes it unnecessary to repeat writing the "+" symbol when adding a long series of numbers. The use of the **ACCUMULATOR** is illustrated in the program below.

Example Write a program that will allow you to enter any number of numbers and then find their sum. Then **RUN** the program to find the sum $-58 + 119 - 73 - 98$.

Computer allows you to choose number of numbers to be added	`10  INPUT "NUMBER  OF NUMBERS TO ADD? ";N`
Clears all previous numbers from memory so they will not be added to the new numbers	`20  LET A = 0`
Allows you to enter N numbers	`30  FOR J = 1 TO N`
You type in the values of X to be added.	`40  INPUT "ENTER A NUMBER ";X`
Replaces each value of X with the new X value added to it; when J = 2, −58 will be replaced with −58 + 119	`50  LET A = A + X`
	`60  NEXT J`
Prints the sum of all numbers entered	`70  PRINT : PRINT "SUM IS "A`
	`80  END`
	`]RUN`
Type 4. (There are 4 numbers to add.)	`NUMBER  OF NUMBERS TO ADD? 4`
Enter the first number; −58.	`ENTER A NUMBER -58`
119 is second number to be added.	`ENTER A NUMBER 119`
Type third number; −73.	`ENTER A NUMBER -73`
Enter the last number; −98.	`ENTER A NUMBER -98`
PRINT: leaves a blank line for readability.	`SUM IS -110`

See the Computer Section beginning on page 420 for more information.

Exercises

Find each sum. Then type and **RUN** the program above to check your answers.

1. $-67 - 89 + 278 - 45$ **2.** $98 - 121 - 453 + 89 - 76$ **3.** $-114 + 139 - 229$

Chapter Review

Compare. Use > or <. [288]

1. 8 __?__ 9
2. −6 __?__ −1
3. −2 __?__ 4
4. 0 __?__ −6

Add. [291]

5. 8 + 3
6. −5 + (−7)
7. 12 + (−3)
8. −6 + 8
9. 9 + (−9)
10. −7 + 7
11. 0 + 2
12. −3 + 0

Copy and complete. [299]

13. −8 + 8 = __?__
14. 17 + __?__ = 0
15. 10 + (−10) = __?__

Subtract. [295, 301]

16. 13 − 5
17. 12 − 26
18. −27 − 52
19. −36 − (−54)
20. 16 − (−10)

Solve these problems. [303]

21. A kite was flying at a height of 700 m. Then it was lowered 300 m, raised 100 m, lowered 200 m, and raised 500 m. At what height was the kite flying then?

22. At 6:00 A.M., the temperature was 9°C. The temperature rose 9°C between 6:00 A.M. and noon and rose 6°C between noon and 4:00 P.M. What was the final temperature?

23. A football team gained 11 yards, lost 13 yards, and lost 7 yards. What was the net result?

24. Louise started with 30 points. Then she won 12 points, lost 8 points, and won 16 points. How many points does she have now?

Solve for x. [305]

25. $x + (-5) = -9$
26. $4 + x = -5$
27. $x - (-6) = -12$
28. $x - (-12) = 34$

Calculator

4! means $1 \cdot 2 \cdot 3 \cdot 4$
↑
└ read *four factorial*
$4! = 1 \cdot 2 \cdot 3 \cdot 4$
$5! = 1 \cdot 2 \cdot 3 \cdot 4 \cdot 5$
Generally, $n! = 1 \cdot 2 \cdot 3 \cdot \ldots \cdot (n-2)(n-1)n$
Compute. Use a calculator.

1. 2!
2. 3!
3. 4!
4. 5!
5. 6!
6. 7!
7. 8!
8. 9!
9. 10!
10. 11!

Chapter Test

Compare. Use > or <.

1. 8 __?__ −10
2. −5 __?__ −3
3. −12 __?__ 0
4. −6 __?__ −11

Add.

5. −2 + 13
6. −8 + (−6)
7. −11 + 4
8. 4 + (−8) + 7
9. −15 + (−4) + 8
10. 9 + (−3) + (−6)

Copy and complete.

11. 5 + (−5) = __?__
12. __?__ + (−12) = 0
13. −6 + __?__ = 0

Subtract.

14. 36 − 42
15. −12 − (−15)
16. 27 − (−14)
17. −16 − 23
18. 6 − (−6)
19. −8 − (−7)

Solve these problems.

20. A kite was flying at a height of 500 m. Then it was lowered 200 m, raised 300 m, lowered 100 m, and raised 400 m. At what height was the kite flying then?

21. At 7:00 A.M. the temperature was 3°C. The temperature rose 6°C between 7:00 A.M. and 11:00 A.M., and rose 5°C between 11:00 A.M. and 3:00 P.M. What was the temperature at 3:00 P.M.?

Solve for x.

22. $x + (-6) = -8$
23. $-9 + x = 12$
24. $x - (-7) = 7$
25. $x - (-5) = -5$

Cumulative Review

Evaluate if $a = -4$, $b = \frac{1}{5}$, and $c = 9$.

1. $\frac{1}{2}c - b$
2. $a - 3b + \frac{2}{3}c$
3. $\frac{a + 2c}{4b}$

Solve.

4. $x + 9 = 4$
5. $5y - 3 = 17$
6. $-4 + z = -13$

Compute.

7. 32% of 362 is what number?
8. What % of 75 is 15?
9. What % of 50 is 250?
10. 4 is what percent of 5?
11. Find the range, mean, median, and mode for these scores.
    72, 89, 94, 99, 70, 68, 92, 94, 96

Answer these exercises about the spinner.

12. How many possible outcomes?
13. $P(4)$
14. $P(2)$
15. $P(3)$
16. $P(2 \text{ or } 3)$
17. If you spin the spinner 200 times, about how many times should it stop on 2?

Solve each proportion.

18. $\frac{3}{7} = \frac{n}{28}$
19. $\frac{4}{y} = \frac{2}{9}$
20. $n : 12 = 9 : 4$

Add or subtract.

21. $7 + (-4) + (-9)$
22. $12 - 19$
23. $-7 - 4 - (-22)$

Solve these problems.

24. A test had 50 questions. Teresa got 94% right. How many questions did she miss?
25. Ms. Todd is earning $31,000. Next year she will get a raise of 7%. What will be her salary next year?
26. The circle graph shows how Gregg spent $60. How much did he spend on each item?
27. There are 5 candidates for president and 4 for vice-president. How many different ways can these offices be filled?

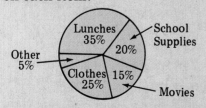

Multiplying and Dividing Integers

13

The Okefenokee Swamp in Georgia is a wildlife refuge covering 293,826 acres.

Multiplying Integers

◇ OBJECTIVES ◇

To multiply two positive integers

To multiply a positive integer and a negative integer

◇ RECALL ◇

$3 \cdot 4 = 12 \qquad 5 \cdot 9 = 45 \qquad 9 \cdot 8 = 72$

$$\begin{array}{r} 26 \\ \times 14 \\ \hline 104 \\ 26 \\ \hline 364 \end{array} \qquad \begin{array}{r} 38 \\ \times 23 \\ \hline 114 \\ 76 \\ \hline 874 \end{array}$$

Example 1

$4 - 1 = 3;\ 20 - 5 = 15$

$3 - 1 = 2;\ 15 - 5 = 10$

$2 - 1 = 1;\ 10 - 5 = 5$

Continue the pattern for two more steps.
$5 \cdot 4 = 20$
$5 \cdot 3 = 15$
↑ ↳ Subtract 5 each time.
↳ Subtract 1 each time.
So, $5 \cdot 2 = 10$,
$5 \cdot 1 = 5$.

The product of two positive integers is a positive integer.

Example 2

Continue the pattern for three more steps.
$5 \cdot 1 = 5$
$5 \cdot 0 = 0$
↑ ↳ Subtract 5 each time.
↳ Subtract 1 each time.

$0 - 1 = -1;\ 0 - 5 = -5$ So, $5 \cdot (-1) = -5$,
$-1 - 1 = -2;\ -5 - 5 = -10$ $5 \cdot (-2) = -10$,
$-2 - 1 = -3;\ -10 - 5 = -15$ $5 \cdot (-3) = -15$.

Example 3

Continue the pattern for three more steps.
$2 \cdot 3 = 6$
$1 \cdot 3 = 3$
$0 \cdot 3 = 0$
↑ ↳ Subtract 3 each time.
↳ Subtract 1 each time.

$0 - 1 = -1;\ 0 - 3 = -3$ So, $-1 \cdot 3 = -3$,
$-1 - 1 = -2;\ -3 - 3 = -6$ $-2 \cdot 3 = -6$,
$-2 - 1 = -3;\ -6 - 3 = -9$ $-3 \cdot 3 = -9$.

The product of a positive integer and a negative integer is a negative integer.

practice ▷ **Multiply.**

1. $6 \cdot 9$ 2. $9 \cdot (-7)$ 3. $-8 \cdot 6$ 4. $-12 \cdot 3$

Example 4

Multiply. $29 \cdot (-48)$
First, multiply as with whole numbers.

$$\begin{array}{r} 48 \\ \times 29 \\ \hline 432 \\ 96 \\ \hline 1{,}392 \end{array}$$

Determine the sign.
(positive) · (negative) = negative   So, $29 \cdot (-48) = -1{,}392$.

practice ▷ **Multiply.**

5. $12 \cdot (-42)$ 6. $29 \cdot (-57)$ 7. $-67 \cdot 98$ 8. $-19 \cdot 46$

Example 5

Multiply. $27 \cdot 0$; $-13 \cdot 0$

$$\begin{array}{cc} 27 \cdot 0 & -13 \cdot 0 \\ 0 & 0 \end{array}$$

So, $27 \cdot 0 = 0$ and $-13 \cdot 0 = 0$.

The product of any integer and 0 is 0.

practice ▷ **Multiply.**

9. $19 \cdot 0$ 10. $0 \cdot 62$ 11. $-28 \cdot 0$ 12. $0 \cdot (-57)$

◇ ORAL EXERCISES ◇

Multiply.

1. $12 \cdot 2$ 2. $13 \cdot 3$ 3. $3 \cdot 15$ 4. $5 \cdot (-7)$ 5. $-9 \cdot 4$
6. $9 \cdot 0$ 7. $11 \cdot 7$ 8. $9 \cdot (-3)$ 9. $0 \cdot (-97)$ 10. $-6 \cdot 7$

◇ EXERCISES ◇

Multiply.

1. $32 \cdot 4$ 2. $5 \cdot 24$ 3. $6 \cdot 13$ 4. $14 \cdot 8$
5. $26 \cdot 71$ 6. $19 \cdot 82$ 7. $29 \cdot 34$ 8. $76 \cdot 41$
9. $26 \cdot (-31)$ 10. $-45 \cdot 45$ 11. $93 \cdot (-91)$ 12. $26 \cdot (-25)$
13. $42 \cdot (-98)$ 14. $-11 \cdot 78$ 15. $62 \cdot (-12)$ 16. $14 \cdot (-17)$
17. $0 \cdot 35$ 18. $-48 \cdot 0$ 19. $0 \cdot 88$ ★ 20. $77 \cdot 0 \cdot (-10)$

MULTIPLYING INTEGERS

Multiplying Negative Integers

◇ OBJECTIVE ◇
To multiply negative integers

◇ RECALL ◇
The product of any integer and 0 is 0.
$$-8 \cdot 0 = 0 \qquad 0 \cdot 7 = 0$$

The product of a positive integer and a negative integer is a negative integer.
$$-3 \cdot 2 = -6 \qquad 4 \cdot (-5) = -20$$

Example 1
Continue the pattern for three more steps.
$$-3 \cdot 3 = -9$$
$$-3 \cdot 2 = -6$$
$$-3 \cdot 1 = -3$$
$$-3 \cdot 0 = 0$$
↑ ↑—Add 3 each time.
└————Subtract 1 each time.

$0 - 1 = -1; 0 + 3 = 3$
$-1 - 1 = -2; 3 + 3 = 6$
$-2 - 1 = -3; 6 + 3 = 9$

So, $-3 \cdot (-1) = 3$,
$-3 \cdot (-2) = 6$,
$-3 \cdot (-3) = 9$.

Example 2
Continue the pattern for three more steps.
$$2 \cdot (-7) = -14$$
$$1 \cdot (-7) = -7$$
$$0 \cdot (-7) = 0$$
↑ ↑—Add 7 each time.
└————Subtract 1 each time.

$0 - 1 = -1; 0 + 7 = 7$
$-1 - 1 = -2; 7 + 7 = 14$
$-2 - 1 = -3; 14 + 7 = 21$

So, $-1 \cdot (-7) = 7$,
$-2 \cdot (-7) = 14$,
$-3 \cdot (-7) = 21$.

The product of two negative integers is a positive integer.

practice ▷ Multiply.

1. $-5 \cdot (-3)$
2. $-7 \cdot (-4)$
3. $-3 \cdot (-9)$
4. $-9 \cdot (-9)$

Example 3

Multiply. $-37 \cdot (-86)$
First multiply as with whole numbers.

$$\begin{array}{r} 86 \\ \times 37 \end{array} \qquad 86 \times 37 = 3{,}182$$

Determine the sign. So, $-37 \cdot (-86) = 3{,}182$.
(negative) · (negative) = positive

practice ▷ Multiply.

5. $-47 \cdot (-81)$ 6. $-63 \cdot (-46)$ 7. $-45 \cdot (-92)$ 8. $-83 \cdot (-200)$

◇ **ORAL EXERCISES** ◇

Multiply.

1. $-1 \cdot (-2)$ 2. $-2 \cdot (-3)$ 3. $-5 \cdot (-1)$ 4. $-6 \cdot (-2)$
5. $-3 \cdot (-3)$ 6. $-4 \cdot (-2)$ 7. $-9 \cdot (-7)$ 8. $-4 \cdot (-8)$

◇ **EXERCISES** ◇

Multiply.

1. $15 \cdot (3)$ 2. $-25 \cdot (-4)$ 3. $-35 \cdot (-2)$
4. $-6 \cdot (-15)$ 5. $-8 \cdot (-25)$ 6. $-5 \cdot (-22)$
7. $-17 \cdot (-15)$ 8. $-36 \cdot (-12)$ 9. $-30 \cdot (-11)$
10. $-78 \cdot (-19)$ 11. $-29 \cdot (-76)$ 12. $-55 \cdot (-55)$
13. $-20 \cdot (-55)$ 14. $-16 \cdot (-16)$ 15. $-100 \cdot (-14)$
16. $-300 \cdot (-11)$ 17. $-21 \cdot (-400)$ 18. $-15 \cdot (-300)$
★ 19. $-300 \cdot (-10)^2$ ★ 20. $-400 \cdot (-5)^2$ ★ 21. $(-4)^2 \cdot (-5)^2$

Challenge

1	2	3	4	5	6	7	8	9	10	11	12	13
A	B	C	D	E	F	G	H	I	J	K	L	M
14	15	16	17	18	19	20	21	22	23	24	25	26
N	O	P	Q	R	S	T	U	V	W	X	Y	Z

Decode the message.
First word: $-13 \cdot (-1)$; $-1 \cdot (-1)$; $-5 \cdot (-5)$
Second word: $-25 + 50$; $30 - 15$; $-7 \cdot (-3)$; $-3 \cdot (-6)$
Third word: $-2 - (-6)$; $-12 - (-13)$; $100 - 75$
Fourth word: $-1 \cdot (-2)$; $-1 \cdot (-5)$
Fifth word: $-20 - (-21)$
Sixth word: $-2 \cdot (-11)$; $2 - (-3)$; $-9 \cdot (-2)$; $-50 - (-75)$
Seventh word: $-2 + 10$; $-4 - (-5)$; $-2 \cdot (-8)$; $20 - 4$; $-26 - (-51)$
Eighth word: $-3 \cdot (-5)$; $-2 \cdot (-7)$; $-10 - (-15)$

MULTIPLYING NEGATIVE INTEGERS

Properties of Multiplication

◇ OBJECTIVE ◇
To use properties of multiplication of integers

◇ RECALL ◇
The product of any whole number and 1 is that whole number.
$$26 \cdot 1 = 26 \qquad 1 \cdot 95 = 95$$

Example 1

Multiply integers as you would whole numbers.

Multiply. $5 \cdot 1;\ 1 \cdot 367;\ -270 \cdot 1$

| $5 \cdot 1$ | $1 \cdot 367$ | $-270 \cdot 1$ |
| 5 | 367 | -270 |

So, $5 \cdot 1 = 5,\ 1 \cdot 367 = 367$, and $-270 \cdot 1 = -270$.

property of 1 for multiplication
The product of any integer and 1 is that integer.
$$1 \cdot n = n \text{ and } n \cdot 1 = n$$

Example 2

Multiply. $7 \cdot (-1);\ -1 \cdot 150;\ -300 \cdot (-1)$

| $7 \cdot (-1)$ | $-1 \cdot 150$ | $-300 \cdot (-1)$ |
| -7 | -150 | 300 |

So, $7 \cdot (-1) = -7,\ -1 \cdot 150 = -150$, and $-300 \cdot (-1) = 300$.

property of −1 for multiplication
The product of any integer and -1 is the opposite of that integer.
$$-1 \cdot n = -n \text{ and } n \cdot (-1) = -n$$

---

practice ▷ **Multiply.**

1. $36 \cdot 1$
2. $1 \cdot (-525)$
3. $18 \cdot (-1)$
4. $-1 \cdot 98$

Example 3

(negative) · (negative) = positive

Show that $-5 \cdot (-2) = -2 \cdot (-5)$.

| $-5 \cdot (-2)$ | $-2 \cdot (-5)$ |
| 10 | 10 |

So, $-5 \cdot (-2) = -2 \cdot (-5)$.

commutative property of multiplication
When multiplying two integers, it does not matter in which order they are multiplied. For all integers a and b, $a \cdot b = b \cdot a$.

practice ▷ Rewrite by using the commutative property. Then multiply.

5. $-5 \cdot (-12)$ 6. $-6 \cdot 8$ 7. $9 \cdot (-7)$ 8. $-11 \cdot 10$

Example 4

Show that $-5 \cdot [-2 \cdot (-11)] = [-5 \cdot (-2)] \cdot (-11)$.

$-5 \cdot [-2 \cdot (-11)]$	$[-5 \cdot (-2)] \cdot (-11)$
$-5 \cdot (22)$	$10 \cdot (-11)$
-110	-110

So, $-5 \cdot [-2 \cdot (-11)] = [-5 \cdot (-2)] \cdot (-11)$.

---

associative property of multiplication
For all integers a, b, and c, $a \cdot (b \cdot c) = (a \cdot b) \cdot c$.

practice ▷ Multiply. Use the associative property.

9. $-6 \cdot 5 \cdot (-2)$ 10. $-9 \cdot 2 \cdot (-10)$ 11. $-1 \cdot (-20) \cdot (-30)$

Example 5

Show that $-5 \cdot [-7 + (-3)] = -5 \cdot (-7) + (-5) \cdot (-3)$.

$-7 + (-3) = -10$

$-5 \cdot [-7 + (-3)]$	$-5 \cdot (-7) + (-5) \cdot (-3)$
$-5 \cdot (-10)$	$35 + 15$
50	50

So, $-5 \cdot [-7 + (-3)] = -5 \cdot (-7) + (-5) \cdot (-3)$.

---

distributive property of multiplication over addition
For all integers a, b, and c, $a \cdot (b + c) = a \cdot b + a \cdot c$.

practice ▷ Multiply. Use the distributive property.

12. $-8 \cdot (-23 + 3)$ 13. $-8 \cdot (-23) + (-8) \cdot 3$

Example 6

Show that $-8 \cdot [-5 - (-7)] = -8 \cdot (-5) - (-8) \cdot (-7)$.

$-5 - (-7) = -5 + 7 = 2$

$-8 \cdot [-5 - (-7)]$	$-8 \cdot (-5) - (-8) \cdot (-7)$
$-8 \cdot 2$	$40 - 56$
-16	-16

So, $-8 \cdot [-5 - (-7)] = -8 \cdot (-5) - (-8) \cdot (-7)$.

---

practice ▷ Multiply. Use the distributive property.

14. $-6 \cdot (7 - 2)$ 15. $-6 \cdot 7 - (-6) \cdot 2$ 16. $-4 \cdot [-7 - (-9)]$

PROPERTIES OF MULTIPLICATION

◇ ORAL EXERCISES ◇

Multiply.

1. $1 \cdot 5$
2. $-6 \cdot 1$
3. $-8 \cdot (-1)$
4. $-1 \cdot (-93)$
5. $-63 \cdot 1$
6. $-1 \cdot 73$
7. $73 \cdot (-1)$
8. $1 \cdot (-963)$
9. $-963 \cdot 1$
10. $-107 \cdot (-1)$

◇ EXERCISES ◇

Multiply.

1. $412 \cdot 1$
2. $1 \cdot 329$
3. $-157 \cdot 1$
4. $1 \cdot (-222)$
5. $1 \cdot 0$
6. $807 \cdot (-1)$
7. $-1 \cdot 584$
8. $-1 \cdot (-396)$
9. $-920 \cdot (-1)$
10. $-1 \cdot 0$

Rewrite by using the commutative property. Then multiply.

11. $-6 \cdot (-4)$
12. $9 \cdot (-9)$
13. $-15 \cdot 6$
14. $-18 \cdot (-7)$
15. $22 \cdot (-15)$
16. $-31 \cdot 31$
17. $-12 \cdot (-48)$
18. $-52 \cdot 11$

Multiply. Use the associative property.

19. $-17 \cdot (-5) \cdot 2$
20. $26 \cdot 25 \cdot (-4)$
21. $-83 \cdot (-34) \cdot 0$
22. $-36 \cdot 27 \cdot (-1)$
23. $-63 \cdot 2 \cdot (-5)$
24. $-20 \cdot 5 \cdot 13$
25. $40 \cdot (-25) \cdot 3$
26. $-12 \cdot (-2) \cdot (-50)$

Multiply. Use the distributive property.

27. $-9 \cdot (-17 + 7)$
28. $-9 \cdot (-17) + (-9) \cdot 7$
29. $13 \cdot 37 + 13 \cdot (-17)$
30. $13 \cdot [37 + (-17)]$
31. $-21 \cdot (5 - 25)$
32. $-21 \cdot 5 - (-21) \cdot 25$
33. $-50 \cdot 51 - (-50) \cdot 1$
34. $-50 \cdot (51 - 1)$

Calculator

1. Compute. Use a calculator.

    $1^3 + 2^3$ $(1 + 2)^2$
    $1^3 + 2^3 + 3^3$ $(1 + 2 + 3)^2$
    $1^3 + 2^3 + 3^3 + 4^3$ $(1 + 2 + 3 + 4)^2$
    $1^3 + 2^3 + 3^3 + 4^3 + 5^3$ $(1 + 2 + 3 + 4 + 5)^2$

2. Do you see a pattern? Use the pattern to compute these.

    $1^3 + 2^3 + 3^3 + 4^3 + 5^3 + 6^3$
    $1^3 + 2^3 + 3^3 + 4^3 + 5^3 + 6^3 + 7^3$
    $1^3 + 2^3 + 3^3 + 4^3 + 5^3 + 6^3 + 7^3 + 8^3$
    $1^3 + 2^3 + 3^3 + 4^3 + 5^3 + 6^3 + 7^3 + 8^3 + 9^3$
    $1^3 + 2^3 + 3^3 + 4^3 + 5^3 + 6^3 + 7^3 + 8^3 + 9^3 + 10^3$

Problem Solving – Applications
Jobs for Teenagers

1. As a camp counselor, Tina earned $240 over $2\frac{1}{2}$ pay periods. How much did she make per pay period?

2. A group of campers went on a 20-kilometer hike. The counselor let them rest after every $3\frac{1}{3}$ kilometers. How many times did they rest?

3. Night patrol at the Green Hills Camp starts at 9:45 P.M. It ends at 12:30 A.M. How long is night patrol?

4. One summer, a camp had 176 campers and 22 counselors. How many campers were there per counselor?

5. A counselor earned $210 in July and $1\frac{1}{7}$ times as much in August. How much did the counselor earn for the two months?

6. Last summer, Kaoni worked in a camp 27 days. He expects to work $1\frac{1}{3}$ as many days next summer. How many days will he work next summer?

7. A group of campers on a canoe trip rowed upstream for $1\frac{1}{2}$ hours. Coming back downstream, they took $\frac{2}{3}$ as long. How long did it take them to come back?

8. A counselor had waterfront duty for $\frac{3}{4}$ hour in the afternoon. Waterfront duty in the morning was $\frac{2}{3}$ as long. How long was waterfront duty in the morning?

PROBLEM SOLVING

Dividing Integers

◇ OBJECTIVE ◇
To divide integers

◇ RECALL ◇
Multiplication and division are related.
$$5 \cdot 2 = 10 \begin{cases} 10 \div 2 = 5 \\ 10 \div 5 = 2 \end{cases}$$

Example 1

Write related division and multiplication sentences.

$7 \cdot 3 = 21$ and $21 \div 3 = 7$.

7 is the *quotient*.

Divide. $21 \div 3$
$21 \div 3 = n$, so $n \cdot 3 = 21$.
Think: What number times 3 is 21?
Answer: $7 \cdot 3 = 21$.
So, $21 \div 3 = 7$.

Example 2

Write related division and multiplication sentences.

$9 \cdot (-2) = -18$ and
$-18 \div (-2) = 9$.

Divide. $-18 \div (-2)$
$-18 \div (-2) = n$, so $n \cdot (-2) = -18$.
Think: What number times -2 is -18?
Answer: $9 \cdot (-2) = -18$.
So, $-18 \div (-2) = 9$.

The quotient of two positive integers or two negative integers is a positive integer.

practice ▷ **Divide.**

1. $35 \div 5$ 2. $42 \div 6$ 3. $-72 \div (-9)$ 4. $-63 \div (-7)$

Example 3

Write related division and multiplication sentences.

$-6 \cdot 3 = -18$ and $-18 \div 3 = -6$.

Divide. $-18 \div 3$
$-18 \div 3 = n$, so $n \cdot 3 = -18$.
Think: What number times 3 is -18?
Answer: $-6 \cdot 3 = -18$.
So, $-18 \div 3 = -6$.

Example 4

Write related division and multiplication sentences.

$-5 \cdot (-4) = 20$ and
$20 \div (-4) = -5$.

Divide. $20 \div -4$
$20 \div (-4) = n$, so $n \cdot (-4) = 20$.
Think: What number times -4 is 20?
Answer: $-5 \cdot (-4) = 20$.
So, $20 \div (-4) = -5$.

The quotient of a positive integer and a negative integer is a negative integer.

practice ▷ Divide.

    5. $-25 \div 5$ **6.** $-30 \div 6$ **7.** $48 \div (-6)$ **8.** $64 \div (-8)$

Example 5

Write related division and multiplication sentences.

Divide. $-17 \div 1$
$-17 \div 1 = n$, so $n \cdot 1 = -17$.
$-17 \cdot 1 = -17$
So, $-17 \div 1 = -17$.

Example 6

Divide. $-25 \div (-1)$
$-25 \div (-1) = n$, so $n \cdot (-1) = -25$.
$25 \cdot (-1) = -25$

25 is the opposite of -25.   So, $-25 \div (-1) = 25$.

The quotient of an integer and 1 is that integer. The quotient of an integer and -1 is the opposite of that integer.

practice ▷ Divide.

    9. $44 \div 1$ **10.** $-36 \div 1$ **11.** $-85 \div (-1)$ **12.** $99 \div (-1)$

Example 7

Write related division and multiplication sentences.

Divide. $0 \div (-8)$
$0 \div (-8) = n$, so $n \cdot (-8) = 0$.
Think: What number times -8 is 0?
Answer: $0 \cdot (-8) = 0$.
So, $0 \div (-8) = 0$.

Example 8

Write related division and multiplication sentences.

Divide. $-5 \div 0$
$-5 \div 0 = n$, so $n \cdot 0 = -5$.
Think: What number times 0 is -5?
There is no such number, since any number times 0 is 0.
So, dividing -5 by 0 has no answer.

The quotient of 0 and an integer is 0. Division by 0 has no answer.

practice ▷ Divide, if possible.

    13. $0 \div 5$ **14.** $0 \div (-25)$ **15.** $-2 \div 0$ **16.** $98 \div 0$

DIVIDING INTEGERS

◇ EXERCISES ◇

Divide, if possible.

1. $36 \div 9$
2. $48 \div 8$
3. $35 \div 7$
4. $56 \div 7$
5. $63 \div 9$
6. $64 \div 8$
7. $30 \div 5$
8. $60 \div 5$
9. $48 \div 4$
10. $51 \div 3$
11. $-27 \div (-3)$
12. $-32 \div (-8)$
13. $-40 \div (-5)$
14. $-81 \div (-9)$
15. $-42 \div (-7)$
16. $-35 \div (-5)$
17. $-45 \div (-9)$
18. $-33 \div (-3)$
19. $-36 \div (-2)$
20. $-80 \div (-5)$
21. $-28 \div 7$
22. $-64 \div 8$
23. $-18 \div 6$
24. $-24 \div 4$
25. $-36 \div 6$
26. $-15 \div 5$
27. $-54 \div 6$
28. $-57 \div 3$
29. $-60 \div 6$
30. $-75 \div 5$
31. $28 \div (-4)$
32. $18 \div (-9)$
33. $49 \div (-7)$
34. $63 \div (-7)$
35. $56 \div (-8)$
36. $20 \div (-5)$
37. $48 \div (-8)$
38. $93 \div (-3)$
39. $84 \div (-4)$
40. $72 \div (-6)$
41. $-8 \div 1$
42. $27 \div 1$
43. $-32 \div 1$
44. $-11 \div 1$
45. $59 \div 1$
46. $40 \div (-1)$
47. $-21 \div (-1)$
48. $-19 \div (-1)$
49. $59 \div (-1)$
50. $-78 \div (-1)$
51. $0 \div 3$
52. $0 \div (-7)$
53. $0 \div 15$
54. $0 \div (-24)$
55. $0 \div (-35)$
56. $4 \div 0$
57. $-8 \div 0$
★ 58. $12^2 \div (-9)$
★ 59. $(-8)^2 \div (-16)$
★ 60. $(-15)^2 \div (-25)$

Reading in Math

For each group of four mathematical terms, identify the term that does not belong with the other three.

Example: integer
fraction
line
percent

Line does not belong; the other three terms refer to numbers.

1. ratio
triangle
proportion
extremes

2. mean
mode
score
median

3. temperature
probability
experiment
event

4. kilometer
gram
centimeter
millimeter

From Integers to Rationals

◇ **OBJECTIVES** ◇

To locate rational numbers on a number line
To divide rational numbers

◇ **RECALL** ◇

$7 \div 3 = n$, so $n \cdot 3 = 7$.
There is no whole number which when multiplied by 3 gives 7.
The quotient of two numbers is not always a whole number.

$$\frac{2}{3} \div \frac{4}{5} = \frac{2}{3} \cdot \frac{5}{4}$$

In general, $\frac{a}{b} \div \frac{c}{d} = \frac{a}{b} \cdot \frac{d}{c}$.

Example 1

What number corresponds to point A? to point B?

A is halfway between 2 and 3. B is two-thirds of the way from -2 to -3.

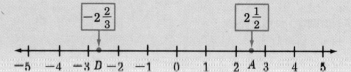

So, $2\frac{1}{2}$ corresponds to point A and $-2\frac{2}{3}$ corresponds to point B.

---

practice ▷ What number corresponds to the point?

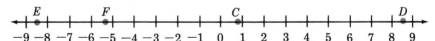

1. C 2. D 3. E 4. F

Positive fractions, negative fractions, positive mixed numbers, and negative mixed numbers are rational numbers:

$\frac{1}{2}, \frac{3}{7}, \frac{17}{5}, 4\frac{5}{8}, -\frac{2}{3}, -\frac{4}{9}, -\frac{12}{5}, -5\frac{2}{3}.$

Decimals that terminate or repeat are also rational numbers: $1.\dot{6}, -0.5, -2.65, 0.007, 0.\overline{6}$.
All integers are also rational numbers:
$4, 21, -3, -15$.

FROM INTEGERS TO RATIONALS

Dividing rational numbers is like dividing fractions.

Example 2   Divide. $\dfrac{2}{3} \div \left(-\dfrac{5}{7}\right)$ $\quad -3\dfrac{3}{4} \div \left(-2\dfrac{1}{2}\right)$

$\dfrac{2}{3} \div \left(-\dfrac{5}{7}\right)$ $\quad -3\dfrac{3}{4} \div \left(-2\dfrac{1}{2}\right)$

$\dfrac{2}{3} \cdot \left(-\dfrac{7}{5}\right)$ $\quad -\dfrac{15}{4} \div \left(-\dfrac{5}{2}\right)$

$\dfrac{2(-7)}{3 \cdot 5}$ $\quad -\dfrac{15}{4} \cdot \left(-\dfrac{2}{5}\right)$

(positive) · (negative) = negative
(negative) · (negative) = positive

$-\dfrac{14}{15}$ $\quad \dfrac{-\overset{3}{\cancel{15}} \cdot (-\cancel{2})}{\underset{2}{\cancel{4}} \cdot \underset{1}{\cancel{5}}} = \dfrac{3}{2},$ or $1\dfrac{1}{2}$

So, $\dfrac{2}{3} \div \left(-\dfrac{5}{7}\right) = -\dfrac{14}{15}$ and $-3\dfrac{3}{4} \div \left(-2\dfrac{1}{2}\right) = 1\dfrac{1}{2}$.

practice ▷   Divide.

5. $\dfrac{4}{5} \div \left(-\dfrac{1}{3}\right)$ 6. $-\dfrac{2}{3} \div \left(-\dfrac{4}{9}\right)$ 7. $-3\dfrac{1}{2} \div \left(-5\dfrac{1}{4}\right)$ 8. $-4 \div 2\dfrac{1}{3}$

◇ **ORAL EXERCISES** ◇

Is the quotient an integer?

1. $13 \div (-1)$ 2. $-21 \div (-7)$ 3. $-15 \div (-6)$ 4. $-17 \div 2$
5. $14 \div (-3)$ 6. $27 \div (-3)$ 7. $-42 \div 6$ 8. $45 \div (-9)$

◇ **EXERCISES** ◇

What number corresponds to the point?

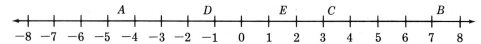

1. A 2. B 3. C 4. D 5. E

Divide.

6. $\dfrac{4}{5} \div \dfrac{3}{7}$ 7. $\dfrac{9}{4} \div \dfrac{3}{10}$ 8. $\dfrac{2}{9} \div \dfrac{3}{5}$ 9. $\dfrac{4}{9} \div \dfrac{8}{13}$

10. $-\dfrac{2}{5} \div \left(-\dfrac{4}{3}\right)$ 11. $-\dfrac{5}{8} \div \left(-\dfrac{9}{10}\right)$ 12. $-\dfrac{3}{7} \div \dfrac{5}{6}$ 13. $-\dfrac{2}{3} \div \dfrac{9}{4}$

14. $-2\dfrac{1}{3} \div \dfrac{3}{7}$ 15. $-5\dfrac{2}{5} \div 1\dfrac{1}{2}$ 16. $-3 \div \left(-2\dfrac{1}{4}\right)$ 17. $4\dfrac{1}{4} \div \left(-5\dfrac{1}{5}\right)$

Problem Solving – Careers
Industrial Machinists

1. Ms. Barkley repairs industrial sewing machines. One week, she repaired 12 machines. The following week, she repaired $2\frac{1}{3}$ times as many. How many did she repair then?

2. Mr. Bing works in a printing plant. One week, 24 pieces of machinery broke down. On the average, he can repair $1\frac{1}{2}$ pieces per day. How many days will it take him to repair the 24 pieces?

3. Ms. Garcia's factory spent $1,500 for new parts in one month. The following month, it spent $1\frac{1}{4}$ times as much. How much was that?

4. Mr. Ferguson runs the repair shop in a food products plant. One month, he ordered 1,200 parts for the shop. The following month he ordered 3 times as many parts. How many parts did he order in the two months?

5. Lucille works in a textile mill. She repaired 78 pieces of machinery during January. During February, she repaired $1\frac{1}{2}$ times as many. How many pieces did she repair in the two months?

6. Raymond's Packaging Plant spent $2,100 for parts one month. The next month the plant spent $\frac{2}{3}$ as much for parts. How much did the plant spend on parts in the two months?

PROBLEM SOLVING

Simplifying Expressions

◇ OBJECTIVES ◇

To simplify expressions like
$3x + (-5x)$, $4 - (-6x)$,
and $-4 \cdot (-3y)$

To simplify expressions like
$3(x - 5)$, $-2(3m - 1)$,
and $-(-3 - 2x)$

◇ RECALL ◇

$5 + (-9) = -4$ | $3[5 + (-1)] = 3(5) + 3(-1)$
$-3 - (-4) = 1$ |
$-30 \div 5 = -6$ | $2(4) + (-3)(4) = [2 + (-3)]4$
$-3 \cdot (-4) = 12$ |
$5 \cdot (-7) = -35$ | $-2 \cdot [-3 \cdot 4] = [-2 \cdot (-3)]4$

Example 1

Use the distributive property.
$5x$ means $5 \cdot x$; $-9x$ means $-9 \cdot x$;
$5 + (-9) = -4$.

Simplify. $5x + (-9x)$
$5x + (-9x)$
$[5 + (-9)]x$
$-4x$
So, $5x + (-9x) = -4x$.

---

practice ▷ Simplify.

1. $3x + (-7x)$ 2. $-5x + 3x$ 3. $-6x + (-2x)$ 4. $8x + (-8x)$

Example 2

Use the distributive property.
$-3 - (-4) = -3 + 4 = 1$
$1x$ means $1 \cdot x$, or x.

Simplify. $-3x - (-4x)$
$-3x - (-4x)$
$[-3 - (-4)]x$
$1x$, or x
So, $-3x - (-4x) = x$.

---

practice ▷ Simplify.

5. $-2x - (-6x)$ 6. $5x - 9x$ 7. $8x - (-3x)$ 8. $-4x - x$

Example 3

$-6y$ means $-6 \cdot y$.
Use the associative property.
$-5 \cdot (-6) = 30$

Simplify. $-5 \cdot (-6y)$
$-5 \cdot (-6y)$
$-5 \cdot (-6) \cdot y$
$[-5 \cdot (-6)]y$
$30y$
So, $-5 \cdot (-6y) = 30y$.

---

practice ▷ Simplify.

9. $3 \cdot 7x$ 10. $-4 \cdot 5y$ 11. $-2 \cdot (-6y)$ 12. $6 \cdot (-3x)$

Example 4

Use the distributive property.

$6x + (-8)$ means $6x - 8$.

Simplify. $2(3x - 4)$

$2(3x - 4) = 2(3x) + 2(-4)$
$ = 6x + (-8)$
$ = 6x - 8$

So, $2(3x - 4) = 6x - 8$.

Example 5

Use the distributive property.
$-3x + (-15)$ means $-3x - 15$.

Simplify. $-3(x + 5)$

$-3(x + 5) = -3(x) + (-3)(5)$
$ = -3x + (-15)$
$ = -3x - 15$

So, $-3(x + 5) = -3x - 15$.

Example 6

$-(6 - 5y)$ means $-1(6 - 5y)$.
Use the distributive property.

Simplify. $-(6 - 5y)$

$-(6 - 5y) = -1(6 - 5y)$
$ = -1(6) + (-1)(-5y)$
$ = -6 + 5y$

So, $-(6 - 5y) = -6 + 5y$.

practice ▷ Simplify.

13. $3(2x - 1)$ 14. $-2(y - 4)$ 15. $-(3 + 4x)$ 16. $-(5 - 3y)$

◇ EXERCISES ◇

Simplify.

1. $3x + (-7x)$
2. $-6y + (-7y)$
3. $4z + 9z$
4. $-10a + 4a$
5. $-6c + (-6c)$
6. $10m + (-14m)$
7. $-3n + (-7n)$
8. $4c + (-2c)$
9. $17t + (-20t)$
10. $-19p + 7p$
11. $-2x - (-3x)$
12. $-5x - 2x$
13. $7u - (-1u)$
14. $-6v - 6v$
15. $-8a - (-8a)$
16. $4y - (-10y)$
17. $-12h - (-30h)$
18. $14k - (-5k)$
19. $r - (-r)$
20. $-d - 9d$
21. $-3 \cdot (-7m)$
22. $5 \cdot 9d$
23. $-7 \cdot 2b$
24. $8 \cdot (-9n)$
25. $-6 \cdot (-6t)$
26. $3(2x + 3)$
27. $2(5y - 1)$
28. $-2(4x + 1)$
29. $-4(3z - 2)$
30. $6(2h + 1)$
31. $4(n - 2)$
32. $3(p + 7)$
33. $-5(x - 3)$
34. $-2(y + 2)$
35. $7(a - 1)$
36. $-(r - 3)$
37. $-(8 - z)$
38. $-(4 - 2b)$
39. $-(4 + 3w)$
40. $-(5y - 3)$

★ 41. $-8(2m - p) + 5$
★ 42. $6(-m - n) - 10$

SIMPLIFYING EXPRESSIONS

Solving Equations

◇ OBJECTIVE ◇

To solve equations like
$-4x = -20$, $\frac{x}{-3} = -2$, and
$-3x + 4 = -8$

◇ RECALL ◇

Solve. $3x = 12$

$x = \frac{12}{3}$, or 4

So, 4 is the solution.

Solve. $\frac{x}{4} = 5$

$x = 5 \cdot 4$, or 20

So, 20 is the solution.

Example 1

Solve and check.

Undo multiplication by division.

$5x = -30$
$\frac{5x}{5} = \frac{-30}{5}$
$x = -6$
Check:
$5x = -30$
$5 \cdot (-6)$ | -30
$-30 = -30$

$-3y = -39$
$\frac{-3y}{-3} = \frac{-39}{-3}$
$y = 13$
Check:
$-3y = -39$
$-3 \cdot 13$ | -39
$-39 = -39$

So, -6 is the solution of $5x = -30$, and 13 is the solution of $-3y = -39$.

practice ▷ Solve and check.

1. $2x = -18$
2. $-4x = 28$
3. $-6y = -42$

Example 2

Solve.

Undo division by multiplication.

$\frac{x}{-3} = 4$
$-3 \cdot \left(\frac{x}{-3}\right) = 4 \cdot (-3)$
$x = -12$

$\frac{a}{-7} = -5$
$-7 \cdot \left(\frac{a}{-7}\right) = -5 \cdot (-7)$
$a = 35$

Check on your own.

So, -12 is the solution of $\frac{x}{-3} = 4$, and 35 is the solution of $\frac{a}{-7} = -5$.

practice ▷ Solve and check.

4. $\frac{x}{-5} = 4$
5. $\frac{n}{-4} = -3$
6. $\frac{w}{6} = -8$

Example 3

Solve and check. $-4x + 7 = -1$

Undo addition by subtraction.	$-4x + 7 = -1$
Subtract 7 from each side.	$-4x + 7 - 7 = -1 - 7$
Undo multiplication by division.	$-4x = -8$
Divide each side by -4.	$\dfrac{-4x}{-4} = \dfrac{-8}{-4}$
	$x = 2$

Check: $-4x + 7 = -1$
$-4 \cdot 2 + 7$
$-8 + 7$
$-1 = -1$

So, 2 is the solution of $-4x + 7 = -1$.

practice ▷ Solve and check.

7. $2x - 8 = -14$
8. $-4x + 7 = 11$
9. $\dfrac{x}{-3} - 6 = 2$

◇ EXERCISES ◇

Solve and check.

1. $3x = -6$
2. $-7y = 28$
3. $-4z = -36$
4. $9a = 81$
5. $6t = -42$
6. $-8m = -64$
7. $\dfrac{x}{-4} = -1$
8. $\dfrac{y}{-2} = 6$
9. $\dfrac{m}{-6} = -9$
10. $\dfrac{r}{9} = -7$
11. $\dfrac{c}{-8} = 7$
12. $\dfrac{a}{-4} = -12$
13. $3x - 5 = 1$
14. $-2y + 7 = -9$
15. $-4a - 2 = -10$
16. $5x + 8 = -7$
17. $\dfrac{x}{-2} + 8 = -1$
18. $\dfrac{y}{3} - 7 = 4$
19. $\dfrac{z}{-5} - 3 = 2$
20. $\dfrac{a}{4} + 6 = 1$

Challenge

1. A farmer has some chickens and some goats. Altogether there are 43 heads and 108 legs. How many chickens does the farmer have? How many goats does the farmer have?

2. You need 5 liters of water. Three containers are available to you. One holds 3 liters; the second, 4 liters; and the third, 7 liters. How can you get 5 liters of water?

SOLVING EQUATIONS

Non-Routine Problems

There is no simple method used to solve these problems. Use whatever methods you can. Do not expect to solve these easily.

The cards below are made in such a way that the number in the lower right corner of each card is obtained by doing something with the other three numbers. The cards in each row follow the same pattern. For each row, find the pattern and figure out what number should replace the question mark to fit the pattern.

Example

2 4		5 6		9 2		7 7
3 5		0 30		8 10		12 ?	←?

Multiply these two numbers and subtract this number from the product: $(7 \times 7) - 12 = 37$.

1.
10 2	3 1	4 2	6 3
5 10	6 9	8 10	4 ?

2.
1 2	4 2	0 5	2 3
2 6	3 18	3 15	4 ?

3.
8 2	10 2	13 1	18 2
2 9	4 9	6 9	9 ?

4.
1 2	5 0	6 2	3 2
4 6	1 1	1 13	0 ?

5.
5 4	3 2	5 1	6 1
8 12	0 6	2 3	2 ?

6.
10 2	15 1	20 8	16 2
3 2	2 5	2 2	2 ?

330   CHAPTER THIRTEEN

If a program calls for an evaluation that results in division by zero, the program will stop **RUN**ning. (Recall that you cannot divide by the number 0.) The computer will print an error message such as **DIVISION BY ZERO ERROR IN 80**. The program below shows how to avoid such a problem.

Example Use **READ/DATA** to write a program to evaluate $\frac{2a-4}{a-2}$, if $a = 1$, 2, -4. **RUN** the program.

```
                                   10   HOME
    There are 3 sets of data.      20   FOR K = 1 TO 3
                                   30   READ A
    The 3 values of A.             40   DATA 1,2,-4
    N represents numerator.        50   LET N = 2 * A - 4
    D represents denominator.      60   LET D = A - 2
    Prints the values of A, N, and D for  70   PRINT "A = "A", N = "N", D = "D
    each RUN.
    Computer tests to see if       80   IF D = 0 THEN 120
    denominator is 0.
    Otherwise, evaluation can be done.  90   LET Q = N / D
                                   100  PRINT "THE VALUE OF Q IS "Q"."
                                   110  GOTO 130
                                   120  PRINT "YOU CANNOT DIVIDE BY ZERO."
    The program will END after 3 sets of  130  NEXT K
    data.                          140  END
                                   ]RUN
    When k = 1, A = 1.             A = 1, N = -2, D = -1
                                   THE VALUE OF Q IS 2.
    When k = 2, A = 2.             A = 2, N = 0, D = 0
                                   YOU CANNOT DIVIDE BY ZERO.
    When k = 3, A = -4.            A = -4, N = -12, D = -6
                                   THE VALUE OF Q IS 2.
```

See the Computer Section beginning on page 420 for more information.

Exercises

Write and **RUN** a program to evaluate each for the given data.

1. $\frac{5x-20}{x-5}$ if $x = 0, 5, 10$

2. $\frac{4x+6}{2x-2}$ if $x = -4, 1, 2, 3$

COMPUTER ACTIVITIES

Chapter Review

Multiply. [312, 314]

1. $11 \cdot 23$
2. $24 \cdot (-18)$
3. $-29 \cdot 71$
4. $48 \cdot 0$
5. $-49 \cdot (-62)$
6. $-68 \cdot (-91)$
7. $-100 \cdot (-12)$
8. $-17 \cdot (-200)$

Multiply. [316]

9. $1 \cdot 275$
10. $-312 \cdot (-1)$
11. $42 \cdot 25 \cdot (-4)$
12. $-61 \cdot (-22) \cdot 0$

Multiply. Use the distributive property. [316]

13. $-7 \cdot (-19 + 9)$
14. $14 \cdot 26 + 14 \cdot (-6)$
15. $-12 \cdot (7 - 27)$
16. $-30 \cdot 41 - (-30) \cdot 1$

Divide, if possible. [320]

17. $-38 \div 1$
18. $0 \div (-5)$
19. $88 \div (-1)$
20. $-42 \div 0$
21. $48 \div 6$
22. $-22 \div 2$
23. $-55 \div (-5)$
24. $57 \div (-3)$
25. $150 \div (-5)$
26. $-200 \div 4$
27. $-500 \div 50$
28. $-600 \div (-30)$

Divide. [323]

29. $\dfrac{3}{7} \div \dfrac{2}{9}$
30. $-\dfrac{3}{5} \div \left(-\dfrac{2}{3}\right)$
31. $-\dfrac{4}{5} \div \dfrac{1}{2}$
32. $-3\dfrac{1}{4} \div \left(-4\dfrac{1}{5}\right)$

Simplify. [326]

33. $3y + (-16y)$
34. $-5t - (-11t)$
35. $-4 \cdot 5x$
36. $-7 \cdot (-8m)$
37. $3(2y - 5)\ 6y$
38. $5(a + 6)$
39. $-(6 + 2x)$
40. $-4(4n - 3)$

Solve. [328]

41. $-9x = 54$
42. $-7y = -63$
43. $\dfrac{t}{-6} = 2$
44. $2k - 7 = -3$

Chapter Test

Multiply.

1. $12 \cdot 47$
2. $54 \cdot (-23)$
3. $-64 \cdot (-93)$
4. $32 \cdot 0$
5. $-1 \cdot 518$
6. $20 \cdot (-5) \cdot 37$

Multiply. Use the distributive property.

7. $-22 \cdot (-9 + 15)$
8. $9 \cdot (-11) - 9 \cdot (-24)$

Divide, if possible.

9. $16 \div (-1)$
10. $-800 \div (-40)$
11. $12 \div 0$
12. $0 \div (-31)$

Divide.

13. $-\frac{5}{7} \div \frac{2}{3}$
14. $-\frac{2}{5} \div \left(-\frac{3}{8}\right)$
15. $-5\frac{1}{2} \div \left(-6\frac{1}{3}\right)$

Simplify.

16. $-5t + 16t$
17. $-9 \cdot (-4m)$
18. $-2(2x - 1)$

Solve.

19. $-7x = 63$
20. $\dfrac{y}{-4} = -8$
21. $-2x + 5 = -7$

Challenge

Replace each letter by a digit so that correct additions will result. Within the same problem, the same letters should be replaced by the same digit, different letters by different digits.

1.   MERRY
   $+$ XMAS
     TO ALL

2.   SEND
   $+$MORE
   MONEY

3.   HOCUS
   $+$POCUS
   PRESTO

4.   AHAHA
   $+$ TEHE
   TEHAW

Equations and Inequalities

14

The mirror in the Mayall telescope at the Kitt Peak National Observatory weighs 14 t.

Solving Equations

◇ OBJECTIVE ◇

To solve an equation like
$4x - 9 = 2x + 5$

◇ RECALL ◇

Solve.	$6x - 5 = 19$
Add 5 to each side.	$6x - 5 + 5 = 19 + 5$
	$6x = 24$
Divide each side by 6.	$\frac{6x}{6} = \frac{24}{6}$
So, 4 is the solution.	$x = 4$

Equation properties can be used to solve equations in which the variable is on each side of the equation.

Example 1

Solve and check. $5x + 4 = x + 16$

	$5x + 4 = x + 16$	Check.	
Subtract 4 from each side.	$5x + 4 - 4 = x + 16 - 4$	$5x + 4$	$= x + 16$
	$5x = x + 12$	$5 \cdot 3 + 4$	$3 + 16$
Subtract x from each side.	$5x - x = x + 12 - x$	$15 + 4$	19
Combine like terms.	$4x = 12$	19	
Divide each side by 4.	$\frac{4x}{4} = \frac{12}{4}$	$19 = 19$	
	$x = 3$	true	

So, 3 is the solution of $5x + 4 = x + 16$.

practice ▷ Solve.

1. $3x + 8 = x + 2$ 2. $6x + 9 = x + 4$

Example 2

Solve and check. $6x - 7 = 4x + 5$

	$6x - 7 = 4x + 5$	Check.	
Add 7 to each side.	$6x - 7 + 7 = 4x + 5 + 7$	$6x - 7$	$= 4x + 5$
	$6x = 4x + 12$	$6 \cdot 6 - 7$	$4 \cdot 6 + 5$
Subtract $4x$ from each side.	$6x - 4x = 4x + 12 - 4x$	$36 - 7$	$24 + 5$
Combine like terms.	$2x = 12$	29	29
Divide each side by 2.	$\frac{2x}{2} = \frac{12}{2}$	$29 = 29$	
	$x = 6$	true	

So, 6 is the solution of $6x - 7 = 4x + 5$.

practice ▷ Solve.

3. $3x + 7 = 5x - 3$ 4. $9x - 2 = 2x + 12$

◇ ORAL EXERCISES ◇

Solve.

1. $2x = 12$
2. $3x = 24$
3. $5x = 30$
4. $12x = 6$
5. $15x = 5$
6. $12x = 3$
7. $x + 3 = 5$
8. $x - 7 = 2$
9. $x - 5 = 20$

◇ EXERCISES ◇

Solve.

1. $7x + 3 = 17$
2. $9x + 4 = 4$
3. $12x + 5 = 17$
4. $2x - 5 = 7$
5. $5x - 4 = 11$
6. $4x - 8 = 12$
7. $4x + 9 = x + 6$
8. $2x + 7 = x + 1$
9. $7x + 8 = x + 2$
10. $3x - 6 = x - 2$
11. $7x - 2 = x - 14$
12. $8x - 1 = x - 8$
13. $4x - 5 = 2x + 1$
14. $7x - 5 = 5x + 5$
15. $12x + 3 = 5x - 18$
16. $3x + 2 = 6x - 13$
17. $4x - 5 = 9x + 15$
18. $9x - 4 = 3x + 2$
★ 19. $2x - 3 = 5x - 4$
★ 20. $12 - 3x = 13 - 5x$
★ 21. $5x - 7 = x - 8$
★ 22. $-7x + 4 = 7 - x$
★ 23. $-5x - 3 = -2x + 2$
★ 24. $4 - 2x = 5 + 7x$

Calculator

You can use a calculator to check whether a given number is a solution of an equation.

Which of the numbers 5 or 7 is the solution of the equation $37x + 9 = 54x - 76$?
Replace x with 5 in $37x + 9$ first.
Press 37 ⊗ 5 ⊕ 9 ⊖
Display 194
Now replace x with 5 in $54x - 76$.
Press 54 ⊗ 5 ⊖ 76 ⊖
Display 194   So, 5 is the solution of $37x + 9 = 54x - 76$.

Which of the given numbers is the solution of the equation?

1. $45x + 27 = 74x - 234$ 8, 9, 10
2. $79x - 56 = 120x - 384$ 6, 7, 8
3. $94x + 63 = 136x - 441$ 9, 12, 16

Equations with Parentheses

◇ OBJECTIVE ◇

To solve an equation like
$2(x - 1) = -(8x - 18)$

◇ RECALL ◇

$$4(x + 5) = 4(x) + 4(5)$$
$$= 4x + 20$$
$$-3(y - 6) = -3(y) + (-3)(-6)$$
$$= -3y + 18$$
$$-(4 + 2x) = -1(4 + 2x)$$
$$= -1(4) + (-1)(2x)$$
$$= -4 - 2x$$

If an equation contains parentheses, first rewrite the equation without parentheses. This can be done by using the distributive property as shown in the recall. Then solve the resulting equation.

Example 1

Solve $5(x - 2) = 20$. Then check your solution.

$5(x - 2) = 5(x) + 5(-2)$

$$5(x - 2) = 20$$
$$5(x) + 5(-2) = 20$$
$$5x - 10 = 20$$

Add 10 to each side. $5x - 10 + 10 = 20 + 10$
Combine like terms. $5x = 30$
Divide each side by 5. $\dfrac{5x}{5} = \dfrac{30}{5}$
$x = 6$ So, 6 is the solution.

Check.
$$5(x - 2) = 20$$
$$5(6 - 2) \mid 20$$
$$5(4)$$
$$20 = 20$$

Example 2

Solve $2(n - 6) = -4(n - 3)$.

$$2(n - 6) = -4(n - 3)$$
$$2(n) + 2(-6) = -4(n) + (-4)(-3)$$
$$2n - 12 = -4n + 12$$

Add $4n$ to each side. $2n - 12 + 4n = -4n + 12 + 4n$
Combine like terms. $6n - 12 = 12$
Add 12 to each side. $6n - 12 + 12 = 12 + 12$
$6n = 24$
Divide each side by 6. $\dfrac{6n}{6} = \dfrac{24}{6}$
$n = 4$

Check on your own.   So, 4 is the solution.

practice Solve.

1. $4(n + 2) = 24$
2. $6(z - 3) = 3(z + 4)$

Example 3 Solve $4x - (-6 - 3x) = -8$.

$-(-6x - 3x) = -1(-6x - 3x)$

$$4x - (-6 - 3x) = -8$$
$$4x - 1(-6 - 3x) = -8$$
$$4x + (-1)(-6) + (-1)(-3x) = -8$$

Combine like terms.
$4x + 3x = 7x$
$$4x + 6 + 3x = -8$$
$$7x + 6 = -8$$

Subtract 6 from each side.
$$7x + 6 - 6 = -8 - 6$$
$$7x = -14$$

Divide each side by 7.
$$\frac{7x}{7} = \frac{-14}{7}$$
$$x = -2$$

So, -2 is the solution.

---

practice Solve.

3. $-(c + 8) = 5$
4. $2x - (4 - 3x) = 11$

◇ EXERCISES ◇

Solve.

1. $3(x + 1) = 15$
2. $-2(x + 3) = -12$
3. $4(x - 5) = 16$
4. $-7(y - 1) = 21$
5. $6(4 - z) = 18$
6. $-20 = -5(a + 7)$
7. $4(x - 3) = 8(x + 1)$
8. $-6(y + 4) = 3(y - 2)$
9. $2(z - 6) = -4(z - 9)$
10. $-5(c + 3) = -2(c - 6)$
11. $-3(x + 7) = -4(x - 1)$
12. $7(x + 1) = -4(x + 12)$
13. $-(c + 6) = 9$
14. $-(y - 4) = 17$
15. $-(7 - z) = 13$
16. $-(6 + a) = 10$
17. $17 = -(y + 4)$
18. $-20 = -(x - 9)$
19. $4z - (7 + 3z) = 5$
20. $-6y - (3 - 4y) = 11$
21. $-(-5a + 8) + 3a = 32$
22. $41 = 3n - (7 - 3n)$
23. $3(x - 4) = -(7 - 2x)$
24. $-(6 + 4x) = -2(x + 9)$
25. $2(y - 1) = -(8y - 18)$
26. $-(12 - 3z) = 6(z + 2)$
★ 27. $2(3x + 1) - 5x = -6$
★ 28. $4(-2x - 3) + 6x = 18$
★ 29. $4(2y - 9) - (5y + 6) = 0$
★ 30. $-3(5 - 2c) - (4c + 7) = 0$
★ 31. $\frac{2}{3}(3y + 6) = -10$
★ 32. $-\frac{1}{2}(6 + 4x) = \frac{1}{3}(9 - 3x)$
★ 33. $5z - \frac{3}{4}(8z - 4) = -1$
★ 34. $\frac{3}{5}(10a - 5) - \frac{3}{2}(8a + 10) = 0$

Number Problems

◇ OBJECTIVES ◇

To write an equation for a word problem
To solve a word problem

◇ RECALL ◇

5 increased by 8
$5 + 8$

7 less than twice a number
$2n - 7$

Example 1   Write an equation for the sentence.
Six more than a number is 9.

Let n = the number.   Six more than a number is 9.
Six more than n
$n + 6$ $n + 6 = 9$
So, the equation is $n + 6 = 9$.

Example 2   Write an equation for the sentence.
A number decreased by 4 is equal to 12.

Let n = the number.   A number decreased by 4 is equal to 12.
n decreased by 4
$n - 4$ $n - 4 = 12$
So, the equation is $n - 4 = 12$.

Example 3   Write an equation for the sentence.
Two less than 3 times a number is 16.

Let n = the number.
2 less than   3 times n   Two less than 3 times a number is 16.
$3n - 2$ $3n - 2 = 16$
So, the equation is $3n - 2 = 16$.

---

practice ▷   **Write an equation for the sentence.**

1. A number increased by 7 is equal to 15.
2. Three less than 5 times a number is 7.

Example 4

A number increased by 5 is equal to 19. Find the number.

Let $n =$ the number.
n increased by 5 is equal to 19.

Write an equation.	$n + 5 = 19$
Subtract 5 from each side.	$n + 5 - 5 = 19 - 5$
	$n = 14$
Now check 14 in the problem.	Check: n increased by 5 is 19.

$$\begin{array}{c|c} 14 + 5 & 19 \\ 19 & \text{true} \end{array}$$

So, the number is 14.

Example 5

Eight less than a number is 17. Find the number.

Let $n =$ the number.
8 less than n is 17.

Write an equation.	
8 less than n	$n - 8 = 17$
$n \leftarrow - \rightarrow 8$	
Add 8 to each side.	$n - 8 + 8 = 17 + 8$
Check 25 in the problem.	$n = 25$
$25 - 8 = 17$	So, the number is 25.

Example 6

Three more than twice a number is equal to -11. Find the number.

Let $n =$ the number.
3 more than twice n is -11.

Write an equation.	$2n + 3 = -11$
Subtract 3 from each side.	$2n + 3 - 3 = -11 - 3$
Divide each side by 2.	$2n = -14$
Check -7 in the problem.	$n = -7$

So, the number is -7.

practice ▷ **Write an equation for the sentence.**

3. Six dollars less than what you paid for a blouse is $11. How much did you pay for the blouse?

4. Five more than 3 times Paul's score is 23 points. What is Paul's score?

◇ EXERCISES ◇

Write an equation for each sentence.

1. Four more than a number is 15.
2. Six less than a number is 9.
3. A number decreased by 8 is 7.
4. Nine increased by a number is 17.
5. A number increased by 12 is 23.
6. Three more than a number is 18.
7. Seven less than twice a number is equal to 19.
8. Eight, decreased by three times a number, is equal to 2.
9. Five more than four times a number is equal to 21.
10. Twice a number, increased by 11, is equal to 25.
11. Four times a number, decreased by 5, is 11.
12. Eight less than 3 times a number is equal to 7.

Solve each problem.

13. Eight less than a number is 15. Find the number.
14. Three more than a number is 18. Find the number.
15. A number increased by 12 is 21. Find the number.
16. Sixteen decreased by a number is 7. Find the number.
17. Twenty-four increased by a number is equal to 31. Find the number.
18. A number decreased by 13 is equal to 7. Find the number.
19. Five more than a number is equal to 7. Find the number.
20. Ten less than a number is equal to 12. Find the number.
21. Four increased by twice the number of tomato plants you planted this year is 30. How many tomato plants did you plant this year?
22. Six less than ten times the number of absences you had this year is 14. How many absences did you have this year?
23. Five times the number of books Maria carries to school each day increased by 11 is 31. How many books does Maria carry to school each day?
24. Four more than 9 times the width of a sidewalk is 49 feet. How wide is the sidewalk?
25. Ten times the number of students on a committee decreased by 7 is 43. What is the number of students on the committee?
26. Seven less than 6 times the number of students that boarded the bus is equal to 35. How many students boarded the bus?
★ 27. Twice the number of good friends you have increased by 3 times the number is equal to 45. How many good friends do you have?
★ 28. Four times the number of days you went to school during the last two weeks decreased by 3 is equal to 3 times that number, increased by 7. How many days did you go to school during the last two weeks?

NUMBER PROBLEMS

Problem Solving – Applications
Jobs for Teenagers

Example

Jane weighed 32 lb in March. That was an increase of 5 lb in a month. How much did she weigh a month ago?

You can write an equation for this problem.
Let w = weight a month ago.
Weight in March $\longrightarrow w + 5$
Equation $\quad\quad w + 5 = 32$

Solve. $\quad w + 5 - 5 = 32 - 5$
$\quad\quad\quad\quad\quad\quad w = 27$

Check 27 lb in the problem.
So, Jane weighed 27 lb a month ago.

practice ▷ In June, Mrs. Sadowski paid $27.50 for diapers. That was an increase of $5.75 over the cost in May. How much was the cost in May?

Solve these problems.

1. Bob weighed 29 lb in May. This was an increase of 6 lb in 3 months. What was Bob's weight 3 months ago?

2. In Exercise 1, assume that Bob's weight increased the same amount each month. What were Bob's weights in February, March, and April?

3. Sharon's age is 2 more than 3 times Tod's age. Tod is now 3 years old. How old is Sharon?

4. Lori charges $1.00 an hour for babysitting. She worked from 8:00 P.M. to 11:00 P.M. How much did she earn?

5. For babysitting, Jerry charges $1.00 per hour before midnight and $5.00 to stay overnight. Saturday night, he arrived at 6:30 P.M. and stayed overnight. How much did he earn?

6. Jackie works during the summer as a mother's helper. She is paid $8 per day and works 5 days a week. How much will she earn in 11 weeks?

7. Julie charges $0.75 per hour before midnight and $1.25 per hour after midnight for babysitting. She babysat from 7:00 P.M. to 2:00 A.M. How much did she earn?

8. Karen took two children to a museum. Bus fare was $0.50 per person each way. Admission to the museum was $0.75 per person. How much did the trip cost for all three?

9. Bill works at a day-care center 5 days a week. He is paid $25 per day. How much will he earn in 2 weeks?

10. When a new baby came along, the Katsaris family had to increase the purchase of baby foods by 25%. They are now spending $44 per week. How much were they spending previously?

11. The Happy Hour Nursery had 36 babies to take care of during the month of May. This was 20% more than were there during April. How many babies were there in April?

12. Mr. Willis pays his babysitter $1.25 per hour and gives a $0.50 tip. How much does he pay a babysitter for 5 hours of babysitting?

13. Mrs. Reed pays her babysitter $1.25 per hour and gives a $0.50 tip. She uses the formula $1.25h + 0.50$, where h is the number of hours. Use the formula to find a babysitter's pay for sitting from 7:00 P.M. to midnight.

14. Pablo gets babysitting jobs through an agency. He is paid $2.00 per hour, but he must pay the agency $3.00 for each job. Thursday he babysat from 8:00 P.M. until 12:30 A.M. How much did he earn after paying the agency?

15. Pablo uses the formula $2.00h - 3.00$ to help figure his earnings. Use the formula to find his earnings for sitting from 7:30 P.M. until 1:00 A.M.

16. Rosemarie earns $1.75 per hour babysitting, but pays an agency $2.50 for each job. How much does she earn for babysitting for 6 hours?

PROBLEM SOLVING

Properties for Inequalities

◇ OBJECTIVE ◇

To write a true inequality after performing a given operation on each side of an inequality like $-2 < 5$

◇ RECALL ◇

$6 < 4$ false	$-3 \geq 2$ false
$6 > 4$ true	$-3 \leq 2$ true
$2 \geq 2$ true	$3 \leq 3$ true
$2 > 2$ false	$3 < 3$ false

Example 1

Add 3 to each side of the inequality $2 < 5$. Then use $<$ or $>$ to write a true inequality.

Different numbers
Add the same numbers.
Different numbers, same order

$2 < 5$ same
$2 + 3 \; ? \; 5 + 3$ order
So, $5 < 8$.

Example 2

Subtract 2 from each side of $7 > -1$. Then write a true inequality.

$7 > -1$ same
$7 - 2 \; ? \; -1 - 2$ order
So, $5 > -3$.

$6 > -1$ ← true inequality
$+2 \quad +2$ ← Add 2 to each side.
$8 > \; 1$ ← same order

addition/subtraction properties for inequalities
Adding or subtracting the same number to or from each side of a true inequality does not change the order.

practice

1. Subtract 5 from each side of $-3 \leq 4$. Then write a true inequality.

2. Add 4 to each side of $-2 \geq -5$. Then write a true inequality.

Example 3

Multiply each side of $3 < 6$ by 2 and write a true inequality. Then multiply each side of $3 < 6$ by -2 and write a true inequality.

Multiplying by a negative number reverses the order.

$3 < 6$ same $3 < 6$ reverse
$3 \cdot 2 \; ? \; 6 \cdot 2$ order $3 \cdot (-2) \; ? \; 6 \cdot (-2)$ order
So, $6 < 12$.   So, $-6 > -12$.

Example 4

Multiply each side of $2 \geq -4$ by 3 and write a true inequality. Then multiply each side of $2 \geq -4$ by -3 and write a true inequality.

Multiplying by a negative number reverses the order.

$2 \geq -4$ same order
$2 \cdot 3 \; ? \; -4 \cdot 3$
So, $6 \geq -12$.

$2 \geq -4$ reverse order
$2 \cdot (-3) \; ? \; -4 \cdot (-3)$
So, $-6 \leq 12$.

$-5 < -2$ ← true inequality
$-5(-3) \; ? \; -2(-3)$ ← Multiply by -3.
$15 > 6$ ← reverse order

multiplication property for inequalities
Multiplying each side of a true inequality by the same positive number does not change the order. Multiplying each side of a true inequality by the same negative number reverses the order.

practice

3. Multiply each side of $-1 < 5$ by 4. Then write a true inequality.

4. Multiply each side of $-1 < 5$ by -4. Then write a true inequality.

Example 5

Divide each side of $-6 < 8$ by 2 and write a true inequality. Then divide each side of $-6 < 8$ by -2 and write a true inequality.

$-6 < 8$
$\dfrac{-6}{2} \; ? \; \dfrac{8}{2}$ same order
So, $-3 < 4$.

$-6 < 8$
$\dfrac{-6}{-2} \; ? \; \dfrac{8}{-2}$ reverse order
So, $3 > -4$.

Reverse the order for a true inequality when multiplying or dividing by a negative number.

division property for inequalities
Dividing each side of a true inequality by the same positive number does not change the order. Dividing each side of a true inequality by the same negative number reverses the order.

practice

5. Divide each side of $-6 \geq -9$ by 3. Then write a true inequality.

6. Divide each side of $-6 \geq -9$ by -3. Then write a true inequality.

PROPERTIES FOR INEQUALITIES

◇ ORAL EXERCISES ◇

If the indicated operation is performed on an inequality, will the order of the inequality change?

1. Add 6.
2. Multiply by 4.
3. Divide by 2.
4. Multiply by −2.
5. Subtract −4.
6. Multiply by 6.
7. Divide by −4.
8. Multiply by −5.
9. Divide by 6.
10. Subtract 5.
11. Add 1.
12. Multiply by 3.
13. Divide by −3.
14. Divide by 4.
15. Subtract −2.
16. Add 0.
17. Multiply by −1.
18. Divide by −1.

◇ EXERCISES ◇

Perform the operation on each side of the inequality. Then write a true inequality.

1. $2 < 5$ Add 2.
2. $3 > 1$ Add −1.
3. $3 < 4$ Multiply by 3.
4. $5 > 2$ Divide by −1.
5. $6 > -1$ Subtract 6.
6. $-4 < 6$ Divide by 2.
7. $1 > -2$ Multiply by −4.
8. $-1 < 4$ Multiply by 5.
9. $6 > -3$ Divide by 3.
10. $-2 > -4$ Add 0.
11. $-1 < 3$ Add 6.
12. $4 > -3$ Multiply by −2.
13. $-9 < 3$ Divide by −3.
14. $-3 < -2$ Multiply by 4.
15. $-2 > -6$ Subtract 5.
16. $-5 > -30$ Divide by −5.
17. $6 \geq -12$ Divide by 6.
18. $4 \geq 4$ Add 5.
19. $-2 \leq 6$ Multiply by 1.
20. $-3 \leq -1$ Add 10.
21. $12 \geq -8$ Divide by −4.
22. $-3 \leq 6$ Multiply by −5.
23. $-6 < 7$ Subtract −8.
24. $4 > -6$ Divide by −2.
25. $3 \geq -1$ Multiply by −1.
26. $2 \leq 5$ Add −7.
27. $14 > -7$ Divide by 7.
28. $-5 < -2$ Multiply by 10.
★ 29. $2 \leq 5$ Multiply by 0.
★ 30. $-4 \geq -1$ Multiply by 0.

Challenge

Using each of the numbers 1 through 9 exactly once, place a number in each circle so that the sum of the three numbers along every straight line is 15.

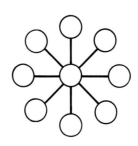

Solving Inequalities

◇ OBJECTIVE ◇

To solve inequalities like
$x + 1 > -4$, $x - 5 \leq 3$,
$-3x \geq 9$, and $\frac{1}{2}x < -5$
using integers

◇ RECALL ◇

Solve. $x - 4 = -13$
$ x - 4 = -13$
$ x - 4 + 4 = -13 + 4$
$ x = -9$
So, the solution is -9.

Solve. $-4y = 24$
$ \dfrac{-4y}{-4} = \dfrac{24}{-4}$
$ y = -6$
So, the solution is -6.

The sum of a number and 3 is greater than -1.
$ x + 3 > -1$

Example 1

Find the integers that make $x + 3 > -2$ true.

$x + 3 > -2$

Subtract 3 from each side.
Subtracting the same number from each side of a true inequality does not change the order.

$x + 3 - 3 > -2 - 3$ same order   {subtraction property for inequalities}
$x + 0 > -5$
$x > -5$

Integers greater than -5:
$-4, -3, -2, -1, 0, \ldots$
Try -4 and 1.

Check.
$\begin{array}{c|c} x + 3 > -2 & -2 \\ -4 + 3 & \\ -1 & \\ -1 > -2 \text{ true} & \end{array}$
$\begin{array}{c|c} x + 3 > -2 & -2 \\ 1 + 3 & \\ 4 & \\ 4 > -2 \text{ true} & \end{array}$

All integers greater than -5 make $x + 3 > -2$ true.

So, $-4, -3, -2, \ldots$ are solutions of $x + 3 > -2$.

Example 2

Solve. $y - 2 \leq -1$

$y - 2 \leq -1$

Add 2 to each side.
$y - 2 + 2 \leq -1 + 2$ same order   {addition property for inequalities}
$y + 0 \leq 1$
$y \leq 1$

Integers less than or equal to 1: 1, 0, $-1, -2, -3, \ldots$ Try 1 and -4.

Check.
$\begin{array}{c|c} y - 2 \leq -1 & -1 \\ 1 - 2 & \\ -1 & \\ -1 \leq -1 \text{ true} & \end{array}$
$\begin{array}{c|c} y - 2 \leq -1 & -1 \\ -4 - 2 & \\ -6 & \\ -6 \leq -1 \text{ true} & \end{array}$

The solutions are $1, 0, -1, -2, \ldots$

So the solutions are integers less than or equal to 1.

practice ▷ Solve.

1. $x + 2 > -5$
2. $y + 1 \le -7$
3. $n - 5 \ge -9$

Let n be the number. $n \cdot 4 > -12$ may also be written $4n > -12$.

A number multiplied by 4 is greater than -12.
↓   ↓   ↓   ↓   ↓
n $\cdot$ 4 $>$ -12

Example 3

Find the integers that make $4n > -12$ true.

$4n > -12$
$\frac{4n}{4} > \frac{-12}{4}$ same order   { division property for inequalities
$1n > -3$
$n > -3$

Divide each side by 4.

Dividing each side of a true inequality by the same positive number does not change the order. Any integer greater than -3 is a solution.

So, the integers that make $4n > -12$ true are $-2, -1, 0, \ldots$

Example 4

Solve. $-5x \le -15$

$-5x \le -15$
$\frac{-5x}{-5} \ge \frac{-15}{-5}$ reverse order   { division property for inequalities
$1x \ge 3$
$x \ge 3$

Divide each side by -5.

Dividing each side of a true inequality by the same negative number reverses the order. The solutions are $3, 4, 5, \ldots$

So, any integer greater than or equal to 3 solves the inequality.

Example 5

Solve. $\frac{2}{3}x < -6$

$\frac{2}{3}x < -6$

$\frac{3}{2}$ is the reciprocal of $\frac{2}{3}$.

$\frac{3}{2} \cdot \frac{2}{3}x < -6 \cdot \frac{3}{2}$

Use the multiplication property for inequalities.

$1x < \frac{-18}{2}$
$x < -9$

So, $-10, -11, -12, \ldots$ are solutions.

Solve. $-\frac{1}{2}n \ge -3$

$-\frac{1}{2}n \ge -3$

$-2 \cdot \left(-\frac{1}{2}n\right) \le -3 \cdot (-2)$

$1n \le 6$
$n \le 6$

So, $6, 5, 4, \ldots$ are solutions.

practice ▷ Solve.

4. $2x < -8$
5. $-3y > 15$
6. $-\frac{3}{4}y \le -6$

◇ EXERCISES ◇

Solve.

1. $x + 1 > 3$
2. $y + 6 \leq 7$
3. $z + 2 \leq 6$
4. $a + 5 > 8$
5. $c + 6 \geq 2$
6. $w + 5 < 3$
7. $d + 7 \geq 2$
8. $x + 1 < 0$
9. $m + 1 < -6$
10. $e + 3 \leq -2$
11. $y + 7 > -1$
12. $n + 2 \leq -4$
13. $r - 3 > 5$
14. $k - 5 \geq 0$
15. $z - 10 \geq 3$
16. $b - 7 < 1$
17. $s - 9 < -5$
18. $n - 6 < -3$
19. $p - 4 \geq -1$
20. $a - 8 \geq -7$
21. $v - 3 \geq -12$
22. $x - 7 \leq -9$
23. $t - 6 < -8$
24. $z - 1 > -5$
25. $2n \geq -6$
26. $3x < 6$
27. $4y \leq -12$
28. $3a > -18$
29. $8k < -16$
30. $3n \geq 3$
31. $5d \leq 15$
32. $7z > -14$
33. $-2m \leq 10$
34. $-4x > -4$
35. $-2c < 8$
36. $-5s \leq 5$
37. $-7t > -14$
38. $-3s \geq -12$
39. $-8v < 16$
40. $-12n \leq -12$
41. $\frac{1}{2}x > -1$
42. $\frac{1}{3}y > 2$
43. $\frac{1}{5}a \geq -2$
44. $\frac{1}{4}m < -7$
45. $\frac{3}{4}a \leq -9$
46. $\frac{2}{5}c \geq 4$
47. $\frac{7}{10}d < -21$
48. $\frac{5}{6}w > 10$
49. $-\frac{1}{3}c < -4$
50. $-\frac{1}{5}s \leq 1$
51. $-\frac{2}{3}y \geq -4$
52. $-\frac{3}{8}z < 15$

Challenge

1. Find a fraction equal to $\frac{2}{3}$ such that the sum of the numerator and denominator is 100.

2. Write a ten-digit numeral so that the first digit on the left tells how many 0's there are in the entire numeral, the second digit tells how many 1's, the third digit how many 2's, and so on.

3. A car traveled for 2 hours and averaged 80 km/h. On the return trip it averaged 70 km/h. What was the average speed for the entire trip?

SOLVING INEQUALITIES

Problem Solving – Careers
Railroad Brakers

1. Mr. Scorsese is a railroad braker. He is stationed at the rear of the train to display warning lights and signals. Last year he earned $24,640. What were his average earnings per month?

2. Ms. Nicholas is a passenger braker. She often helps the conductor by collecting or selling tickets. A passenger bought four one-way tickets at $3.40 each. He gave Ms. Nicholas a $20 bill. How much change should she give him?

3. John bought a one-way ticket for $2.55. He paid with a $10 bill. How much change did he get?

4. A round-trip ticket cost $7.60. How much did Shelly pay for three round-trip tickets?

5. As a yard braker, Mr. Martinez earns about $2,250 per month. As a passenger braker, Ms. Rosen earns about $2,517 per month. How much more does Ms. Rosen earn in a year than Mr. Martinez?

6. Sometimes passenger brakers regulate car lighting and temperature. The braker saw that the car thermostat read 32°C. Was the car too hot or too cold?

7. In 1980, there were nearly 72,000 brakers in the U.S.A. By 1990 that number may decrease by about 5%. If so, how many brakers will there be in the U.S.A. in 1990?

Exponents

◆ OBJECTIVES ◆

To evaluate expressions like
$-3xy^2$ if x is 4 and y is -1

To evaluate expressions like
$(-2a)^2$ if a is 3

◆ RECALL ◆

Evaluate $4x^3$ if x is 2.
$$4x^3$$
$$\downarrow\downarrow$$
$$4 \cdot 2^3$$
$$4 \cdot 2 \cdot 2 \cdot 2$$
$$4 \cdot 8$$
$$32$$

Example 1

Evaluate x^3 if x is -3.
$$x^3$$
$$\downarrow$$
Substitute -3 for x. $\quad (-3)^3$
$$\underbrace{(-3)(-3)(-3)}$$
$(-3)(-3) = 9 \qquad\qquad (9)(-3)$
$(9)(-3) = -27 \qquad\qquad -27$

So, the value of x^3 is -27 if $x = -3$.

Example 2

Evaluate $-4y^2$ if $y = -5$.
$$-4y^2$$
$$\downarrow\downarrow$$
Substitute -5 for y. $\quad -4(-5)^2$
$$-4\underbrace{(-5)(-5)}$$
$(-5)(-5) = 25 \qquad\qquad -4 \cdot 25$
$$-100$$

So, the value of $-4y^2$ is -100 if $y = -5$.

Example 3

Evaluate $-a^2b^3$ if $a = 6$ and $b = -1$.
$$-a^2b^3$$
$$(-1)a^2b^3$$
Substitute 6 for a and -1 for b. $\quad (-1)(6)^2(-1)^3$
$$(-1)(6)(6)\underbrace{(-1)(-1)(-1)}$$
$$\downarrow\qquad\qquad\downarrow$$
$(-1)(-1)(-1) = -1 \qquad (-1)\ (36)(-1)$
$$36$$

So, the value of $-a^2b^3$ is 36 if $a = 6$ and $b = -1$.

practice ▷ Evaluate.

1. y^2 if $y = -3$
2. $-3x^3$ if $x = -2$
3. $-mn^3$ if $m = 4$ and $n = -2$

Example 4

Evaluate $(3x)^2$ if $x = -4$.
$(3x)^2$

Substitute -4 for x. $[(3)(-4)]^2$
$(3)(-4) = -12$ $(-12)^2$
$(-12)^2 = (-12)(-12)$ $(-12)(-12)$
$(-12)(-12) = 144$ 144

So, the value of $(3x)^2$ is 144 if $x = -4$.

practice ▷ Evaluate.

4. $(4y)^2$ if $y = -2$
5. $(-2x)^3$ if $x = 2$
6. $(-6a)^4$ if $a = -1$

◇ **EXERCISES** ◇

Evaluate.

1. x^2 if $x = 7$
2. y^3 if $y = -1$
3. z^3 if $z = -4$
4. a^5 if $a = 2$
5. b^4 if $b = -3$
6. c^5 if $c = -1$
7. $2x^2$ if $x = 5$
8. $-3n^2$ if $n = -2$
9. $5a^3$ if $a = -1$
10. $-4c^4$ if $c = 1$
11. $-2x^4$ if $x = -3$
12. $3y^5$ if $y = -1$
13. $-n^2$ if $n = 7$
14. $-z^2$ if $z = -8$
15. $-x^3$ if $x = 3$
16. $-a^3$ if $a = -4$
17. $-b^4$ if $b = 2$
18. $-c^5$ if $c = -1$
19. $2xy^2$ if $x = 3, y = 2$
20. $3a^2b$ if $a = -1, b = 8$
21. $-4m^2n^2$ if $m = 2, n = 1$
22. $-3ab^4$ if $a = -7, b = 1$
23. $-xy^3$ if $x = 5, y = 2$
24. $-r^2s^3$ if $r = 9, s = -1$
25. $(6x)^2$ if $x = -1$
26. $(3a)^2$ if $a = 2$
27. $(4n)^2$ if $n = -5$
28. $(-2c)^2$ if $c = 5$
29. $(-4n)^2$ if $n = -2$
30. $(3x)^3$ if $x = -2$
31. $(-y)^3$ if $y = -7$
32. $(-4d)^3$ if $d = 1$
33. $(-5z)^3$ if $z = -1$
★ 34. x^3y^2z if $x = 2, y = 3, z = 1$
★ 35. ab^3c if $a = 3, b = -2, c = -1$
★ 36. d^2ef^3 if $d = 1, e = 4, f = -2$
★ 37. $-xy^3z$ if $x = 3, y = 2, z = -1$
★ 38. $-r^2st^2$ if $r = 5, s = -1, t = 2$
★ 39. $-a^3b^2c$ if $a = 1, b = 3, c = -4$
★ 40. $(xy)^2$ if $x = 4, y = 2$
★ 41. $(ab)^3$ if $a = 2, b = -1$
★ 42. $(x^2y)^3$ if $x = 3, y = -1$
★ 43. $(rs)^3$ $r = 4, s = -2$
★ 44. $(p^2q^2)^2$ if $p = -1, q = 3$
★ 45. $(-x^2y^3)^2$ if $x = 2, y = -1$
★ 46. $-(pq)^2$ if $p = -3, q = 1$
★ 47. $-(3r^2s)^2$ if $r = -1, s = 2$

Properties of Exponents

 OBJECTIVE  RECALL

To simplify expressions like $(-2x^3)(5x^2)$, $(x^4)^3$, and $(-3a)^4$

$x^3 = x \cdot x \cdot x \leftarrow$ 3 factors
x^3 is the third power of x

$x^1 = x$

In this lesson, you will learn some shortcuts for multiplying expressions with exponents.

Example 1   Show that $y^2 \cdot y^3 = y^5$.

5 factors, y
$$\frac{y^2 \cdot y^3}{y \cdot y \cdot y \cdot y \cdot y} \Big| \frac{y^5}{y^5}$$
$$y^5 \qquad \text{So, } y^2 \cdot y^3 = y^5.$$

In Example 1, notice that the factors y^2 and y^3 have y as a base; the product y^5 also has y as a base; the exponent of the product is the sum of the exponents of the factors. This suggests the following property.

When multiplying with the same base, add the exponents.

product of powers
$x^m \cdot x^n = x^{m+n}$

Example 2   Simplify. $a^3 \cdot a^4$ | Simplify. $x^3 \cdot x$
$a^3 \cdot a^4$ | $x^3 \cdot x$
a^{3+4} | $x^3 \cdot x^1$
a^7 | x^{3+1}
So, $a^3 \cdot a^4 = a^7$. | x^4
| So, $x^3 \cdot x = x^4$.

Example 3   Simplify. $(7c^4)(-3c^2)$

Group numbers together. $(7c^4)(-3c^2)$
Add the exponents. $7 \cdot (-3) \cdot c^4 \cdot c^2$
$-21c^{4+2}$
$-21c^6$ So, $(7c^4)(-3c^2) = -21c^6$.

practice ▷ Simplify.

1. $b^2 \cdot b^2$
2. $(-2y^2)(6y^5)$
3. $(-5r)(-7r^8)$

PROPERTIES OF EXPONENTS

Example 4

Show that $(x^2)^3 = x^{2\cdot 3}$.

	$(x^2)^3$	$x^{2\cdot 3}$
3 factors, x^2	$x^2 \cdot x^2 \cdot x^2$	x^6
	x^{2+2+2}	
	x^6	So, $(x^2)^3 = x^{2\cdot 3}$.

Example 4 suggests that a power is raised to a power by multiplying the exponents.

Multiply the exponents.

power of a power
$(x^m)^n = x^{m\cdot n}$

Example 5

Simplify $(a^4)^2$. | Simplify $(y^3)^3$.

Multiply the exponents.

$(a^4)^2$ | $(y^3)^3$
$a^{4\cdot 2}$ | $y^{3\cdot 3}$
a^8 | y^9
So, $(a^4)^2 = a^8$. | So, $(y^3)^3 = y^9$.

practice ▷ Simplify.

4. $(z^5)^2$ 5. $(x^4)^3$ 6. $(c^2)^4$

Example 6

Show that $(x \cdot y)^3 = x^3 \cdot y^3$.

	$(x \cdot y)^3$	$x^3 \cdot y^3$
3 factors, $(x \cdot y)$	$(x \cdot y)(x \cdot y)(x \cdot y)$	$x^3 \cdot y^3$
Group x's and y's together.	$x \cdot x \cdot x \cdot y \cdot y \cdot y$	
	$x^3 \cdot y^3$	
$x^3 \cdot y^3$ can be written x^3y^3.	So, $(x \cdot y)^3 = x^3 \cdot y^3$.	

Example 6 suggests that a product can be raised to a power by distributing the exponent.

Distribute the exponent.

power of a product
$(x \cdot y)^n = x^n \cdot y^n$

Example 7

Simplify $(2a)^3$. | Simplify $(-3c)^2$.

$(2a)^3 = (2^1 \cdot a^1)^3 = 2^{1\cdot 3} \cdot a^{1\cdot 3}$

$(2a)^3$ | $(-3c)^2$
$2^3 \cdot a^3$ | $(-3)^2 \cdot c^2$
$8a^3$ | $9c^2$
So, $(2a)^3 = 8a^3$. | So, $(-3c)^2 = 9c^2$.

practice ▷ Simplify.

7. $(ab)^4$ 8. $(4y)^2$ 9. $(-2s)^3$

◇ ORAL EXERCISES ◇

Should you add, multiply, or distribute exponents?

1. $a^3 \cdot a^5$ 2. $(x^2)^5$ 3. $(3a)^2$ 4. $(x^3)^2$ 5. $c^4 \cdot c$
6. $(x \cdot y)^3$ 7. $(4x)^3$ 8. $y \cdot y^5$ 9. $(-3y)^3$ 10. $(z^5)^3$
11. $(a \cdot b)^5$ 12. $x^4 \cdot x^3$ 13. $(-2c)^6$ 14. $(b^2)^4$ 15. $x \cdot x^3$
16. $(-4y)^4$ 17. $(3c)^4$ 18. $c^2 \cdot c^6$ 19. $(x^5)^2$ 20. $y^6 \cdot y$

◇ EXERCISES ◇

Simplify.

1. $x^2 \cdot x^2$ 2. $y^3 \cdot y$ 3. $z^4 \cdot z^3$ 4. $c^5 \cdot c$
5. $r^4 \cdot r^5$ 6. $b \cdot b^6$ 7. $x^7 \cdot x$ 8. $y \cdot y^4$
9. $(x^2)(2x^3)$ 10. $(3y^2)(2y^2)$ 11. $(-4c^3)(c^4)$ 12. $(7x^2)(x^5)$
13. $(5y)(3y^2)$ 14. $(-4x)(2x^3)$ 15. $(5d^2)(\ 2d)$ 16. $(3r^4)(-5r)$
17. $(-7x^2)(-2x^4)$ 18. $(-3y)(-7y^4)$ 19. $(-4a)(-3a^5)$ 20. $(2y^5)(-7y)$
21. $(r^5)^2$ 22. $(x^3)^2$ 23. $(y^4)^3$ 24. $(x^4)^2$
25. $(z^4)^3$ 26. $(m^6)^2$ 27. $(c^2)^5$ 28. $(r^3)^4$
29. $(x^2)^2$ 30. $(y^3)^3$ 31. $(a^2)^4$ 32. $(c^4)^4$
33. $(xy)^3$ 34. $(2a)^2$ 35. $(ab)^4$ 36. $(-3x)^3$
37. $(4y)^2$ 38. $(-2z)^3$ 39. $(-3y)^2$ 40. $(rs)^5$
41. $(-2x)^4$ 42. $(cd)^2$ 43. $(5p)^2$ 44. $(-2y)^5$
★ 45. $(5x^3)^2$ ★ 46. $(-3c^2)^2$ ★ 47. $(2x^3)^4$ ★ 48. $(-2r^3)^3$

Calculator

A calculator can be used to evaluate $3xy^2$ if x is 9 and y is 7.

    3   x y^2

Press 3 ⊗ 9 ⊗ 7 ⊗ 7 ⊜
Display 1323
So, the value of $3xy^2$ is 1,323 when x is replaced with 9 and y with 7.

Evaluate.

1. z^5 if z is 7 2. $4x^3$ if x is 8
3. $5a^2b$ if a is 4, b is 11 4. $7c^3d^4$ if c is 5, d is 3
5. $(6y)^4$ if y is 5 6. $(rs)^3$ if r is 6, s is 7

PROPERTIES OF EXPONENTS

Polynomials

◇ OBJECTIVES ◇
To classify polynomials
To simplify polynomials by combining like terms

◇ RECALL ◇
Simplify. $4x + 2x$
            ↑   ↑
          like terms

So, $4x + 2x = 6x$.

Thus far, you have been working with polynomials of one term. These are called *monomials*. The prefix *mono* means *one*.
Following are examples of polynomials.

$3x^2 + 2x - 1$ $7y + 4$ $4a^2$

3 terms   2 terms   1 term
Trinomial   Binomial   Monomial

Tri means *three* and *bi* means *two*.

Example 1   Classify each polynomial.

$4x^2 - 3$ $9y$ $6a^2 - 5a + 1$

2 terms   1 term   3 terms
Binomial   Monomial   Trinomial

---

practice ▷ Classify each polynomial.

1. $2y^2 - 3y + 1$ 2. $5x^2$ 3. $3c - 2$

Example 2   Simplify. $6x^2 - 4x + 3 - 4x^2 - 3x - 5$

$6x^2 - 4x + 3 - 4x^2 - 3x - 5$
Group like terms together. $(6x^2 - 4x^2) + (-4x - 3x) + (3 - 5)$
Combine like terms. $2x^2 - 7x - 2$

So, $6x^2 - 4x + 3 - 4x^2 - 3x - 5 = 2x^2 - 7x - 2$.

Example 3   Simplify and write the result in descending order of exponents. $3x + 6x^2 - 4 - 8x^2 + 7x + 3$

Descending means in order from highest to lowest.
Group like terms together, with highest exponents first.
Combine like terms.

$3x + 6x^2 - 4 - 8x^2 + 7x + 3$
$(6x^2 - 8x^2) + (3x + 7x) + (-4 + 3)$
$-2x^2 + 10x - 1$

$3x + 6x^2 - 4 - 8x^2 + 7x + 3 = -2x^2 + 10x - 1$.

practice ▷ **Simplify.**

4. $7 + 6y^2 - 3y + 8y - 9 - 2y^2$

Example 4

Simplify. $x^2 + 3x^3 - 4 + 6x - 2x^2 - 3x$

$\ 1x^2 + 3x^3 - 4 + 6x - 2x^2 - 3x$

$x^2 = 1x^2$ $\quad 3x^3 + (1x^2 - 2x^2) + (6x - 3x) - 4$
Group like terms in descending order.
Combine like terms. $\quad 3x^3 - 1x^2 + 3x - 4$

$-1x^2 = -x^2$ $\quad$ So, $x^2 + 3x^3 - 4 + 6x - 2x^2 - 3x =$
$\quad 3x^3 - x^2 + 3x - 4.$

---

practice ▷ **Simplify.**

5. $5a^2 - 3a^4 + 6 - 2a + a^2 - 7$

◇ **ORAL EXERCISES** ◇

Arrange each polynomial in descending order of exponents.

1. $6x + 3x^2 - 1$
2. $1 - 4y$
3. $7 + 2a - 3a^2$
4. $2 - x^2 + 3x$
5. $6x^9 - 4 + 2x$
6. $3 + 5z^2$
7. $3x^2 - 2x + 1 + 3x^3$
8. $6y - 5y^3 + 4 - 3y^2$
9. $7a - 3 - a^3 + 4a^2$

◇ **EXERCISES** ◇

Classify each polynomial.

1. $4y^2 - 1$
2. $3x^2 - 2x + 1$
3. $3x$
4. $-7 + a$
5. $-2x^3 + 3x - 2$
6. 6
7. $-5x^2$
8. $4y^2 + 3y - 1$
9. $3c^2 + 4c$
10. $a^2 + a + 2$
11. $6x - 1$
12. -2

Simplify.

13. $5x^2 - 3x + 2 - 2x^2 - 4x + 1$
14. $6y^2 + 2y - 1 + y^2 - 3y + 5$
15. $4a^2 - 2a - 1 - 3a^2 + 5a - 4$
16. $n^2 + 5n + 8 + 4n^2 - 2n - 9$
17. $-y^2 + 3y + 5 - 2y^2 - 3y + 2$
18. $3x^2 + 4x - 1 - 3x^2 - 2x + 8$
19. $5y + 4y^2 - 3 - 8y^2 + 6y + 1$
20. $3 + 2a - 5a^2 + a^2 - 4 - 6a$
21. $7x^2 - 3 + x - 4x^2 - 6x + 2$
22. $3c + 2c^2 - 4c + 1 + c^2 - 5$
23. $9 + 8z^2 - 6z - 5 + 4z - 3z^2$
24. $7x^2 - 8 + 4x - 1 - 5x^2 - 4x$
25. $2y - 2y^2 - 2 + 5 - 3y - 4y^2$
26. $-6n^2 + 2 - 4n + 3 + 6n^2 - 5n$
27. $y^2 + 4y^3 - 3 + 7y - 3y^2 - 2y$
28. $6r^2 - 2r^4 + 5 - 3r + r^2 - 8$
29. $3x - 2x^2 + 5x^3 + 2 - x - 4x^2$
30. $3x^4 - 2x^2 - 5x + 2x^2 - 5 + 9$
31. $7c^3 - 5 + 6c^2 - 6c + 4c - 6c^2$
32. $5a - 2a^4 + 4a^2 - 7 + 4a^2 - 2$
33. $6z - 8z^3 + 3 - 2z^2 + z^3 - 3$
34. $6a^3 - 5a^4 + 2a - 1 + 5a^4 - a^3$
35. $x^2 + x^4 - 3x^3 - 5x + 2 + 3x^4 - 5x^3 + 7x - 5 + x^2$
36. $5 - 8a^2 + 3a^4 - 2a^3 + 7 - 4a - 3a^4 + 6a - a^2 + 5a^3$

POLYNOMIALS

Simplifying Polynomials

◇ OBJECTIVES ◇
To multiply a monomial by a polynomial
To simplify polynomials

◇ RECALL ◇

$2(5x - 3)$

$2(5x) + 2(-3)$
$10x - 6$

$-(6y - 1)$
$-1(6y - 1)$

$(-1)(6y) + (-1)(-1)$
$-6y + 1$

The recall shows that the distributive property can be used to multiply a monomial by a polynomial when the monomial is an integer. The distributive property can also be used when the monomial contains a variable.

Example 1

By commutative property of multiplication, $x(4x^2) = (4x^2)x$.

Multiply. $x(4x^2 + x)$
$x(4x^2 + x)$
$x(4x^2) + x(x)$
$4 \cdot x^2 \cdot x + x \cdot x$
$4x^3 + x^2$
So, $x(4x^2 + x) = 4x^3 + x^2$.

Example 2

Distribute $3y^2$.

Multiply. $3y^2(2y^2 - 4y + 1)$
$3y^2(2y^2 - 4y + 1)$
$3y^2(2y^2) + 3y^2(-4y) + 3y^2(1)$
$3 \cdot 2 \cdot y^2 \cdot y^2 + 3 \cdot (-4) \cdot y^2 \cdot y + 3 \cdot 1 \cdot y^2$
$6y^4 - 12y^3 + 3y^2$
So, $3y^2(2y^2 - 4y + 1) = 6y^4 - 12y^3 + 3y^2$.

---

practice ▷ **Multiply.**

1. $a(a^3 - 5a)$

2. $-3c^2(4c^2 + 6c - 2)$

Example 3

$a^2 = 1a^2; -a = -1 \cdot a$
$3(1a^2 - 1a + 2)$
$= 3(1a^2) + 3(-1a) + 3(2)$
$= 3a^2 - 3a + 6$

Simplify. $a^2 + 5a - 4 + 3(a^2 - a + 2)$
$a^2 + 5a - 4 + 3(a^2 - a + 2)$
$1a^2 + 5a - 4 + 3(1a^2 - 1a + 2)$
$1a^2 + 5a - 4 + 3a^2 - 3a + 6$
$(1a^2 + 3a^2) + (5a - 3a) + (-4 + 6)$
$4a^2 \quad + \quad 2a \quad + \quad 2$
So, $a^2 + 5a - 4 + 3(a^2 - a + 2) = 4a^2 + 2a + 2$.

practice ▷ **Simplify.**

  3. $x^2 - 2x + 6 - 2(3x^2 - x + 5)$

Example 4

Simplify. $-(y^2 - 5y + 4)$

$-y = -1 \cdot y$ $-(y^2 - 5y + 4)$
Distribute -1. $-1(y^2 - 5y + 4)$
$-1(y^2) + (-1)(-5y) + (-1)(4)$
$-y^2 + 5y - 4$

So, $-(y^2 - 5y + 4) = -y^2 + 5y - 4$.

Example 5

Simplify. $4x^2 + 6x - 3 - (x^2 - 5x + 2)$

$-1(x^2 - 5x + 2) =$ $4x^2 + 6x - 3 - (x^2 - 5x + 2)$
$(-1)(x^2) + (-1)(-5x) + (-1)(2)$ $4x^2 + 6x - 3 - 1(x^2 - 5x + 2)$
$4x^2 + 6x - 3 - 1x^2 + 5x - 2$
Group like terms. $(4x^2 - 1x^2) + (6x + 5x) + (-3 - 2)$
$3x^2 + 11x - 5$

So, $4x^2 + 6x - 3 - (x^2 - 5x + 2) = 3x^2 + 11x - 5$.

---

practice ▷ **Simplify.**

  4. $-(a^2 + 7a - 4)$ 5. $3c^2 + 2c - 1 - (c^2 + 5c + 2)$

◇ **EXERCISES** ◇

Multiply.

1. $x(3x^2 + 2)$
2. $y(4y - 1)$
3. $a(-5a^2 + a)$
4. $2x(x^2 + 1)$
5. $3c(2c - 2)$
6. $-4y(2y^2 - 3y)$
7. $3c(5c^2 + 2c - 3)$
8. $-4x(x^2 + 6x + 2)$
9. $5a(a^2 + 5a - 2)$
10. $2z^2(z^2 - 4z + 2)$
11. $y^2(3y^2 + 4y - 5)$
12. $-3b^2(-b^2 + 2b - 1)$

Simplify.

13. $-(x^2 + 3x)$
14. $-(r^2 - 6r + 2)$
15. $-(3c^2 + 2c + 5)$
16. $-(-2a^2 + 4a - 3)$
17. $-(6x^2 - 5x - 4)$
18. $-(-4y^2 - y + 1)$
19. $x^2 + 2x - 5 + 4(x^2 - 3x - 2)$
20. $y^2 - 3y - 2 + 3(y^2 - 2y + 1)$
21. $2z^2 - 6z - 3 + 2(z^2 - 5z + 4)$
22. $3c^2 + 4c - 1 - 3(c^2 - 6c - 7)$
23. $5x^2 - 3x + 4 - 4(2x^2 + 6x - 1)$
24. $-4y^2 + 6y - 2 - 5(-y^2 + 2y + 3)$
25. $3a^2 + 2a - 1 - (a^2 - 4a + 2)$
26. $c^2 - 5c + 6 - (2c^2 + 5c - 4)$
27. $z^2 - 3z + 4 - (4z^2 - 2z + 1)$
28. $-y^2 + 3y - 1 - (4y^2 + 3y - 2)$
29. $-6x^2 + 2x - 1 - (-x^2 + 4x - 3)$
30. $-z^2 - 8z - 1 - (-3z^2 + 2z - 2)$
★ 31. $2x^2(3x^2 - 6x + 4) - x^2(x^2 + 4x - 2)$ ★ 32. $7y^2(3y^2 - y + 1) - 2y^2(4y^2 + 3y - 1)$
★ 33. $5x(2x^3 - 3x^2 + 1) - 4x^2(2x + 5)$ ★ 34. $6a^2(5a^2 + 7a - 1) - a^2(a^2 + 4a - 2)$

Common Factors

 **OBJECTIVE**

To factor a common monomial from a polynomial

 **RECALL**

$3(x^2 + 1)$ ← distributive property
$3(x^2) + 3(1)$
$3x^2 + 3$

We can rewrite $2x + 6$ as $2(x + 3)$ by using the distributive property in reverse.
$2x + 6 = 2(x + 3)$
     ↑ 2 is a common monomial factor.

Example 1

Factor a common monomial factor from $3y + 9$.

$3y + 9$
$3(y) + 3(3)$
$3(y + 3)$

3 is a common factor of $3y$ and 9.
Distributive property in reverse.

So, $3y + 9 = 3(y + 3)$.

---

practice ▷ Factor a common monomial factor from each.

1. $5c - 10$
2. $21 + 7z$
3. $12 - 6a$

Example 2

Factor the greatest common factor (GCF) from $8c^2 - 12c + 4$.

The GCF is the greatest number that is a common factor of all three terms.

$8c^2 - 12c + 4$
$4(2c^2) + 4(-3c) + 4(1)$
$4(2c^2 - 3c + 1)$

4 is the GCF.
Factor out 4.

So, $8c^2 - 12c + 4 = 4(2c^2 - 3c + 1)$.

---

practice ▷ Factor the GCF from each.

4. $6x^2 - 18$
5. $2y^2 - 6y + 10$
6. $4n^2 + 6n - 12$

Example 3

Factor the GCF from $x^2 + 5x$.

x is the GCF of x^2 and $5x$.
Factor out x.

$$x^2 + 5x$$
$$\underline{x} \cdot x + \underline{x} \cdot 5$$
$$x(x + 5)$$

So, $x^2 + 5x = x(x + 5)$.

practice ▷ Factor the GCF from each.

7. $y^2 + 4y$ 8. $c^3 - 2c^2$ 9. $n^4 + 3n^2$

Example 4

Factor $4a^2 + 6a$.

To factor $4a^2 + 6a$ means to factor the GCF from $4a^2 + 6a$.

First, look for the greatest common whole number factor.

2 is the greatest common whole number factor.
Factor out 2.

$$4a^2 + 6a$$
$$\underline{2}(2a^2) + \underline{2}(3a)$$
$$2(2a^2 + 3a)$$

Now look for the greatest common variable factor.

a is the greatest common variable factor.
Factor out a.

$$2(2a^2 + 3a)$$
$$2(\underline{a} \cdot 2a + \underline{a} \cdot 3)$$
$$2 \cdot a(2a + 3)$$

So, $4a^2 + 6a = 2a(2a + 3)$.

Example 5

Factor $6x^3 - 8x^2 + 2x$.

2 is the greatest common whole number factor.

x is the greatest common variable factor.

$$6x^3 - 8x^2 + 2x$$
$$\underline{2}(3x^3) + \underline{2}(-4x^2) + \underline{2}(x)$$
$$2(3x^3 - 4x^2 + x)$$
$$2(\underline{x} \cdot 3x^2 + \underline{x} \cdot -4x + \underline{x} \cdot 1)$$
$$2 \cdot x(3x^2 - 4x + 1)$$

So, $6x^3 - 8x^2 + 2x = 2x(3x^2 - 4x + 1)$.

practice ▷ Factor.

10. $3y^2 - 12y$ 11. $10y^3 + 15y^2 - 5y$ 12. $3y^4 - 6y^2 + 9y$

◇ ORAL EXERCISES ◇

What is the GCF of each?

1. $2x + 4$
2. $3y - 9$
3. $4a + 16$
4. $10c^2 - 5$
5. $3z^2 + 9z - 6$
6. $4x^2 - 2x + 10$
7. $5y^2 + 15y - 20$
8. $8a^2 - 4a + 12$
9. $x^2 + 4x$
10. $y^2 - 7y$
11. $2a^3 + 6a^2 + a$
12. $5c^3 - 3c^2 + 2c$
13. $3y^2 + 6y$
14. $8b^2 - 4b$
15. $2x^3 - 4x^2 + 6x$
16. $10a^3 - 15a^2 + 5a$

◇ EXERCISES ◇

Factor.

1. $4n + 8$
2. $6x - 9$
3. $3c + 12$
4. $8y - 2$
5. $2d - 10$
6. $15y + 5$
7. $7x^2 - 14$
8. $16c^2 + 8$
9. $2x^2 + 4x - 8$
10. $3y^2 - 9y + 6$
11. $5x^2 - 10x + 20$
12. $4a^2 - 12a - 20$
13. $6z^2 + 12z + 3$
14. $8x^2 - 6x + 2$
15. $10n^2 + 5n - 15$
16. $12y^2 - 8y - 4$
17. $y^2 + 2y$
18. $x^2 - 3x$
19. $c^3 + 6c^2$
20. $n^3 - 5n$
21. $3z^3 - 2z^2 + 4z$
22. $7y^3 + 2y^2 - 3y$
23. $6n^3 - 2n^2 + n$
24. $5x^3 - 4x^2 - x$
25. $2x^2 + 4x$
26. $3y^2 + 6y$
27. $8a^2 - 2a$
28. $10d^2 - 20d$
29. $14n^2 + 7n$
30. $8y^3 + 4y^2$
31. $6n^3 - 12n^2$
32. $15b^3 - 5b^2$
33. $4x^3 + 2x^2 - 6x$
34. $3c^3 - 6c^2 + 12c$
35. $5r^3 + 15r^2 - 10r$
36. $4y^3 - 8y^2 - 12y$
37. $10c^3 - 4c^2 + 2c$
38. $20n^3 - 30n^2 + 10n$
39. $6a^3 + 18a^2 - 6a$
40. $4x^4 - 8x^3 + 6x^2 - 2x$
★ 41. $3y^4 - 9y^3 + 6y^2 + 3y$
42. $10c^4 - 20c^3 + 30c^2 - 5c$
★ 43. $12n^4 - 6n^3 - 18n^2 + 6n$

Algebra Maintenance

Solve.

1. $x - 3 = -10$
2. $-12 + x = -17$
3. $x + 9 = -4$
4. $-8x = 40$
5. $\dfrac{u}{-6} = -5$
6. $-4k + 6 = -14$

Solve for x.

7. $x + u = t$
8. $t - x = r$
9. $e + x = k$

Solve.

10. A number added to twice the number is -36. What is the number?
11. A number diminished by 3 times the number is -8. What is the number?

Mathematics Aptitude Test

This test will prepare you for taking standardized tests.
Choose the best answer for each question.

1. The average of 4 numbers is 6. If two of the numbers are 2 and 6, which of the following cannot be one of the other two numbers?
   a. 8   **b.** 10   **c.** 16   **d.** 18

2. What is the value of $1^n + 1^{2n} + 1^{3n}$ for any whole number n?
   a. 1   **b.** 1^{6n} **c.** 3^{6n} **d.** 3

3. A student has grades of 90, 78, and 82. What grade must she get on her fourth test to get an average of 85?
   a. 85   **b.** 90   **c.** 95   **d.** 100

4. Which of the following is a list of numbers in decreasing order?
   a. 8, 4/5, −0.2, −2
   b. −2, −0.2, 4/5, 8
   c. 8, 4/5, −2, −0.2
   d. −0.2, −2, 4/5, 8

5. If $n = 3^2$, then $n^2 =$ _____?
   a. 12   **b.** 27   **c.** 36   **d.** 81

6. If the sum of two numbers is divisible by 5, then which of the following is true?
   a. The difference of the two is divisible by 5.
   b. At least one of the numbers is divisible by 5.
   c. None of the two numbers is divisible by 5.
   d. none of the above

7. A micromillimeter is defined as one millionth of a millimeter. What is the length of 12 micromillimeters?
   a. 12,000,000 mm   **b.** 0.00000012 mm   **c.** 0.0000012 mm   **d.** 0.000012 mm

8. The original price of an item is $6.80. It is reduced by 20%. What is the new price?
   a. $1.36   **b.** $8.16   **c.** $5.44   **d.** $4.80

9. $m \overline{)n}$ with quotient q and remainder r over s indicates division of n by m. Which of the following is true?
   a. $n = m \cdot q + s$ **b.** $m \cdot q \cdot n = s$ **c.** $m \cdot s + q = n$ **d.** $q = m \cdot n - s$

COMPUTER ACTIVITIES

You have learned to solve equations of the type $-2x + 10 = 4x - 8$. You can write a program to solve any equation of this type. First, it is necessary to write a general equation.

Specific equation: $-2x + 10 = 4x + (-8)$
General equation: $Ax + B = Cx + D$

Second, you must write a formula for solving the general equation $Ax + B = Cx + D$ for x. The solution is shown below.

$$Ax + B = Cx + D$$

Subtract Cx from each side.
$$Ax - Cx + B = Cx - Cx + D$$
$$Ax - Cx + B = D$$

Subtract B from each side.
$$Ax - Cx + B - B = D - B$$
$$Ax - Cx = D - B$$

Factor out common factor x.
$$(A - C)x = D - B$$

Divide each side by $(A - C)$.
$$x = \frac{D - B}{A - C}$$

Example Write a program to solve any equation of the form $AX + B = CX + D$.

```
10  INPUT "TYPE IN A VALUE FOR A ";A
20  INPUT "TYPE IN A VALUE FOR B ";B
30  INPUT "TYPE IN A VALUE FOR C ";C
40  INPUT "TYPE IN A VALUE FOR D ";D
50  LET X = (D - B) / (A - C)
60  PRINT "THE SOLUTION IS X = "X
70  END
```

See the Computer Section beginning on page 420 for more information.

Exercises

1. **RUN** the program above to solve $-2x + 10 = 4x - 8$. (*Hint:* $A = -2$, $B = 10$, $C = 4$, $D = -8$)
2. Modify the program above with a loop to allow you to solve any 3 equations without having to retype.

Use the program from Exercise 2 to solve each of these equations.

3. $3x - 2 = 7x + 10$
4. $-5x + 7 = 10x + 8$
5. $6x - 4 = 14$
(*Hint:* What is the value of D?)

Chapter Review

Solve. [335]

1. $4x + 9 = 2x + 3$
2. $8x - 5 = 2x - 17$
3. $7x + 2 = 2x - 23$

Solve. [337]

4. $2(x - 3) = 18$
5. $-4(z + 7) = 24$
6. $-(2 + 2y) = 4(y - 2)$

Write in mathematical terms. [339]

7. 5 increased by 3
8. 8 less than x
9. 4 more than 7 times a number

Solve. [339]

10. Three more than twice a number is equal to 36. Find the number.

Perform the operation on each side of the inequality. Then write a true inequality. [344]

11. $5 < 6$ Subtract -3.
12. $-8 > -12$ Divide by -4.

Solve. [347]

13. $5n \leq -25$
14. $-\frac{1}{7}t > -5$

Evaluate. [351]

15. $4xy^2$ if $x = 8$ and $y = -1$
16. $(-3b)^2$ if $b = -4$

Simplify. [353]

17. $y^4 \cdot y$
18. $(3c^2)(-8c^5)$
19. $(-3z)^3$

Classify each polynomial. [356]

20. $3 - 2x$
21. $-6y$
22. $3x^2 - 5 + 9x$

Simplify. [356]

23. $2c - 3c^2 + 6 - 8 - 3c + c^2$
24. $3d - 4d^4 + 5d^2 - 6 + 3d^2 + 5$

Multiply. [358]

25. $m(-3m^2 + m)$
26. $-2y(y^2 + 3y - 1)$

Simplify. [358]

27. $-(a^2 + 3a)$
28. $-5c^2 + 3c - 2 - (4c^2 + 2c - 1)$

Factor. [360]

29. $3y^2 - 6y + 24$
30. $5a^3 + 10a^2 - 25a$

Chapter Test

Solve.

1. $3x + 5 = 2x + 1$
2. $9x - 8 = 4x - 23$
3. $3x + 7 = 5x - 7$
4. $3(z - 8) = 9$
5. $3(r - 1) = -(6 - 2r)$

Write in mathematical terms.

6. 4 increased by x
7. 6 less than 5 times a number

Solve.

8. Seven less than 4 times a number is equal to -3. Find the number.

Perform the operation on each side of the inequality. Then write a true inequality.

9. $-4 < 3$ Add 6.
10. $6 \geq -6$ Divide by -3.

Solve.

11. $x - 6 > -10$
12. $-9x \geq 18$

Evaluate.

13. x^5 if $x = -2$
14. $-4y^2$ if $y = -3$
15. $3a^2b$ if $a = -1, b = 9$
16. $(-5n)^2$ if $n = -2$

Simplify.

17. $x^5 \cdot x$
18. $(4d^3)(-7d^4)$
19. $(m^4)^2$
20. $(-2y)^3$

Classify each polynomial.

21. $-5d^3$
22. $4x^2 - 2x + 1$

Simplify.

23. $2x^2 + 3x - 1 - 2x^2 - 2x + 4$
24. $6r^3 - 3 + 4r^2 - 5r + 3r - 2r^2$

Multiply.

25. $z(3z^2 - 1)$
26. $-2x(-x^2 + 5x - 4)$

Simplify.

27. $-(3x^2 - 4x - 1)$
28. $d^2 + 3d - 2 - (5d^2 - 6d + 1)$

Factor.

29. $9x - 18$
30. $t^3 + 6t^2$
31. $4b^3 - 8b^2 + 20b$

Some Ideas from Geometry

15

George Washington's head on Mt. Rushmore is as tall as a five-story building.

Points, Lines, and Planes

◇ OBJECTIVES ◇

To show that two points determine a unique line
To show that three noncollinear points determine a plane

◇ RECALL ◇

 point A

 line containing points X and Y

In mathematics some key words are not defined. They are used to define other words. Three key geometric words that are not defined are *point*, *line*, and *plane*.
Think of a point as being a location in space.

Example 1

Draw and name a point, a line, and a line segment.

Use a dot on paper to represent a point.

A
• point A

Choose any two points. Name them. Draw a line through the points.

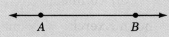 Line AB, or $\overleftrightarrow{AB}$

A line continues without end in both directions. The arrowheads are used to suggest that the line has no end.

A part of a line made up of two points and all the points between them is called a *segment*.

Choose any two points. Name them. Draw a line segment between the points.

•————————•
A B

Line segment AB, or $\overline{AB}$

Points A and B are called *endpoints*. A line segment has a definite length.

Points that are on the same line are called *collinear points*.

Example 2

Name the collinear points.

K, M, and T are on the same line.

So, K, M, and T are collinear points.

practice ▷ **Draw a picture of**

1. a line   2. a point   3. a segment

Name the collinear points.

Three points can be noncollinear.
Noncollinear means *not on the same line*. Three noncollinear points determine a *unique plane*.

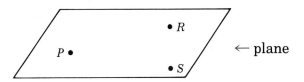

◇ EXERCISES ◇

1. Draw a picture of a line. Mark on it four points. Name them A, B, C, and D. Why are the four points collinear?
2. Choose a point *not* on the line in Exercise 1. Name it E. How many planes are determined by points A, B, and E?
3. How many points are there in space?
★ 4. Can four points in space be coplanar (in one plane)? Can four points be noncoplanar?

Algebra Maintenance

Write in mathematical terms.

1. -8 increased by t
2. 7 less than 8 times a number

Solve.

3. $5x + 3 = 2x + 15$
4. $4(y - 3) = 12$
5. $4(m - 2) = -(5 - 2m)$
6. $x - 4 > -6$
7. $3x \leq 12$
8. $-5x \geq 15$

Evaluate.

9. x^4 if $x = -2$
10. $-5z^2$ if $z = -4$
11. $4c^2g^2$ if $c = -2$, $g = 3$
12. $(-2m)^2$ if $m = -3$

POINTS, LINES, AND PLANES

Angles

◇ OBJECTIVES ◇

To classify angles as acute, right, or obtuse
To define complementary and supplementary angles
To solve problems involving complementary and supplementary angles

◇ RECALL ◇

$\overline{AB}$ has two endpoints and has a definite length.

A *ray* has one endpoint and continues indefinitely in one direction. A ray is part of a line.

   ray AB, or $\overrightarrow{AB}$

An *angle* is made up of two rays that have the same endpoint.

vertex ⟶ A   angle BAC, or ∠BAC

Example 1

Use a protractor to measure angle COB.

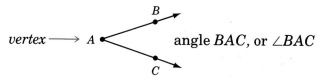

Place the center of the protractor at the vertex O. Line up the bottom with $\overrightarrow{OB}$. Read the number of degrees where $\overrightarrow{OC}$ falls on your protractor.

m∠COB = 55°   So angle COB measures 55 degrees, or 55°.

Angles are classified according to their measures.

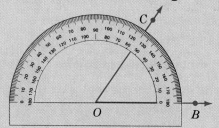

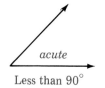
acute
Less than 90°

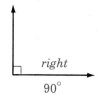
right
90°

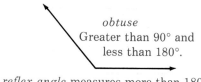
obtuse
Greater than 90° and less than 180°.

A *reflex angle* measures more than 180°.

practice ▷ Measure each angle and classify it as acute, right, or obtuse.

1.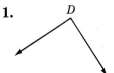

2.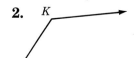

3.

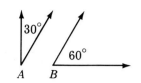

∠A is the complement of ∠B.

complementary angles
Two angles are said to be *complementary* if the sum of their measures is 90°.

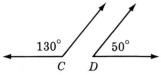

∠C is the supplement of ∠D.

supplementary angles
Two angles are said to be *supplementary* if the sum of their measures is 180°.

Example 2

Tell whether the angles are complementary or supplementary.

$120° + 60° = 180°$ 120°, 60° | 50°, 40° $50° + 40° = 90°$

supplementary | complementary

practice ▷ Complementary or supplementary?

4. 25°, 65°   5. 98°, 82°   6. 1°, 179°   7. 5°, 85°

Example 3

One of two complementary angles is twice the other. What are the measures of the two angles?
Let x = measure of one angle
$2x$ = measure of the other angle

Write an equation.
The two angles are complementary.
Solve.

$x + 2x = 90$
$x = 30$
So, one angle measures 30° and the other 60°.

◇ EXERCISES ◇

Measure each angle and classify it as acute, right, or obtuse.

1.   2. 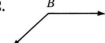   3.   4.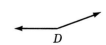

Complementary or supplementary?

5. 17°, 163°   6. 45°, 45°   7. 90°, 90°   8. 101°, 79°

9. One of two supplementary angles is 5 times the size of the other. What are the measures of the two angles?

Problem Solving – Applications
Jobs for Teenagers

1. Jim works as a stockperson at the Josco Supermarket. He works 7 hours a day Monday through Friday and 5 hours on Saturday. He is paid $4.50 per hour. How much does he earn in a week?

2. In the stockroom at Josco's, Jim counted 10 crates of avocados. There are 48 avocados in each crate, and avocados sell for 49¢ apiece. What is the value of the 10 crates of avocados?

3. Josco's usually sells 3 cases of canned peaches during the week and 2 cases over the weekend. How many cases should Jim recommend that Josco's buy to last for 4 weeks?

4. During one week, Josco's sold $68.40 worth of potatoes. They made 20% profit on potatoes. How much profit does Jim figure that Josco's made on potatoes that week?

5. There is an 8¢ profit on each can of tomatoes. What is the profit on 10 cases of tomatoes if there are 12 cans in a case?

6. Maria has worked at the Josco Supermarket for a year. She works 40 hours a week and is paid $4.50 per hour. When her pay is raised to $4.75 per hour, how much more will she earn each week?

7. Maria unpacked 8 cases of peaches. There are 24 cans in a case. How many cans did she unpack?

8. The store makes a profit of 9¢ on each can of corn sold. There are 24 cans in a case of corn. How much profit does Maria figure the store will make on 8 cases of corn?

9. Josco's paid $5.88 for a crate of 12 melons. At what price should Maria mark each melon so that Josco's will make a profit of $4.80 on a crate of melons?

10. Maria allows 40% of the canned vegetable shelf to be stock by Pixie brand products. The shelf is 15 meters long. How many meters of the shelf will she fill with Pixie products?

Parallel Lines and Angles

◇ OBJECTIVES ◇

To identify pairs of alternate interior angles, corresponding angles, and vertical angles

To determine measures of angles on the basis of congruence of special pairs of angles

is parallel to

◇ RECALL ◇

A line continues without end in both directions.

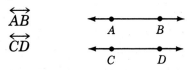

Two lines in the same plane which do not intersect are called *parallel lines*.

When lines intersect, angles are formed.

Vertical angles are opposite angles formed by two intersecting lines.

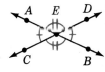

Vertical angles are congruent. They have the same measure. The symbol for congruence is ≅.

∠AEC ≅ ∠DEB and ∠AED ≅ ∠CEB

Example 1

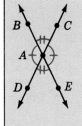

← Note the convenient way of marking congruent angles.

$\overleftrightarrow{BE}$ and $\overleftrightarrow{CD}$ intersect at point A.
Name the vertical angles.

∠BAC and ∠DAE are opposite angles.
∠BAD and ∠CAE are opposite angles.

So, ∠BAC and ∠DAE are vertical angles and ∠BAD and ∠CAE are vertical angles.

Line a ∥ Line b

A *transversal* is a line that intersects two or more lines.

When a transversal intersects parallel lines, pairs of congruent angles are formed.

∠2 ≅ ∠7   ∠4 ≅ ∠5   ∠2 and ∠7, ∠4 and ∠5 are *alternate interior angles*.
∠1 ≅ ∠5   ∠2 ≅ ∠6   ∠3 ≅ ∠7   ∠1 and ∠5, ∠2 and ∠6, ∠3 and ∠7, ∠4 and ∠8
∠4 ≅ ∠8   are *corresponding angles*.

PARALLEL LINES AND ANGLES

practice ▷ In the figure at the right, identify all pairs of vertical angles, alternate interior angles, and corresponding angles.

Line r ∥ Line t

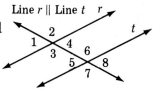

If the sides of an angle lie on a straight line, then that angle is called a *straight angle*. A straight angle measures 180°.

Example 2

Determine the measures of the angles labeled 1 through 7.

The 65° angle and angle 1 form a straight angle.
65° + 115° = 180°
So, m∠1 = 115°.

Line c ∥ Line d

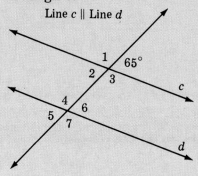

angle 1: 115°
angle 2: 65°
angle 3: 115°
angle 4: 115°
angle 5: 65°
angle 6: 65°
angle 7: 115°

◇ EXERCISES ◇

1. In the figure at the right, name two pairs of congruent
   a. vertical angles
   b. alternate interior angles
   c. corresponding angles.

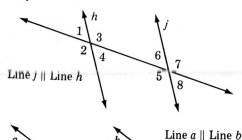

Line j ∥ Line h

2. Determine the measures of angles labeled 1 through 7.

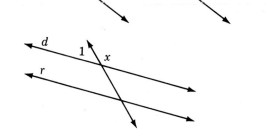

Line a ∥ Line b

★ 3. Line d ∥ line r. The measure of the marked angle is x. Express the measure of angle 1 in terms of x.

374   CHAPTER FIFTEEN

Perpendicular Lines

◇ OBJECTIVE ◇

To construct a line perpendicular to a given line and passing through a point not on the given line

◇ RECALL ◇

A right angle is an angle that has a measure of 90°.

If two lines intersect in such a way that they form right angles, then they are *perpendicular lines*.

$k \perp m$
↑ is perpendicular to

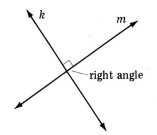
right angle

Example

Construct a line perpendicular to line t that passes through point P.

Step 1 Draw line t and mark the point P.

Step 2 Place the tip of your compass at point P and make two arcs that cut line t at points A and B.

Step 3 Place the tip of the compass at A and make an arc below the line. Do the same with the same opening of the compass but now with the tip of the compass on the point B.

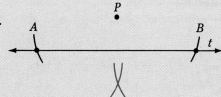

Step 4 Name the point of intersection of the two arcs below line t, R. Draw a line through R and P.

$\overleftrightarrow{RP}$ is perpendicular to line t and passes through point P not on line t.

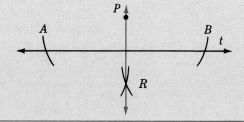

PERPENDICULAR LINES

◇ ORAL EXERCISES ◇

1. What is the measure of a right angle?
2. Define perpendicular lines.

◇ EXERCISES ◇

1. Construct a line perpendicular to line k that passes through point D. First duplicate the picture on your paper.

★ 2. Construct a line perpendicular to line n that passes through point R on the line. First duplicate the picture on your paper.

★ 3. Construct a line that is parallel to line z and passes through point R. First duplicate the picture on your paper. (*Hint:* First construct a line perpendicular to line z that passes through point R; then construct a line through point R that is perpendicular to the line you constructed.)

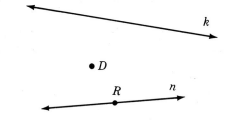
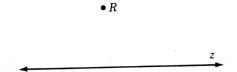

Calculator

Solve each equation. Round the answer to the nearest hundredth.

$$2.6x + 1.9 = 0.3$$
$$2.6x + 1.9 = 8.3$$
$$2.6x + 1.9 - 1.9 = 8.3 - 1.9$$
$$2.6x = 6.4$$
$$x = \frac{6.4}{2.6}$$
$$= 2.462$$

The solution to the nearest hundredth is 2.46.

1. $3.1x + 2.7 = 2.3x + 1.6$
2. $4.8(x - 2.9) = 1.4$
3. $9.3(x - 0.4) = -2.7(3.3 - 5.6x)$
4. $2.5(5.4 - 2.5x) = 1.3(0.5 - 10.7x)$

Triangles

◇ OBJECTIVES ◇

To classify triangles according to measures of their angles
To classify triangles according to measures of their sides
To solve problems involving angle measures in triangles

◇ RECALL ◇

An acute angle measures less than 90°.
A right angle measures 90°.
An obtuse angle measures more than 90° and less than 180°.

Triangles can be classified according to the measures of their angles.

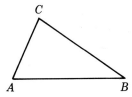

△ ABC is acute, because each of its three angles is acute.

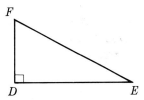

△ DEF is a right triangle, because it has one right angle.

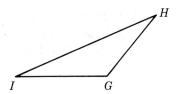

△ GHI is obtuse, because it has one obtuse angle.

practice ▷ Classify these triangles according to the measures of their angles.

1.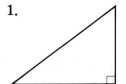
2.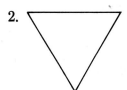
3.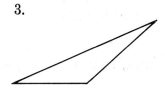

The sum of the measures of the three angles in a triangle is 180°.

Example

Find the measure of the third angle, given the measures of two angles of a triangle. Let x = the measure of the third angle.

$$90°, 12°$$
$$90 + 12 + x = 180$$
$$102 + x = 180$$
$$x = 78°$$

$$130°, 11°$$
$$130 + 11 + x = 180$$
$$141 + x = 180$$
$$x = 39°$$

TRIANGLES

Triangles can be classified according to the relationships among their sides.

No two sides are congruent.

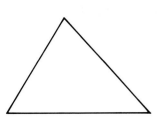

Scalene triangle

Two sides are congruent.

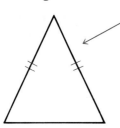

Sides are marked with the same symbol to indicate that they are congruent.

Isosceles triangle

All three sides are congruent.

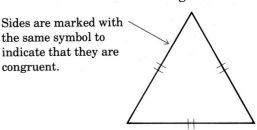

Equilateral triangle

◇ **EXERCISES** ◇

Classify the triangles according to the measures of their angles.

1.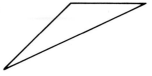
2.
3.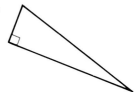

4. 35°, 37°, 108°
5. 40°, 90°, 50°
6. 70°, 70°, 40°

Find the measure of the third angle, given the measures of two angles of a triangle.

7. 47°, 92°
8. 50°, 65°
★ 9. 179°, $\frac{7}{8}$°
★ 10. 178°, $1\frac{1}{2}$°
★ 11. 90°, $89\frac{3}{8}$°
★ 12. 89°, 90.6°

Classify the triangles according to the measures of their sides.

13.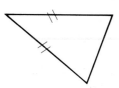
14.
15.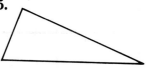

16. 5 in., 3 in., 4 in.
17. 4 in., 4 in., 4 in.
18. 7 in., 7 in., $8\frac{1}{2}$ in.
19. 7 ft, 7ft, 12 ft
★ 20. 3 ft, 30 in., 4 ft
★ 21. 13 in., 15 in., 2 ft

CHAPTER FIFTEEN

Polygons

◇ OBJECTIVES ◇

To classify polygons according to the number of their sides
To classify quadrilaterals
To determine measures of sides and angles in special polygons

◇ RECALL ◇

Segments are parts of lines; they have definite lengths.

A *polygon* is made up of three or more segments that intersect only at their endpoints, and each endpoint is shared by exactly two segments.

Polygons are named according to the number of sides they have.

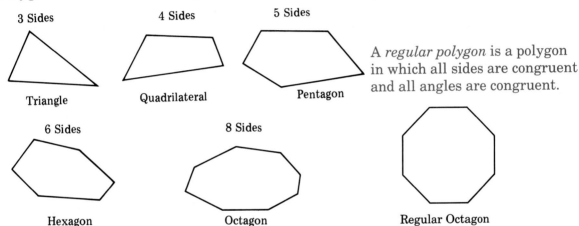

3 Sides — Triangle
4 Sides — Quadrilateral
5 Sides — Pentagon
6 Sides — Hexagon
8 Sides — Octagon
Regular Octagon

A *regular polygon* is a polygon in which all sides are congruent and all angles are congruent.

Example 1

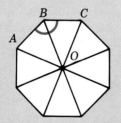

Find the measure of each angle and central angle in a regular octagon.

$\angle ABC$ is one angle of the regular octagon.
Angles like $\angle AOB$ are called *central angles*.

The octagon is divided into 8 triangles.

Sum of the measures of angles in the 8 triangles: $8 \times 180°$ or $1{,}440°$.
Sum of the measures of all central angles: $360°$.
To find the sum of the measures of the angles of the octagon, subtract:
$1{,}440° - 360° = 1{,}080°$.

Measure of each angle: $\frac{1080°}{8} = 135°$

So, each angle in a regular octagon measures $135°$.

Measure of each central angle: $\frac{360°}{8} = 45°$.

So, each central angle in a regular octagon measures $45.°$

POLYGONS

Quadrilaterals can be classified according to their special properties.

Parallelogram

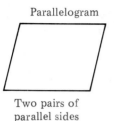

Two pairs of parallel sides

Trapezoid

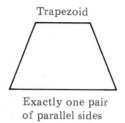

Exactly one pair of parallel sides

Rectangle

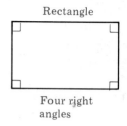

Four right angles

Rhombus

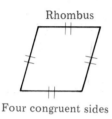

Four congruent sides

Square

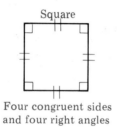

Four congruent sides and four right angles

◇ EXERCISES ◇

1. How many sides does each one of the following polygons have?

   a. pentagon
   b. octagon
   c. quadrilateral
   d. triangle
   e. hexagon
   ★ f. decagon

2. Find the measure of each angle of a regular pentagon.

Name each of the special quadrilaterals.

3.   4.   5.    6.

7. In what sense is a square a special rhombus?

8. ABCD is a square. The length of $\overline{CB}$ is 5 cm. What is the length of $\overline{CD}$? Why is this true?

9. Find the measure of each angle in a regular hexagon.

Problem Solving – Careers
Tile Setters

1. Mr. Rodriguez is a marble setter. He has been hired to set a marble wall. The wall is 12 m long and 3 m high. How many marble slabs each measuring 1 m by 1.5 m will he need to cover the wall?

2. Ms. Ostfeld is a tile setter who will set a tile floor in a kitchen that measures 4 m by 6 m. Each tile is 20 cm by 20 cm. How many tiles will she need?

3. Mrs. Aaron is a terrazzo worker. Terrazzo is tinted concrete with marble chips, and it is used for decorative floors. Mrs. Aaron has been hired to set a terrazzo star in the floor of a new building. She will use metal strips 100 cm long and 173 cm long to separate the parts of the design. How many strips of each length will she need?

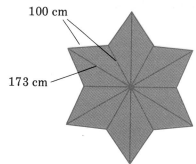

4. As a marble setter, Mr. Alexander earns $8.30 per hour. Mr. Blake works as a marble setter in a different city and earns $8.75 per hour. How much more does Mr. Blake earn in a 40-hour week?

5. A room measures 3 m by 4 m. How many tiles are needed to cover the floor if 36 tiles cover 1 m^2?

PROBLEM SOLVING: CAREERS

Square Roots

 **OBJECTIVES**  **RECALL**

To identify the perfect squares from 1 to 100

To find approximations of square roots from a table

To solve area problems involving square roots

$3 \cdot 3 = 9 \qquad\qquad 8 \cdot 8 = 64$
$-3 \cdot (-3) = 9 \qquad -8 \cdot (-8) = 64$
$(1.2)(1.2) = 1.44$
$(-1.2)(-1.2) = 1.44$

Definition of square root If $x \cdot x = n$, then x is a *square root* of n.

Every positive number has two square roots, one positive and one negative.

Number	Positive Square Root	Negative Square Root
9	3	-3
64	8	-8
1.44	1.2	-1.2

The two square roots of 49 are 7 and -7 because $7 \cdot 7 = 49$ and $-7 \cdot (-7) = 49$. The two square roots of 100 are 10 and -10 because $10 \cdot 10 = 100$ and $-10 \cdot (-10) = 100$. In most instances, especially in problems dealing with lengths of sides, only the positive square root is used. This has a special name, as shown in the definition below.

principal square root
The positive square root of a number is called the *principal square root*. It is symbolized by $\sqrt{}$.

The principal square root of 9 is 3. $\sqrt{9} = 3$ $\sqrt{64} = 8$ $\sqrt{1.44} = 1.2$

perfect square

Definition of perfect square A *perfect square* is a number whose principal square root is a whole number.

9 is a perfect square.

Example 1

$6 \cdot 6 = 36$
$7 \cdot 7 = 49$ and $8 \cdot 8 = 64$
52 is between 49 and 64.

Are 36 and 52 perfect squares?
$\sqrt{36} = 6$, and 6 is a whole number.
$\sqrt{52}$ is not a whole number.
So, 36 is a perfect square, but 52 is not.

practice ▷ Which are perfect squares?

1. 48
2. 25
3. 81
4. 66

$\doteq$ means *is approximately equal to*.

We can use the table on page 473 to find an approximation for $\sqrt{52}$ correct to three decimal places.
$$\sqrt{52} \doteq 7.211$$
$$7.211 \times 7.211 = 51.998521$$
Note that $(7.211)^2$ is not exactly equal to 52. We are not able to find an exact decimal which when squared will equal 52. For this reason, $\sqrt{52}$ is called an *irrational number*.

Example 2

Round to the nearest tenth.

Approximate $\sqrt{48}$ and $\sqrt{82}$ to the nearest tenth. Use the table on page 473.
$\sqrt{48} \doteq 6.9\underline{2}8$ $\sqrt{82} \doteq 9.0\underline{5}5$
    $\doteq 6.9$ ↖ $\doteq 9.1$ ↑
       less than 5   5 or greater
So, $\sqrt{48} \doteq 6.9$ and $\sqrt{82} \doteq 9.1$

practice ▷ Approximate to the nearest tenth. Use the table on page 473.

5. $\sqrt{8}$
6. $\sqrt{67}$
7. $\sqrt{33}$
8. $\sqrt{75}$

Example 3

The area of a square is 62 cm². Find the length of a side, to the nearest tenth of a cm.

Formula for area of a square. $s^2 = A$
Substitute 62 for A. $s^2 = 62$
   $s = \sqrt{62}$
From the table, $\sqrt{62} \doteq 7.874$. $s \doteq 7.874$
Round to the nearest tenth. $s \doteq 7.9$
So, the length of a side is 7.9 cm.

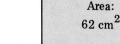

Area: 62 cm²

practice ▷ 9. The area of a square is 54 m². Find the length of a side, to the nearest tenth of a m.

SQUARE ROOTS

◇ ORAL EXERCISES ◇

Give two square roots of each number.

1. 25
2. 9
3. 64
4. 16
5. 81
6. 36
7. 1
8. 4
9. 144
10. 121

◇ EXERCISES ◇

Which are perfect squares?

1. 25
2. 42
3. 1
4. 9
5. 87
6. 8
7. 64
8. 18
9. 79
10. 45
11. 36
12. 28
13. 56
14. 81
15. 125

Approximate to the nearest tenth. Use the table on page 473.

16. $\sqrt{80}$
17. $\sqrt{3}$
18. $\sqrt{90}$
19. $\sqrt{35}$
20. $\sqrt{53}$
21. $\sqrt{12}$
22. $\sqrt{62}$
23. $\sqrt{14}$
24. $\sqrt{79}$
25. $\sqrt{37}$
26. $\sqrt{44}$
27. $\sqrt{58}$
28. $\sqrt{46}$
29. $\sqrt{54}$
30. $\sqrt{2}$

Find *s* to the nearest tenth of a unit. Use the table on page 473.

31. Area: 24 cm²
32. Area: 14 m²
33. Area: 10 cm²
34. Area: 92 m²

Give answers to the nearest tenth of a unit.

35. The area of a square is 42 cm². Find the length of a side.
36. The area of a square room is 38 m². Find the length of a wall.
★ 37. The area of a rectangle is 56 m². The length is twice the width. Find the length and the width.
★ 38. The area of a rectangle is 68 cm². The length is 4 times the width. Find the length and the width.

Challenge

There is a pattern when adding the cubes of consecutive numbers. Observe this pattern in the following:

$$1^3 + 2^3 = 9 \quad \text{also} \quad (1 + 2)^2 = 9$$
$$1^3 + 2^3 + 3^3 = 36 \quad \text{also} \quad (1 + 2 + 3)^2 = 36$$

Give the next 3 lines in this pattern. Using the pattern,

1. give quickly the sum of the cubes of the first 10 counting numbers.
2. give the formula for the sum of the cubes of the first *k* natural numbers.

The Pythagorean Theorem

◇ **OBJECTIVE** ◇   ◇ **RECALL** ◇

To find the length of a side of a right triangle by applying the Pythagorean theorem

Right Triangle ABC

Right Angle

Square $DEFG$

$A = s \cdot s$, or s^2

Example 1

Triangle RST is a right triangle. Figures I, II, and III are squares. Find the area of each square. Add the areas of Squares I and II. Then compare that sum with the area of Square III.

Formula for area of a square

$A = s \cdot s$
Area of I $= 6 \cdot 6 = 36$
Area of II $= 8 \cdot 8 = 64$
Area of III $= 10 \cdot 10 = 100$

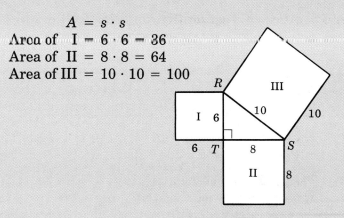

Add the areas of Squares I and II.

This is true for all right triangles.

So, $\underset{\text{Square I}}{36} + \underset{\text{Square II}}{64} = \underset{\text{Square III}}{100}$

$\underset{\text{Square I}}{\text{Area of}} + \underset{\text{Square II}}{\text{Area of}} = \underset{\text{Square III}}{\text{Area of}}$

Pythagorean theorem

In any right triangle, the square of the length of the hypotenuse equals the sum of the squares of the lengths of the other two sides.

c is the length of the hypotenuse. a and b are the lengths of the other two sides.

If $\triangle ABC$ is a right triangle, then $a^2 + b^2 = c^2$.

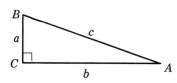

Example 2

The lengths of two sides of a right triangle are 4 cm and 6 cm. Find the length of the hypotenuse, to the nearest tenth of a cm.

Pythagorean theorem. $\quad a^2 + b^2 = c^2$
Replace a with 4 and b with 6. $\quad 4^2 + 6^2 = c^2$
$\qquad 16 + 36 = c^2$
$\qquad\qquad\quad 52 = c^2$
Use the table on page 473. $\quad \sqrt{52} = c$
$\sqrt{52} \doteq 7.211 \doteq 7.2 \qquad 7.2 \doteq c$

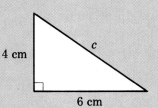

So, the length of the hypotenuse is about 7.2 cm.

practice ▷ Find c to the nearest tenth of a unit.

1.
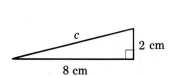

2.
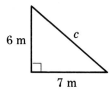

Example 3

The length of the hypotenuse of a right triangle is 12 meters, and the length of one side is 7 meters. Find the length of the other side, to the nearest tenth of a meter.

Pythagorean theorem. $\quad a^2 + b^2 = c^2$
Replace b with 7 and c with 12. $\quad a^2 + 7^2 = 12^2$
$\qquad a^2 + 49 = 144$
Subtract 49 from each side. $\quad a^2 + 49 - 49 = 144 - 49$
$\qquad\qquad a^2 = 95$
Use the table on page 473. $\quad a = \sqrt{95}$
$\sqrt{95} \doteq 9.747 \doteq 9.7 \qquad a \doteq 9.7$

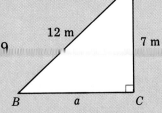

So, the length of the other side is about 9.7 m.

practice ▷ Find a or b to the nearest tenth of a unit.

3.
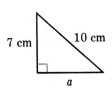

4.
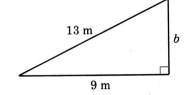

◇ EXERCISES ◇

For each right triangle, find the missing length, to the nearest tenth.

1. $a = 2, b = 3$
2. $a = 3, b = 4$
3. $a = 5, b = 5$
4. $a = 3, b = 6$
5. $a = 4, b = 5$
6. $a = 1, b = 9$
7. $a = 8, c = 10$
8. $b = 12, c = 15$
9. $a = 4, c = 8$
10. $b = 2, c = 6$
11. $b = 10, c = 14$
12. $a = 21, c = 25$

13.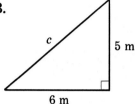

14.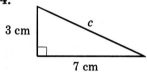

15.

16.

17.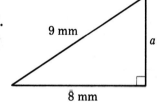

18.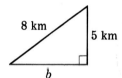

19.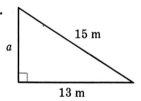

20.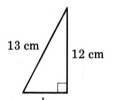

21.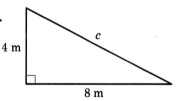

22.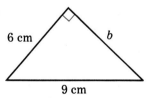

23.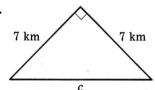

24.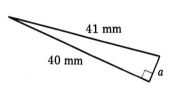

25. The lengths of two sides of a right triangle are 4 m and 7 m. Find the length of the hypotenuse.

26. The length of the hypotenuse of a right triangle is 10 cm, and the length of one side is 5 cm. Find the length of the other side.

★ 27. Bill's room measures 3 m by 5 m. Find the length of the diagonal connecting two opposite corners.

★ 28. Jessie walked 2 km south and 5 km west. How far was she from her starting point?

★ 29. A 4-m ladder is 1 m from the base of a house. At what height does the top of the ladder touch the house?

★ 30. The diagonal of a square measures 8 cm. Find the length of a side of the square.

THE PYTHAGOREAN THEOREM

Non-Routine Problems

Finish drawing the figure on the right so that it is congruent to the figure on the left.

1.
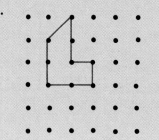

2.
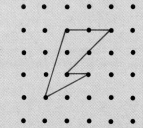

3.
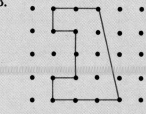

4.
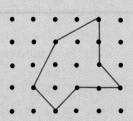
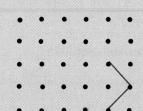

COMPUTER ACTIVITIES

Recall that for a right triangle, the formula for finding the hypotenuse if you know the lengths of the legs is $c = \sqrt{a^2 + b^2}$. You can write a program for finding c if you are given a and b.

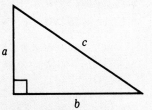

First, you need to know the **BASIC** symbol for square root. The square root of 25 is symbolized by

SQR(25)   no space between SQR and (

Example 1   Compute SQR(3 ∧ 2 + 4 ∧ 2)

3 ∧ 2 means $3^2 = 9$ SQR(3 ∧ 2 + 4 ∧ 2)
4 ∧ 2 means $4^2 = 16$ SQR(9 + 16)
                           SQR(25)
SQR(25) = $\sqrt{25}$ = 5   5

Example 2   Write a program to find C in the formula $C = \sqrt{A^2 + B^2}$. Allow for 2 sets of values for A and B.

First write the formula in **BASIC**: $C = \sqrt{A^2 + B^2}$
C = SQR(A ∧ 2 + B ∧ 2)

```
                Two sets of data.   10  FOR M = 1 TO 2
                                    20  INPUT "TYPE IN LENGTH OF A LEG ";A
                                    30  INPUT "TYPE IN LENGTH OF OTHER LEG ";B
Formula in BASIC for finding the    40  LET C = SQR (A ^ 2 + B ^ 2)
hypotenuse of a right triangle.     50  PRINT "C = "C
                                    60  PRINT : PRINT
                                    70  NEXT M
                                    80  END
```

See the Computer Section beginning on page 420 for more information.

Exercises

RUN the program above to find C for each of the following values of A and B.

1. A = 6, B = 8
2. A = 5, B = 12
3. A = 6, B = 6

★ 4. Write a program to find the leg of a right triangle given the hypotenuse and one of the legs. **RUN** the program to find B if C = 26 and A = 10.

Chapter Review

1. You are given two points. How many lines can contain these two points? [368]

2. You are given four points. Describe all possible relationships that can exist among these four points. [368]

3. Measure each angle and classify as acute, right, or obtuse. [370]

 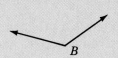

4. Complementary or supplementary? [370]
   a. 24°, 156°   b. 89°, 1°   c. 120°, 60°   d. 45°, 45°

5. [373]
   a. In the figure at the right, name two pairs of congruent vertical angles, alternate interior angles, and corresponding angles.
   b. If m∠1 = 37°, determine the measures of all labeled angles.

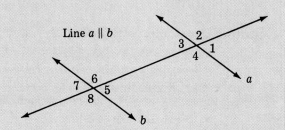

6. Construct a line perpendicular to line t and passing through point A. [375]

7. Name three special parallelograms. Describe their properties. [380]

8. Classify each of the triangles in the picture according to its angles and according to its sides. Tell why you classified each triangle as you did. [377]

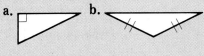

9. Which are perfect squares? [382]   a. 81   b. 93   c. 105   d. 169

10. Approximate to the nearest tenth. Use the table on page 473. [382]
    a. $\sqrt{67}$ b. $\sqrt{19}$ c. $\sqrt{83}$ d. $\sqrt{5}$

11. For each right triangle, find the missing length, to the nearest tenth. a and b are legs. c is the hypotenuse. [385]
    a. $a = 11, b = 15$ b. $a = 8, c = 17$

12. Find the missing lengths. [385]   a.

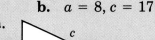

b.

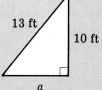

390   CHAPTER FIFTEEN

Chapter Test

1. What does it mean to say that two points determine a unique line?

2. There are two ways in which three points can be arranged. Describe these two ways.

3. Measure each angle and classify as acute, right, or obtuse.

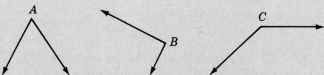

4. Complementary or supplementary?
   a. 36°, 144°   b. 72°, 18°   c. 150°, 30°   d. 90°, 90°

5. a. In the figure at the right, name two pairs of congruent vertical angles, alternate interior angles, and corresponding angles.
   b. If $m\angle 1 = 83°$, determine the measures of all labeled angles.

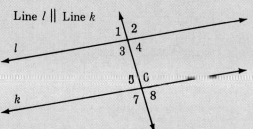

Line l ∥ Line k

6. Construct a line perpendicular to line m and passing through point X.

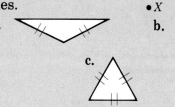

7. Name a special rhombus. Describe its properties.

8. Classify each of the triangles in the picture according to its angles and according to its sides. Tell why you classified each triangle as you did.

9. Which are perfect squares?   a. 49   b. 74   c. 196   d. 110

10. Approximate to the nearest tenth. Use the table on page 473.
    a. $\sqrt{59}$ b. $\sqrt{27}$ c. $\sqrt{91}$ d. $\sqrt{8}$

11. For each right triangle, find the missing length, to the nearest tenth. a and b are legs. c is the hypotenuse.
    a. $a = 14, b = 19$ b. $a = 5, c = 21$

12. Find the missing lengths.   a. 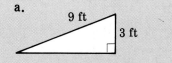   b.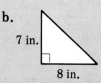

CHAPTER TEST

Coordinate Geometry

16

Dams built by the TVA help control flood waters along the Tennessee River.

Graphing on a Number Line

◇ OBJECTIVES ◇

To graph the integers between two given integers
To graph equations like $3x - 2 = 4$
To graph inequalities like $x < 2$ and $-1 \leq y$

◇ RECALL ◇

Horizontal Number Line

Negative numbers go on forever in the negative direction.   Positive numbers go on forever in the positive direction.

The integers between 2 and 6 are 3, 4, 5. 2 and 6 are not included. There are no integers between 1 and 2. 1 and 2 are not included. The integers greater than 3 are 4, 5, 6, 7, ...
. . . means *goes on forever*.

Example 1

–2 and 4 are not included. Draw a horizontal or vertical number line. Use a heavy dot to graph each number.

Graph the integers between –2 and 4.
The integers between –2 and 4 are –1, 0, 1, 2, 3.

Example 2

2 and 3 are not included.

The graph is blank.

Graph the integers between 2 and 3.
There are no integers between 2 and 3.

practice ▷ **Graph.**

1. the integers between –3 and 2
2. the integers between –1 and 0

Example 3

First find the solution.
Add 2 to each side.

Divide each side by 6.

Graph $6y - 2 = 16$.
$6y - 2 = 16$
$6y - 2 + 2 = 16 + 2$
$6y = 18$
$\dfrac{6y}{6} = \dfrac{18}{6}$
$y = 3$
So, the solution is 3.

Place a heavy dot at 3.

GRAPHING ON A NUMBER LINE

To graph an equation means to graph its solution(s).

practice ▷ **3.** Graph $4x + 3 = -9$. **4.** Graph $a - 3 = a + 5$.

Example 4

Read: x is less than 3.
Substitute numbers for x.
Many numbers make $x < 3$ true.

Graph $x < 3$.

$$x < 3$$
$$2 < 3$$
$$\sqrt{2} < 3$$
$$0 < 3$$
$$-1 < 3$$
$$-2 < 3$$
$$-2\tfrac{1}{2} < 3$$

All rational and irrational numbers less than 3 make $x < 3$ true.

All points that correspond to rational and irrational numbers less than 3 should be graphed. You can show this on a number line. We draw an arrow to show all such points.

The circle around 3 means 3 is not a solution.

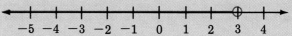

Example 5

Graph all points that correspond to numbers greater than -4.
-4 is not a solution.

Graph $y > -4$.

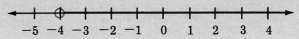

The rational numbers and the irrational numbers together form the *real numbers*. The graph of the real numbers is the real number line.

practice ▷ **5.** Graph $a > 2$. **6.** Graph $x < -1$.

Example 6

Graph $x \leq 2$.
$x \leq 2$ is read x is less than 2 or x is equal to 2.

Any number that is less than 2 or equal to 2 makes $x \leq 2$ true. The graph includes 2 and all points in the negative direction from 2.

The dot at 2 means 2 is a solution.

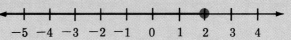

CHAPTER SIXTEEN

practice ▷ 7. Graph $y \le 4$. 8. Graph $c \ge -2$.

◇ ORAL EXERCISES ◇

Which symbol, < or >, will make the inequality true?

1. $4 \underline{\ ?\ } 7$ 2. $9 \underline{\ ?\ } 8$ 3. $-1 \underline{\ ?\ } -3$ 4. $-4 \underline{\ ?\ } -2$ 5. $0 \underline{\ ?\ } -3$
6. $-6 \underline{\ ?\ } 2$ 7. $3 \underline{\ ?\ } -4$ 8. $-2 \underline{\ ?\ } 0$ 9. $-1 \underline{\ ?\ } 1$ 10. $5 \underline{\ ?\ } -5$

Read each inequality. (*Hint:* Read the variable first.) **Then replace x with a number to make the inequality true.**

11. $x < 3$ 12. $x > -1$ 13. $x \le -2$ 14. $x \ge 4$ 15. $x < -3$
16. $4 < x$ 17. $-1 > x$ 18. $2 \ge x$ 19. $-3 \le x$ 20. $-6 \ge x$

◇ EXERCISES ◇

Graph.

1. the integers between 2 and 5
2. the integers between 0 and 6
3. the integers between -5 and -1
4. the integers between -4 and -2
5. the integers between 3 and 4
6. the integers between -2 and -1
7. the integers between -3 and 4
8. the integers between -6 and 2
★ 9. the positive even integers
★ 10. the positive odd integers

11. $4x - 5 = 3$ 12. $2y + 7 = 15$ 13. $3a - 4 = -10$
14. $-5c + 6 = -4$ 15. $-6x - 9 = -3$ 16. $-x + 8 = 8$
17. $7 = n + 3$ 18. $-4 = 3x - 13$ 19. $5 = -2z + 9$
20. $x - 8 = x + 2$ 21. $3r - 4 = 2r - 1$ 22. $2x + 6 = 2x - 4$
23. $4x - 1 = 2 + 5x$ 24. $2x - 1 = 5 + 3x$ 25. $3y - 7 = 4 + 3y$
26. $5a - 7 = 3a + 3$ 27. $4 - 5x = 6 - 5x$ 28. $4c - 2 = 13 - c$
★ 29. $-2x + 2 = x - 8 - 3x$ ★ 30. $5x + 7 + x = -5 + 6x$

Graph.

31. $x < 2$ 32. $y > 3$ 33. $a \le -1$ 34. $c \ge 4$
35. $r \le 0$ 36. $x \ge -3$ 37. $z < -2$ 38. $n > -5$
39. $c > 1$ 40. $y > 0$ 41. $n \le -4$ 42. $x > -1$
43. $-2 \ge y$ 44. $1 > r$ 45. $a > -4$ 46. $x \le -2$

Graph. (*Hint:* ≠ means is not equal to.)

★ 47. $x \ne 2$ ★ 48. $y \ne -3$ ★ 49. $0 \ne n$ ★ 50. $a \ne 1\frac{1}{2}$

GRAPHING ON A NUMBER LINE

More on Graphing Inequalities

◇ **OBJECTIVE** ◇

To graph an inequality like
$4x - 9 > 2x + 5$

◇ **RECALL** ◇

Solve. $x - 3 \geq 4$
$$x - 3 \geq 4$$
$$x - 3 + 3 \geq 4 + 3$$
$$x + 0 \geq 7$$
$$x \geq 7$$
So, all real numbers ≥ 7 are solutions.

Solve. $3x < 15$
$$\frac{3x}{3} < \frac{15}{3}$$
$$1x < 5$$
$$x < 5$$
So, all real numbers < 5 are solutions.

You know how to graph an inequality like $x < 5$. To graph an inequality like $2x - 5 < 3$, first solve the inequality for x.

Example 1

Graph. $2x - 5 < 3$
$$2x - 5 < 3$$

Add 5 to each side.
The order is the same.
$2x - 5 + 5 < 3 + 5$ ← addition property for inequalities

Divide each side by 2.
We divided by a positive number, so the order does not change.
$\frac{2x}{2} < \frac{8}{2}$ ← division property for inequalities
$$x < 4$$

Any number less than 4 should make $2x - 5 < 3$ true.

Check two numbers that are less than 4.

Try 3.
$2x - 5 < 3$	
$2 \cdot 3 - 5$	3
$6 - 5$	
1	
$1 < 3$	True

Try -2.
$2x - 5 < 3$	
$2 \cdot (-2) - 5$	3
$-4 - 5$	
-9	
$-9 < 3$	True

Check two numbers that are not less than 4.

Try 4.
$2x - 5 < 3$	
$2 \cdot 4 - 5$	3
$8 - 5$	
3	
$3 < 3$	False

Try 6.
$2x - 5 < 3$	
$2 \cdot 6 - 5$	3
$12 - 5$	
7	
$7 < 3$	False

Any number less than 4 makes $2x - 5 < 3$ true.

practice ▷ Graph the inequality.

1. $4x + 3 > -9$

2. $2y - 7 < -5$

Example 2

Graph. $-3a + 7 > 13$

$-3a + 7 > 13$

Subtract 7 from each side. $-3a + 7 - 7 > 13 - 7$
Same order
Divide each side by -3. $\frac{-3a}{-3} > \frac{6}{-3}$
Reverse the order.

$a < -2$

Any number less than -2 makes $-3a + 7 > 13$ true.

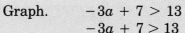

practice ▷ Graph the inequality.

3. $-2y - 7 > 3$ 4. $-3a + 5 < 14$

Example 3

Graph. $5y - 6 \geq 4y - 1$

$5y - 6 \geq 4y - 1$

Subtract $4y$ from each side. $5y - 4y - 6 \geq 4y - 4y - 1$
Same order
Add 6 to each side. $y - 6 \geq -1$
Same order $y - 6 + 6 \geq -1 + 6$

$y \geq 5$

Any number greater than or equal to 5 makes $5y - 6 \geq 4y - 1$ true.

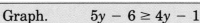

practice ▷ Graph.

5. $4a + 1 \geq 3a - 2$ 6. $3x - 3 \leq 2x - 9$

Example 4

Graph. $4x + 10 \geq 7x + 1$

$4x + 10 \geq 7x + 1$

$4x - 7x + 10 \geq 7x - 7x + 1$

$-3x + 10 \geq 1$

$-3x + 10 - 10 \geq 1 - 10$

$\frac{-3x}{-3} \geq \frac{-9}{-3}$

Divide each side by -3.
Reverse the order.

$x \leq 3$

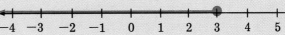

practice ▷ Graph.

7. $2c - 7 \leq 6c + 13$ 8. $3y + 8 \geq 5y - 10$

MORE ON GRAPHING INEQUALITIES

◆ EXERCISES ◆

Graph.

1. $2x - 5 > 5$
2. $3x + 7 < -2$
3. $4a + 6 \leq 10$
4. $5z - 10 > -20$
5. $-3y + 2 > -7$
6. $-6x - 8 \leq 10$
7. $2c - 3 \geq -3$
8. $-4a - 3 < -7$
9. $-x + 9 \leq 7$
10. $-c - 4 > 3$
11. $4 + 2n > 10$
12. $7 - 3x \leq -5$
13. $6 + x \geq 9$
14. $5 - r < 2$
15. $3z - 7 < 8$
16. $4a - 7 > -11$
17. $10y - 32 \leq 8$
18. $2 + 11c \geq 24$
19. $4x - 2 > 3x + 1$
20. $2y + 9 < y + 5$
21. $7a + 1 \leq 6a - 3$
22. $5x - 2 \geq 4x - 2$
23. $3c - 8 \leq 2c - 3$
24. $2a + 7 > a + 4$
25. $4x + 6 < x + 15$
26. $6y - 8 \geq 2y + 28$
27. $5a + 11 > 3a + 7$
28. $4x - 9 < 2x - 1$
29. $8n - 7 \leq 5n + 11$
30. $5z + 2 \geq 3z + 10$
31. $3y - 16 < 7y + 8$
32. $2a - 9 > 3a - 2$
33. $d - 3 \geq -4d + 22$
34. $8x - 13 \leq 21x - 26$
35. $-9y + 18 < 2y + 7$
36. $-4c + 22 > c - 3$
★ 37. $3x - 2x - 8 \geq 5x + 12$
★ 38. $7a + 2 + 4a < 3a - 14$
★ 39. $-8 - 6c > 2 + 8c - 4c$
★ 40. $-y \leq 7 - 8y + 4 - 4y$

★ 41. Five more than twice a number is less than 13. What numbers make this true?

★ 42. Four less than 3 times a number is greater than or equal to 11. What numbers make this true?

Calculator

Replace __?__ with >, <, or =. $(3 + 5)^2$ __?__ $3^2 + 5^2$ >

$(3 + 5)^2$
Press 3 ⊕ 5 ⊖
Display 8
Press ⊗ 8 ⊖
Display 64

$3^2 + 5^2$
Press 3 ⊗ 3 ⊖
Display 9   ← Write down and clear calculator.
Press 5 ⊗ 5 ⊖
Display 25
Press ⊕ 9 ⊖
Display 34

So, $(3 + 5)^2 > 3^2 + 5^2$ since 64 > 34.

Replace __?__ with >, <, or =.
1. $(1 + 1)^2$ __?__ $1^2 + 1^2$
2. $(7 + 9)^2$ __?__ $7^2 + 9^2$
3. $(8 + 15)^2$ __?__ $8^2 + 15^2$
4. $(31 + 24)^2$ __?__ $31^2 + 24^2$
5. For all whole numbers a and b, it appears that $(a + b)^2$ __?__ $a^2 + b^2$.

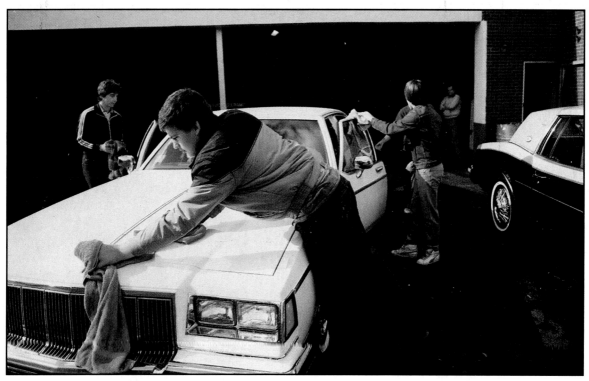

Problem Solving – Applications
Jobs for Teenagers

Example	Bill bought 12 gallons of gas at $1.15 per gallon, and he had a hot-wax treatment for $1.25. How much did he pay?
Identify what is to be found.	Bought: 12 gallons Cost: $1.15 per gallon Other cost: $1.25 for wax
Analyze the information.	Multiply 12 × $1.15 to find the cost of the gas. Add $1.25 for the wax.
Calculate.	$1.15 ← Multiply 12 × $1.15 to find the cost × 12 of the gas. 230 115 $13.80 + 1.25 ← Add $1.25 for the wax. $15.05 Total cost. So, he paid $15.05.

practice ▷ Mary bought 9 gal of gas for $1.25/gal. What was her change from a 20-dollar bill?

PROBLEM SOLVING

Solve these problems.

1. Jim works at the Sparkle Car Wash from 3:00 P.M. until 6:00 P.M. Tuesday through Friday. On Saturday he works from 8:00 A.M. until 4:30 P.M. with a half-hour lunch break. He earns $3.75/h. How much does he earn per week?

2. Nan bought 15 gal of gas and had a polished wax job. If gas costs $1.14/gal, what was the total cost?

3. Mr. Stapleton had his car washed on Friday. He asked for a hot-wax treatment but bought no gas. How much did he pay?

4. Mrs. Schmidt had her car washed on a Thursday. She bought no gas but had a polished wax treatment. How much did Jim charge?

5. Ms. Helinski bought 15 gal of gas and had her car washed. If gas costs $1.29/gal, how much should Jim charge?

6. During the summer, Jan works full time at the Sparkle Car Wash. She earns $3.30 per hour plus time and a half for every hour over 40 hours a week. One week she worked 45 hours. How much did she earn?

7. In one hour on a Tuesday, 3 customers had their cars washed. One bought 5 gal of gas; another bought no gas but had a hot-wax treatment; the third bought 15 gal of gas. How much was taken in that hour if gas costs $1.20/gal?

8. Each of 5 customers bought 13 gal of gas and had their cars washed. How much money was taken in if gas costs $1.15/gal?

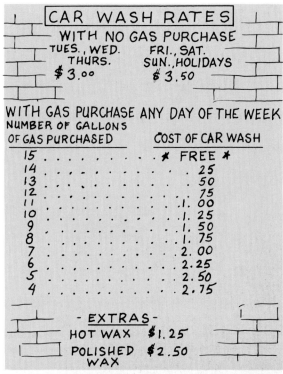

Graphing Points

◇ OBJECTIVES ◇

To give the ordered pair for a point in a plane
To graph points in a plane

◇ RECALL ◇

Perpendicular lines form right angles.

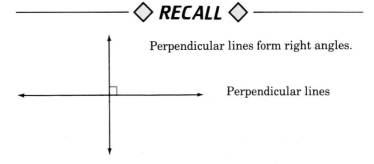

Perpendicular lines

Two perpendicular number lines determine a coordinate plane. Each line is called an *axis*.

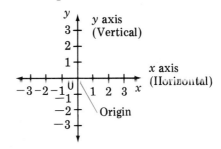

Example 1

Point P is in a coordinate plane.

Describe the location of point P.

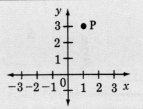

The position of point P is described by an ordered pair of numbers.

x-coordinate y-coordinate
    (1, 3)

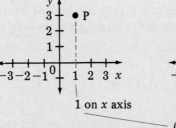

1 on x axis   3 on y axis
         (1, 3)

The x-coordinate is always first.
The y-coordinate is always second.

So, the ordered pair (1, 3) describes point P.

GRAPHING POINTS

practice ▷ Give the ordered pair for each point.

1. P
2. Q
3. R
4. S
5. T
6. U

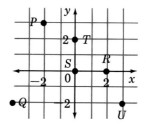

Example 2

Graph (3, 2). Then graph (2, −1).

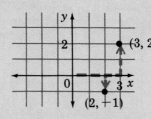

Start at the origin.
(3, 2)
Right 3   Up 2

(2, −1)
Right 2   Down 1

Example 3

Graph (−3, 1). Then graph (−4, −2).

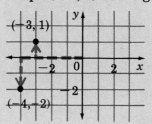

(−3, 1)
Left 3   Up 1

(−4, −2)
Left 4   Down 2

Summary

To graph a point, the signs of the coordinates tell in which directions to move from the origin.

(3, 2)	(−3, 1)	(−4, −2)	(2, −1)
(+, +)	(−, +)	(−, −)	(+, −)
right up	left up	left down	right down

practice ▷ Graph.

7. (1, 3)   8. (−2, 1)   9. (4, −2)   10. (−3, −3)

402   CHAPTER SIXTEEN

Example 4

Start at the origin.

Graph (−3, 0). Graph (0, −1).

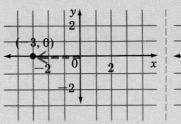
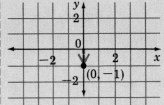

(−3, 0) lies on the x axis.
(0, −1) lies on the y axis.

(−3, 0): left 3, neither up nor down

(0, −1): neither right nor left, down 1

practice ▷ Graph.

11. (−4, 0)   12. (0, 2)   13. (1, 0)   14. (0, −3)

◇ EXERCISES ◇

Give the ordered pair for each point.

1. A   2. B   3. C
4. D   5. E   6. F
7. G   8. H   9. I
10. J   11. K   12. L
13. M   14. N   15. P
★ 16. Q   ★ 17. R   ★ 18. S

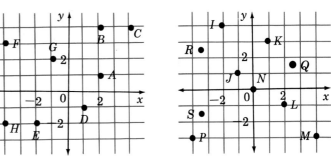

Graph each point. Use a different set of axes for every four

19. $A(-1, 4)$ 20. $B(-3, 0)$ 21. $C(2, 3)$ 22. $D(3, -2)$
23. $E(-2, -3)$ 24. $F(-5, 1)$ 25. $G(3, 0)$ 26. $H(-2, -1)$
27. $I(0, 0)$ 28. $J(4, -2)$ 29. $K(-3, 3)$ 30. $L(4, 1)$
31. $M(6, -3)$ 32. $N(0, -1)$ 33. $D(-4, -4)$ 34. $Q(3, -5)$
★ 35. $E\left(2\frac{1}{2}, 1\frac{1}{2}\right)$ ★ 36. $F\left(-3\frac{1}{2}, 4\frac{1}{2}\right)$ ★ 37. $G\left(-5\frac{1}{2}, 0\right)$ ★ 38. $H\left(-1\frac{1}{2}, -4\frac{1}{2}\right)$

GRAPHING POINTS

Horizontal and Vertical Lines

◇ OBJECTIVE ◇

To determine whether a line is horizontal or vertical, given the coordinates of two points on the line

◇ RECALL ◇

Give the coordinates of points A, B, C, and D.

$A(-2, 1)$
$B(3, 1)$

$C(2, 3)$
$D(2, -1)$

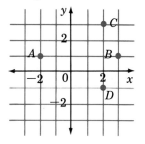

A *horizontal* line is parallel to the x-axis.
A *vertical* line is parallel to the y-axis.

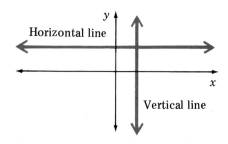

Example 1

For the horizontal line, give the coordinates of points P, Q, and R. For the vertical line, give the coordinates of points S, T, and U.

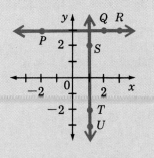

$P(-2, 3)$, $Q(2, 3)$, $R(3, 3)$

same y-coordinate

$S(1, 2)$, $T(1, -2)$, $U(1, -3)$

same x-coordinate

Example 1 suggests that

1. every point on a horizontal line has the same y-coordinate, and
2. every point on a vertical line has the same x-coordinate.

Example 2

Is $\overleftrightarrow{EF}$ a horizontal line or a vertical line?

$E(2, -1), F(2, 3)$	$E(-3, -4), F(1, -4)$
$E(2, -1) \quad F(2, 3)$	$E(-3, -4) \quad F(1, -4)$
same *x*-coordinate	same *y*-coordinate
So, $\overleftrightarrow{EF}$ is vertical.	So, $\overleftrightarrow{EF}$ is horizontal.

practice ▷ Is $\overleftrightarrow{KL}$ a horizontal line or a vertical line?

1. $K(2, -1), L(-3, -1)$
2. $K(1, -4), L(1, 5)$

◇ EXERCISES ◇

Is $\overleftrightarrow{PQ}$ a horizontal line or a vertical line?

1. $P(2, 3), \quad Q(2, 1)$
2. $P(4, 3), \quad Q(1, 3)$
3. $P(1, 4), \quad Q(1, -2)$
4. $P(-2, 1), Q(3, 1)$
5. $P(1, -4), Q(-3, -4)$
6. $P(4, 2), \quad Q(5, 2)$
7. $P(3, -5), Q(3, -3)$
8. $P(-1, 6), Q(-1, -4)$
9. $P(5, -2), Q(-4, -2)$
10. $P(-2, 5), Q(-2, -1)$

For what value of *b* will the line through *R* and *S* be horizontal or vertical as indicated?

★ 11. $R(b, 3), \quad S(-3, -1)$; vertical
★ 12. $R(2, b), \quad S(5, -2)$; horizontal
★ 13. $R(0, -1), S(1, b)$; horizontal
★ 14. $R(4, -4), S(b, 4)$; vertical

Challenge

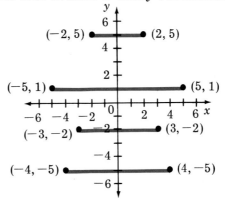

Each pair of points is symmetrical with respect to the *y*-axis.
1. Graph and label four pairs of points that are symmetrical with respect to the *x*-axis.
2. What is true of their *x*- and *y*-coordinates?

HORIZONTAL AND VERTICAL LINES

Graphing Equations and Inequalities

◇ OBJECTIVES ◇

To graph an equation like
$y = 2x - 1$
To graph an inequality like
$y < -3x + 1$

◇ RECALL ◇

Evaluate $3x - 5$ if x is 2; if x is -1.

$3x - 5$	$3x - 5$
$3 \cdot 2 - 5$	$3 \cdot (-1) - 5$
$6 - 5$	$-3 - 5$
1	-8

Example 1

Find five ordered pairs that make the equation $y = x + 2$ true. Then graph the ordered pairs.

Select 5 values for x.	Evaluate $x + 2$ for each value of x.	For each value of x, $y = x + 2$.	Resulting ordered pairs
x	$x + 2$	y	(x, y)
0	$0 + 2$	2	$(0, 2)$
1	$1 + 2$	3	$(1, 3)$
2	$2 + 2$	4	$(2, 4)$
-1	$-1 + 2$	1	$(-1, 1)$
-2	$-2 + 2$	0	$(-2, 0)$

Graph the ordered pairs.

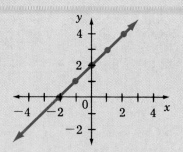

The five points appear to lie on a straight line.

practice ▷

Graph five ordered pairs that make each equation true.

1. $y = x - 1$
2. $y = -x + 3$
3. $y = -x - 2$

Example 2

Graph $y = -3x + 1$.

Find at least three ordered pairs that make the equation true.

x	$-3x + 1$	y	(x,y)
0	$-3 \cdot 0 + 1$	1	$(0, 1)$
2	$-3 \cdot 2 + 1$	-5	$(2, -5)$
-1	$-3 \cdot (-1) + 1$	4	$(-1, 4)$

All ordered pairs that make the equation true lie on the line.

Graph the ordered pairs. Draw a straight line through them.

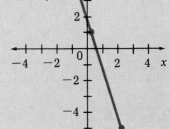

Example 3

Graph $y < -3x + 1$

Find ordered pairs that make $y < -3x + 1$ true.

Try $(-1, 4)$.   Try $(-3, -1)$.

Ordered pairs on the graph of the line $y = -3x + 1$ do *not* make $y < -3x + 1$ true. Ordered pairs *below* the graph of the line make $y < -3x + 1$ true.

y	$< -3x + 1$
4	$-3 \cdot (-1) + 1$
	$3 \quad + 1$
	4

$4 < 4$ false

y	$< -3x + 1$
-1	$-3 \cdot (-3) + 1$
	$9 \quad + 1$
	10

$-1 < 10$ true

The dotted line means the line is *not* part of the graph of $y < -3x + 1$.

Draw a dotted line for $y = -3x + 1$. Shade the region below the line.

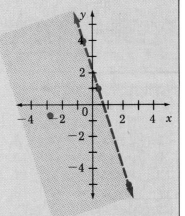

The shaded region is the graph of $y < -3x + 1$.

practice ▷ Graph.

4. $y = -3x$ 5. $y < 4x - 3$ 6. $y > 3x + 4$

GRAPHING EQUATIONS AND INEQUALITIES

Summary

To *graph*
1. an equation like $y = -3x + 1$, graph ordered pairs that make the equation true and draw a straight line through the points.
2. an inequality like $y < -3x + 1$, find ordered pairs that make the inequality true and shade the region where the points lie.

◆ ORAL EXERCISES ◆

Evaluate for each value of x.

1.
x	$x + 3$
0	
3	
-1	

2.
x	$-4x$
0	
2	
-3	

3.
x	$-x$
0	
1	
-2	

4.
x	$-x + 5$
0	
2	
-1	

5.
x	$2x + 1$
0	
1	
-2	

6.
x	$3x - 2$
0	
2	
-1	

7.
x	$-2x + 2$
0	
3	
-2	

8.
x	$-4x - 1$
0	
2	
-3	

◆ EXERCISES ◆

Graph.

1. $y = x + 1$
2. $y = x - 2$
3. $y = x + 3$
4. $y = x - 3$
5. $y = x$
6. $y = -x$
7. $y = -x + 2$
8. $y = -x - 1$
9. $y = 3x$
10. $y = -2x$
11. $y = 4x$
12. $y = -4x$
13. $y = 3x + 1$
14. $y = 2x + 1$
15. $y = -3x + 2$
16. $y = -2x + 2$
17. $y < 2x - 1$
18. $y < 3x - 2$
19. $y < 4x + 1$
20. $y < 5x - 3$
21. $y < 4x - 2$
22. $y < -5x + 2$
23. $y < -4x - 1$
24. $y < 5x + 4$
25. $y > -2x + 3$
26. $y > -4x + 3$
27. $y > 5x - 1$
28. $y > 6x - 5$
★ 29. $x + y = 1$
★ 30. $x - y = 3$
★ 31. $4x + 2y = 6$
★ 32. $6x - 3y = 9$
★ 33. $y = \frac{1}{2}x$
★ 34. $y = -\frac{1}{3}x$
★ 35. $y = \frac{1}{2}x + 1$
★ 36. $y = -\frac{1}{3}x - 2$
★ 37. $y = \frac{2}{3}x$
★ 38. $y = -\frac{4}{3}x$
★ 39. $y = \frac{3}{4}x - 2$
★ 40. $y = -\frac{5}{2}x + 1$

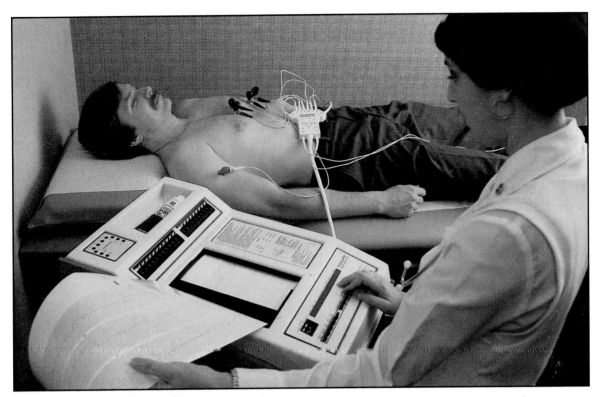

Problem Solving – Careers
Hospital Technicians

1. Ms. Ballin is an EKG technician. She operates equipment that monitors a patient's heart action. Her salary is $980 per month. What is her yearly salary?

2. Phyllis works 5 days a week. She gives 6 EKG's a day. How many EKG's does she give in 4 weeks?

3. Francis earns $210 a week as an EKG technician. He gets 2 weeks paid vacation. What is his yearly salary?

4. As a high school graduate, Mr. Sanders is entering a 3-month training program to become an EKG technician. As a trainee, his salary is $600 per month. After the training period, his salary will be $720 per month. How much more will he earn per year after training?

5. In 1974, there were about 10,000 EKG technicians working in the U.S.A. Due to increased reliance on EKG's in diagnosing heart conditions, the number of EKG technicians could increase by 25% by 1990. If it does, how many EKG technicians would be working in the U.S.A. by 1990?

PROBLEM SOLVING

Slope and Y-Intercept

◇ OBJECTIVES ◇

To find the slope of a line, given two points on the line

To find the slope and y-intercept of a line, given its equation

◇ RECALL ◇

Find three ordered pairs that make the equation $y = 2x + 1$ true.

x	$2\ x + 1$	y	(x, y)
0	$2 \cdot 0 + 1$	1	(0, 1)
2	$2 \cdot 2 + 1$	5	(2, 5)
3	$2 \cdot 3 + 1$	7	(3, 7)

The steepness of a line is called the *slope* of the line. The greater the steepness, the greater the slope.

$\overleftrightarrow{OA}$ is steeper than $\overleftrightarrow{OB}$. So, the slope of $\overleftrightarrow{OA}$ is greater than the slope of $\overleftrightarrow{OB}$.

Example 1

Find the slope of $\overleftrightarrow{AB}$. Find the slope of $\overleftrightarrow{CD}$.

Rise is vertical distance.
Run is horizontal distance.

The *rise* is the difference of the y-coordinates.
The *run* is the difference of the x-coordinates.

y-coordinates
$A(3, 1), B(6, 3)$
x-coordinates

Slope can be determined by dividing the amount of rise by the amount of run.

slope of $\overleftrightarrow{AB} = \dfrac{\text{rise}}{\text{run}}$

$= \dfrac{3 - 1}{6 - 3}$, or $\dfrac{2}{3}$

slope of $\overleftrightarrow{CD} = \dfrac{\text{rise}}{\text{run}}$

$= \dfrac{7 - 2}{3 - 1}$, or $\dfrac{5}{2}$

Slope of a line $= \dfrac{\text{Difference of the } y\text{-coordinates}}{\text{Difference of the } x\text{-coordinates}}$

Example 2

Find the slope of $\overleftrightarrow{MN}$ for $M(4, 2)$ and $N(7, 3)$.

$M(4, 2), N(7, 3)$ with y-coordinates and x-coordinates labeled.

slope of $\overleftrightarrow{MN} = \dfrac{\text{Difference of the y-coordinates}}{\text{Difference of the x-coordinates}}$

$= \dfrac{3 - 2}{7 - 4}$, or $\dfrac{1}{3}$

practice Find the slope of $\overleftrightarrow{AB}$ for the given pairs of points.

1. $A(3, 1), B(6, 5)$
2. $A(5, 7), B(3, 9)$
3. $A(4, -3), B(2, -1)$

The slope of a line can be determined directly from the equation of the line.

Example 3

Graph $y = 2x + 1$. Determine the slope of the line and the y-coordinate of the point of intersection of the line with the y-axis.

x	$2x + 1$	y	(x, y)
1	$2 \cdot 1 + 1$	3	$(1, 3)$
2	$2 \cdot 2 + 1$	5	$(2, 5)$
3	$2 \cdot 3 + 1$	7	$(3, 7)$

Slope $= \dfrac{7 - 3}{3 - 1} = \dfrac{4}{2}$, or 2.

Graph the ordered pairs.

The line intersects the y-axis at $(0, 1)$.
↑
y intercept

So, the slope is 2 and the y-intercept is 1.

The *y-intercept* is the y-coordinate of the point of intersection of a line with the y-axis.

Example 4

Graph $y = \dfrac{2}{3}x - 2$. Determine the slope and the y-intercept.

x	$\dfrac{2}{3}x - 2$	y	(x, y)
0	$\dfrac{2}{3} \cdot 0 - 2$	-2	$(0, -2)$
3	$\dfrac{2}{3} \cdot 3 - 2$	0	$(3, 0)$
6	$\dfrac{2}{3} \cdot 6 - 2$	2	$(6, 2)$

The line intersects the y-axis at $(0, -2)$.

slope $= \dfrac{-2 - 2}{0 - 6} = \dfrac{-4}{-6}$, or $\dfrac{2}{3}$.

So, the slope is $\dfrac{2}{3}$ and the y-intercept is -2.

SLOPE AND Y-INTERCEPT

Slope-intercept form of an equation $y = mx + b$ is an equation of a line.
m is the slope; b is the y-intercept.

Example 5   Find the slope and the y-intercept of the line whose equation is $y = -\frac{4}{5}x - 3$.

$$y = -\frac{4}{5}x - 3$$
$$y = mx + b$$

So, the slope is $-\frac{4}{5}$ and the y-intercept is -3.

practice   Give the slope and the y-intercept.

4. $y = 5x + 3$ 5. $y = \frac{1}{3}x - 2$ 6. $y = -\frac{5}{7}x + 8$

◇ EXERCISES ◇

Find the slope of $\overleftrightarrow{AB}$ for the given pairs of points.

1. $A(4, 1), B(6, 3)$
2. $A(2, 2), B(6, 5)$
3. $A(-1, 6), B(5, 9)$
4. $A(4, 3), B(-6, 4)$
5. $A(7, 3), B(2, -8)$
6. $A(1, -6) B(7, 4)$
7. $A(5, 1), B(8, -3)$
8. $A(4, 1), B(-6, 2)$
9. $A(6, -2), B(7, -3)$

Find the slope and the y-intercept.

10. $y = \frac{3}{2}x + 5$
11. $y = \frac{1}{2}x - 4$
12. $y = -\frac{2}{3}x + 1$
13. $y = \frac{4}{5}x + 7$
14. $y = \frac{2}{5}x - 2$
15. $y = -\frac{1}{3}x + 5$

The slope of a line can be written as $\frac{y_2 - y_1}{x_2 - x_1}$, where y_2 and y_1 are the y-coordinates and x_2 and x_1 are the x-coordinates of two points on the line. That is, the two points on the line are (x_1, y_1) and (x_2, y_2). Recall that for a horizontal line, the y-coordinates of its points are the same.

16. What is $\frac{y_2 - y_1}{x_2 - x_1}$ equal to if $y_1 = y_2$?   17. What is the slope of a horizontal line?

Recall that for a vertical line the x-coordinates are the same.

18. Does $\frac{y_2 - y_1}{x_2 - x_1}$ have a value if $x_1 = x_2$? Why or why not?
19. Why does the slope of a vertical line not exist?

Find the slope and the y-intercept.

★ 20. $2y - 3x = 8$ ★ 21. $x - 3y - 7 = 0$ ★ 22. $4x + 2y = 6$
★ 23. $3y + 5x = 6$ ★ 24. $2y - x = 6$ ★ 25. $5x - 3y = 5$

Solving Systems of Equations

◊ OBJECTIVES ◊

To determine whether coordinates of a point satisfy an equation

To solve a system of two equations in two variables by graphing

◊ RECALL ◊

Graph $y = 2x - 1$.

x	$2x - 1$	y	(x, y)
0	$2 \cdot 0 - 1$	-1	$(0, -1)$
1	$2 \cdot 1 - 1$	1	$(1, 1)$
2	$2 \cdot 2 - 1$	3	$(2, 3)$

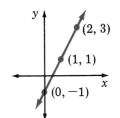

$y = -x + 3$	$y = 3x - 1$
2 | $-(1) + 3$	2 | $3 \cdot 1 - 1$
$-1 + 3$	$3 - 1$
2	2
$2 = 2$ true	$2 = 2$ true

The graphs of $y = -x + 3$ and $y = 3x - 1$ are shown. Point $P(1, 2)$ lies on both lines. It is the point of intersection of the two lines. The coordinates of P satisfy both equations.

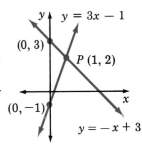

Example 1

Determine whether the coordinates of point $P(2, 3)$ satisfy both equations.
$$y = 3x - 3$$
$$y = -2x + 7$$
Substitute 2 for x and 3 for y in each equation, and determine if the result is true.

$y = 3x - 3$	$y = -2x + 7$
3 | $3 \cdot 2 - 3$	3 | $-2 \cdot 2 + 7$
$6 - 3$	$-4 + 7$
3	3
$3 = 3$ true	$3 = 3$ true

So, $(2, 3)$ satisfies both equations.

The coordinates of $P(2, 3)$ are solutions of both equations. So, $(2, 3)$ is the solution of the system of equations $y = 3x - 3$ and $y = -2x + 7$

practice ▷ Determine whether the coordinates of $P(1, 2)$ satisfy both equations.

1. $y = -2x + 4$
   $y = 3x - 1$

2. $y = -x + 3$
   $y = 2x - 1$

Example 2

Solve the system by graphing.
$y = 2x + 5$
$y = -2x + 1$

Graph each equation on the same set of axes. Find at least three ordered pairs that satisfy each equation.

x	$2x + 5$	y	(x, y)
0	$2 \cdot 0 + 5$	5	$(0, 5)$
1	$2 \cdot 1 + 5$	7	$(1, 7)$
-1	$2 \cdot (-1) + 5$	3	$(-1, 3)$

x	$-2x + 1$	y	(x, y)
0	$-2 \cdot 0 + 1$	1	$(0, 1)$
1	$-2 \cdot 1 + 1$	-1	$(1, -1)$
2	$-2 \cdot 2 + 1$	-3	$(2, -3)$

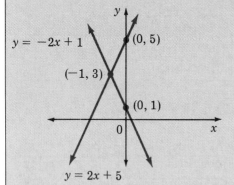

The two lines intersect at $(-1, 3)$.
So, $(-1, 3)$ is the solution.

solving a system of equations by graphing
1. Graph each equation on the same set of axes.
2. Locate the coordinates of the point of intersection of the two lines.

practice ▷ Solve each system by graphing.

3. $y = -2x + 4$
   $y = 3x - 1$

4. $y = 2x - 5$
   $y = -3x + 5$

◇ EXERCISES ◇

Determine whether the coordinates of $P(-1, 2)$ satisfy both equations.

1. $y = -x + 1$
   $y = 2x + 4$
2. $y = 3x + 5$
   $y = -2x + 3$
3. $y = 2x + 4$
   $y = -3x - 1$
4. $y = 3x - 7$
   $y = -x + 1$
5. $y = 4x$
   $y = 2x$
6. $y = -6x - 4$
   $y = 9x + 11$

Solve each system by graphing.

7. $y = -2x + 3$
   $y = x + 3$
8. $y = x + 3$
   $y = 2x + 4$
9. $y = 3x - 5$
   $y = -x + 3$
10. $y = 4x + 4$
    $y = 3x + 2$
11. $y = -2x$
    $y = 3x - 5$
12. $y = -x + 3$
    $y = -2x + 7$

Recall that the slope of a line equals $\frac{\text{Difference of } y\text{-coordinates}}{\text{Difference of } x\text{-coordinates}}$.

You can write a program to find the slope of a line given the coordinates of two points. Before you can write the program you must know the formula for finding the slope.

Given the coordinates of any two points (x_1, y_1) and (x_2, y_2)

$$\text{slope} = \frac{y_2 - y_1}{x_2 - x_1}.$$ The formula in **BASIC:** $S = (Y2 - Y1)/(X2 - X1)$.

Example Write a program to find the slope of a line given the coordinates of two points. Allow for three sets of data.

```
10  FOR I = 1 TO 3
20  INPUT "TYPE COORD OF FIRST POINT X1
    AND Y1 ";X1,Y1
30  INPUT "TYPE COORD OF SECOND POINT X2
    AND Y2 ";X2,Y2
40  LET S = (Y2 - Y1) / (X2 - X1)
50  PRINT "SLOPE IS ";S
60  NEXT I
70  END
```

This program will find the slope of any line that is NOT vertical. Consider the line through the points A(3, 4) and B(3, 11). This will be a vertical line; the x-coordinates are the same, 4.

The slope is $\frac{y_2 - y_1}{x_2 - x_1} = \frac{11 - 4}{3 - 3} = \frac{7}{0}$, which is *undefined*.

You can alter the program so that it will recognize vertical lines.
See the Computer Section beginning on page 420 for more information.

Exercises

1. Type and **RUN** the program above to find the slope of the line through the points (2, 5), (3, 7); (4, 6), (8, 8); (6, 1), (14, 7).
2. **RUN** the program for the following three sets of points: (2, 4), (7, 9); (4, 5), (4, 12); (9, 1), (7,0).
   Why does the computer come up with an error message on the second **RUN**?
3. Use an **IF-THEN** statement to modify the program to avoid the error message above. (*Hint:* $(Y2 - Y1)/(X2 - X1)$ is undefined if the denominator $X2 - X1$ is?)

COMPUTER ACTIVITIES

Chapter Review

Graph. [393]

1. the integers between -2 and 3
2. $-2x - 8 = -6$

Graph. [393]

3. $x < -3$
4. $y \geq -1$
5. $c \leq 2$
6. $r > 3$

Graph. [396]

7. $2x - 3 > 7$
8. $-4y + 7 \leq -9$
9. $5a - 2 < 4a - 7$
10. $3c + 5 \geq 11 + 5c$

Give the ordered pair for each point. [401]

11. A
12. B
13. C
14. D
15. E
16. F

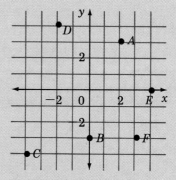

Graph each point. [401]

17. $G(4, -1)$
18. $H(-3, -3)$
19. $J(-2, 0)$
20. $K(-4, 2)$

Is $\overleftrightarrow{RS}$ a horizontal line or a vertical line? [404]

21. $R(3, 1), S(3, -2)$
22. $R(5, -2), S(-1, -2)$
23. $R(4, -4), S(3, -4)$
24. $R(1, 0), S(-1, 0)$

Graph. [406]

25. $y = x + 2$
26. $y = -x + 1$
27. $y = 3x$
28. $y < -2x - 1$

Find the slope of $\overleftrightarrow{AB}$ for the given pairs of points. [410]

29. $A(3, 2), B(5, 3)$
30. $A(2, -3), B(6, 7)$

Find the slope and y-intercept. [410]

31. $y = \frac{2}{3}x + 5$
32. $y = -\frac{5}{6}x - 3$

Solve the system by graphing. [413]

33. $y = 2x - 5$
    $y = -2x + 7$

CHAPTER SIXTEEEN

Chapter Test

Graph.

1. the integers between 0 and -4
2. $-3 = 3x - 12$

Graph.

3. $y > -2$
4. $x \leq 0$
5. $c \geq 4$
6. $m < 1$

Graph.

7. $3x - 6 > 9$
8. $-y + 2 \leq 7$
9. $4r - 2 \geq 5r + 1$

Give the ordered pair for each point.

10. R
11. S
12. T
13. U
14. V
15. W

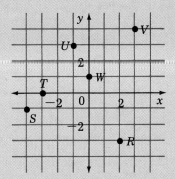

Graph each point.

16. $L(2, 4)$
17. $M(0, -3)$
18. $N(-2, -3)$
19. $P(-3, 2)$

Is $\overleftrightarrow{JK}$ a horizontal line or a vertical line?

20. $J(-1, 4), K(3, 4)$
21. $J(-2, -1), K(-2, 0)$

Graph.

22. $y = x - 1$
23. $y = -2x$
24. $y < 3x + 1$

Find the slope of $\overleftrightarrow{AB}$ for the given points.

25. $A(3, 5), B(6, 9)$
26. $A(-1, 4), B(2, -7)$

Find the slope and y-intercept.

27. $y = \frac{4}{5}x + 2$

Solve the system by graphing.

28. $y = -x + 3$
    $y = 3x - 1$

Cumulative Review

Perform the indicated operation.

1. $-22 + 17$
2. $12 - (-16)$
3. $-13(-6 + 9)$
4. $-\frac{3}{4} \div \left(-\frac{5}{8}\right)$

Simplify.

5. $3x^2 - 2 + 2x - 5x^2 - 7 - 5x$
6. $(3a^2) \cdot (-4a^4)$
7. $(a^2)^3 \cdot (-2a)^2$

Write in mathematical terms.

8. 3 less than 3 times a number
9. three times $-x$
10. r divided by -6

Factor.

11. $-2x^3 - 6x^2 + 6x$

Solve.

12. $-4x + 6 = -2$
13. $-5x \geq 15$

Solve the system by graphing.

14. $y = -x + 6$
    $y = 5x - 6$

Graph on a number line.

15. $x > -3$
16. $x \leq 4$

Evaluate.

17. $(-3x)^2$ if $x = \frac{1}{9}$
18. $-5x^2y^2$ if $x = 2$ and $y = -\frac{1}{4}$

19. In a right triangle, one leg is 3 in. long and the hypotenuse is 7 in. long. How long is the other leg?

20. In a triangle, two sides are 8 in. long each and the third side is 3 in. long. What kind of a triangle is it?

21. Name three kinds of angles.

22. $m\angle A = 25°$. What is its complement? Its supplement?

23. What is the slope of $\overleftrightarrow{XY}$ with $X(1, 4)$ and $Y(5, 7)$?

24. $A = \frac{bh}{2}$. Find A if $b = 7$ in. and $h = 9$ in.

25. The base of an isosceles triangle is 7 cm and the perimeter is 19 cm. Find the length of each leg.

Solve these problems.

26. Martha bought 3 items at these prices: $3.78, $2.45, and $6.15. How much change will she receive from a $20 bill?

27. You worked $19\frac{3}{4}$ h Monday through Thursday. How many more hours must you work on Friday to work 26 h altogether?

Appendix

Computer Section	**420**
Extra Practice	**436**
Answers to Practice Exercises	**455**
Odd-numbered Answers to Extra Practice Exercises	**466**
Table of Measures	**471**
Symbol and Formula Lists	**472**
Table of Squares and Square Roots	**473**
Glossary	**474**
Index	**479**

Computer Section

INPUT

OBJECTIVE
To write programs using the **INPUT** statement

Suppose you want to calculate $7 + 2y$ for any value of y.

First assign some variable (letter) to the value of $7 + 2y$, say S.

Write a formula. $\qquad\qquad\qquad\qquad\qquad$ $S = 7 + 2 \cdot Y$

Now express the formula in **BASIC:** $\qquad$ **LET S = 7 + 2 ∗ Y**

You are now ready to write a program to evaluate $7 + 2 * Y$ for any value of Y.

Example 1 $\quad$ Write a program to evaluate $7 + 2 * Y$ for any value of Y.
$\qquad\qquad\qquad$ Run the program to find the value of $7 + 2 * Y$ for $Y = 3$.
$\qquad\qquad\qquad$ Remember to press **RETURN** at the end of each line.

Tells the computer that you will be assigning a value to Y. The computer will ask the user for a value of Y by printing a "?".	10 INPUT Y
This formula tells how to find S.	20 LET S = 7 + 2 ∗ Y
	30 PRINT S
Tells the computer that the program is ENDed.	40 END
Type **RUN**. (No line number needed.)	]RUN
The computer is asking for a value for Y.	?
The value for Y.	Type **3**. Press **RETURN**.
The computer displays the results.	13

You can now use the program in **EXAMPLE 1** to evaluate $7 + 2 * Y$ for any other value of Y by **RUN**ning the program for each new value of Y.

Example 2 The formula for the perimeter of a rectangle is $P = 2l + 2w$. Write a program to find the perimeter of any rectangle. Then type and **RUN** the program to find P for $l = 8$ and $w = 6$.

First write the formula in **BASIC**.

$P = 2 \cdot l + 2 \cdot w$ becomes **P = 2 * L + 2 * W**.

Computer will expect a value for L.	10 INPUT L
Computer will expect a value for W.	20 INPUT W
Formula in **BASIC**.	30 LET P = 2 * L + 2 * W
	40 PRINT P
	50 END

Now **RUN** the program to find the value of P for L = 8 and W = 6.

Press **RETURN**.

Type **RUN**.

Press **RETURN**.

The computer asks you for a value for L.	**?** is displayed on screen.
The value of L.	You type **8**.

Press **RETURN**.

The computer now wants a value for W.	**?** is displayed again.
The value of W.	You type **6**.

Press **RETURN**.

The value of P for L = 8, W = 6	**28** is displayed.

Example 3 The following is a copy of a program. Use your knowledge of algebra to fill in the blank in the screen display.

Formula for finding T.	10 INPUT A
	20 LET T = (4 * A + 10) / (2 * A - 2)
	30 PRINT T
	40 END
	]RUN
	?2
Programmer types in 2 as value of A. What will be displayed on the screen?	☐

The program gives the value of T in the formula. T = 9

INPUT

Exercises

Fill in the missing blanks. Then type and RUN the program to evaluate for the given variable.

1. Evaluate $3A + 4$: $A = 7$

   10 INPUT A
   20 LET S = _____
   30 PRINT _____
   40 END _____

2. Evaluate $\frac{2X + 6}{4}$: $X = 9$

   10 INPUT _____
   20 LET T = (2 * X + 6) / 4
   30 PRINT _____
   40 END _____

3. Evaluate x^2: $X = 19$

   10 INPUT _____
   20 LET M = _____
   30 _____
   40 _____

4. Evaluate $4X + 9Y$: $X = 5$, $Y = 2$

   10 INPUT _____
   20 INPUT _____
   30 LET G = _____
   40 _____
   50 _____

5. Evaluate $\frac{A + 4}{B}$: $A = 6$, $B = 2$

   10 INPUT _____
   20 INPUT _____
   30 LET K = (A + 4) / B
   40 _____
   50 _____

6. Evaluate $\frac{A + B + C}{3}$: $A = 6$, $B = 9$, $C = 12$

   10 INPUT _____
   20 INPUT _____
   30 _____
   40 LET _____
   50 _____
   60 _____

Use your knowledge of algebra to predict the final screen display.

7. ```
   10  INPUT A
   20  LET S = 3 * A + 4
   30  PRINT S
   40  END

   ]RUN
   ?5
   ☐
   ```

8. ```
   10  INPUT M
   20  LET G = 24 - 2 * M
   30  PRINT G
   40  END

   ]RUN
   ?5
   ☐
   ```

Evaluate. Then write and RUN a program for each evaluation.

9. $5A + 9$ for $A = 6$

10. X^3 for $X = 8$

11. $\frac{3B + 6}{A}$ for $B = 2$, $A = 4$

12. The formula for the area of a square is $A = s^2$ where s is the length of a side. Write a program to find the area of any square. **RUN** the program to find A for $s = 6$.

13. The formula for the braking distance in meters for a car traveling at a given rate in kilometers per hour is $B = 0.006s^2$. Write a program to find the distance for any speed. **RUN** the program to find B for $s = 45$ km/h.

14. For the program in **Exercise 12** above, find out what happens to the area when a side is doubled. Is the area doubled? (*Hint:* **RUN** for $s = 12, 24, 48$).

LOOPS

OBJECTIVE
To write programs using **FOR-NEXT** loops

The formula for finding a person's salary if you know the wage and the hours worked is $S = W \cdot H$. Suppose you want to write a program to find S for different values of W and H. Once the program is written and **RUN** you can get the result by merely typing in the values of W and H. However, for each new set of values of W and H you must retype **RUN**. Programmers use a technique that avoids the repetition of typing **RUN** over and over. The technique involves using a *loop* as shown below.

Example 1 Write a program to find the value of S in the formula $S = W \cdot H$. Allow for executing the program for three different sets of values of W and H. Then **RUN** the program to find S for: $W = \$5.95$, $H = 42$; $W = \$7.00$, $H = 39$; $W = \$5.75$, $H = 40$.

First write the formula in **BASIC**.

$S = W \cdot H$ becomes **S = W * H.**

Use I or any letter other than S, W, or H.	10 FOR I = 1 TO 3
	20 INPUT W
	30 INPUT H
The program calculates S 3 times.	40 LET S = W * H
The computer loops back 3 times for 3 different sets of values for W and H.	50 PRINT S
	60 NEXT I
	70 END
Type 5.95 for the first wage.	]RUN
Type 42 for the number of hours.	?5.95
The value of S	?42
The computer asks for second set of values for W and H.	249.9
	?7
	?39
The value of S for W = 7 and H = 39.	273
Type in the third set of values.	?5.75
	?40
The value of S for the third set of values of W and H corresponds to I = 3.	230

FOR I = 1 TO 3 instructed the computer to calculate S for 3 sets of values of W and H and then stop or **END**.

LOOPS

In the program found in Example 1, the statements between lines 10 and 60 are processed 3 times because the **FOR** statement in line 10 tells the computer to do so. Each time the computer reaches line 60, it goes back to line 10 for the next value of I. After using I = 3, the computer goes back with a value of I = 4. It checks and finds that I cannot be bigger than 3. The computer then drops out of the **FOR-NEXT** loop and goes to line 70 **END**. In any **FOR-NEXT** loop, there must be agreement between the letter used in the **FOR** statement and the letter used in the **NEXT** statement.

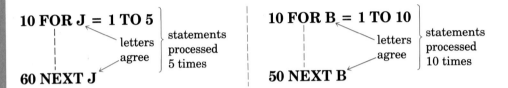

Example 2 A rocket is launched with an initial velocity of 128 ft/s. The formula for the height of the rocket at the end of T seconds is $H = 128T - 16T^2$. Write a program to find the value of H for any 2 values of T. Then **RUN** the program to find H for T = 2, T = 5.

First, write the formula in **BASIC**.

$H = 128T - 16T^2$ becomes **H = 128 * T − 16 * T ∧ 2.**

Use K or any letter other than T or H.	10 FOR K = 1 TO 2 20 INPUT T 30 LET H = 128 * T - 16 * T ^ 2 40 PRINT H 50 NEXT K 60 END
Computer asks for first value of T; you type 2. Screen displays the answer; height of 192. Computer loops back to line 10, asks for second and last value of T; you type 5. Computer **PRINTS** the answer, 240.	]RUN ?2 192 ?5 240

Since the computer is told that there are only 2 values of T (**FOR K = 1 TO 2**), the looping process is now complete. The computer goes to the next line of the program: 60 **END**. The program is **END**ed.

Exercises

Complete the program by filling in the missing blanks.

1. Write a program to find the value of Y for 6 different values of X.
   $Y = 4X^3 + 5X$
   10 FOR I = 1 to _____
   20 INPUT X
   30 LET Y = _____
   40 PRINT Y
   50 NEXT _____
   60 END

2. Write a program to find the value of A for 8 different sets of values of B and H.
   $A = \frac{BH}{2}$
   10 FOR K = _____
   20 INPUT _____
   30 INPUT H
   40 LET _____
   50 PRINT _____
   60 _____
   70 _____

Use the following program to answer Exercises 3 through 5.

```
10  FOR M - 1 TO 4
20  INPUT X
30  LET Y = 2 * X ^ 3
40  PRINT Y
50  NEXT M
60  END
```

3. What is the significance of line 10?

4. What will cause the program to go to line 60?

5. Predict, without **RUN**ning the program, the output for the following values of X: X = 2, X = 4, X = 5, X = 7. Then **RUN** the program to see if the computer agrees with your algebraic interpretation.

6. The formula for simple interest is I = PRT.
   Write a program to evaluate I for any 2 different sets of values of P, R, and T. **RUN** the program for P = $2000, R = 12%, T = 4 years, and for P = $4500, R = 23%, T = 3 years. (*Hints:* Type values of P without "$". Type 12% as .12, 23% as .23. Do not use commas in numbers like 2000 or 4500.)

7. $Y = 4X^2 + 3X$.
   Write a program to evaluate Y for any 5 different values of X. Compute the values of Y for the following values of X: 5, 4, 6, 7, 2. Then **RUN** the program. Check whether or not the computer agrees with your computations.

LOOPS

MORE ON PRINT AND INPUT

OBJECTIVES

To use the print command with expressions in quotation marks
To write programs using *enhanced* **INPUT**

Example 1 Type and **RUN** the following program. What is the difference in results among the 3 **PRINT** statements?

Line 20 instructs the computer to print exactly what is in quotation marks, the letter P, not the value of P. Line 30 results in the **PRINT**ing of the value of P, 5. Line 40 combines the two types of **PRINT** command.	10 LET P = 5 20 PRINT "P" 30 PRINT P 40 PRINT "P="P 50 END]RUN P 5 P=5

Example 2 What will be the output of the following program when **RUN**?

When the program is **RUN**, line 20 will produce this output. The value of L is 24 since 8 ∗ 3 = 24.	10 LET L = 8 * 3 20 PRINT "L="L 30 END]RUN L=24

Recall that when a program using **INPUT** is RUN a "?" appears on the screen. If someone else were **RUN**ning the program the meaning of the "?" might not be clear. In this lesson you will learn how to *enhance* the **INPUT** statement. It will be clear to anyone **RUN**ning your program what is expected when the "?" appears. This is illustrated in **Example 3**.

Example 3 Write a program to find the product of any two numbers. Allow for **RUN**ning the program for any two pairs of numbers. When **RUN**, the screen should display an actual instruction, not merely a **"?"**. The output should also explain what the answer represents. Then **RUN** the program to find the products 5 * 8 and 4 * 6.

First write a formula in **BASIC**.
Product means multiply.
Let P = A * B, where A and B are any two numbers.

```
                                    10   PRINT "PROGRAM FINDS PRODUCTS."
Tells the operator what the program
                will do              20   FOR I = 1 TO 2
INPUT is split into two separate     30   INPUT "TYPE IN A NUMBER ";A
              statements.
                                     40   INPUT "TYPE IN A SECOND NUMBER ";B
                                     50   LET P = A * B
                                     60   PRINT "THEIR PRODUCT IS "P
Allows one line of space before next 70   PRINT
             question
                                     80   NEXT I
                                     90   END

                                     ]RUN
                                     PROGRAM FINDS PRODUCTS.
                                     TYPE IN A NUMBER 5
                     Type 5.         TYPE IN A SECOND NUMBER 8
                     Type 8.         THEIR PRODUCT IS 40
            Screen displays result.  TYPE IN A NUMBER 4
        This corresponds to I = 2.   TYPE IN A SECOND NUMBER 6
                                     THEIR PRODUCT IS 24
```

Leave 1 space after **IS** so the computer will print a space before the answer.

Notice the structure and role of the *enhanced* **INPUT** statement in lines 30 and 40 of the program.

30 INPUT "TYPE IN A NUMBER "; A

Instruction in quotation marks — space — A semicolon is needed.

There is a space between number and " to avoid screen display **TYPE IN A NUMBER 5.** The result is that when the program is **RUN**, the **INPUT** is clarified or *enhanced* by an actual instruction, **TYPE IN A NUMBER.**

MORE ON PRINT AND INPUT

Exercises

What will be the output of each program when it is RUN?

1. ```
   10  LET W = 4
   20  PRINT "W"
   30  PRINT W
   40  PRINT "W="W
   50  END
   ```

2. ```
   10  LET M = 3 * 14
   20  PRINT "M="M
   30  END
   ```

3. ```
   10  LET S = 12 / 4
   20  PRINT "S="S
   30  END
   ```

4. When the following program is **RUN** what will be the output: if B = 8?; if B = 5?

   ```
   10  PRINT "PROGRAM EVALUATES 6+4B FOR ANY TWO DIFFERENT VALUES OF B"
   20  FOR I = 1 TO 2
   30  INPUT "TYPE IN A VALUE FOR B ";B
   40  LET Y = 6 + 4 * B
   50  PRINT "THE VALUE OF Y IS ";Y
   60  NEXT I
   70  END
   ```

Find the error in each statement. Then rewrite it correctly.

5. 10 INPUT TYPE IN A VALUE FOR A; A

6. 10 INPUT "WHAT IS THE VALUE OF B?" B

7. 10 INPUT "WHAT IS THE VALUE OF M?; M

8. 10 PRINT THE ANSWER IS W

9. 10 PRINT "THE ANSWER IS W" (Numerival value of W is desired output.)

For each of the following:
(a) Write a program using *enhanced* **INPUT** and the new **PRINT** technique of this lesson;
(b) predict the output algebraically before **RUN**ning the program;
(c) type and **RUN** the program for the given value(s) of the variable(s).

10. Evaluate $7X + 4$ if: $X = 8$; $X = 3$; $X = 23$

11. Evaluate $17 - 2M$ if: $M = 4$; $M = 5$

12. Evaluate $2L + 2W$ if: $L = 8$, $W = 6$; $L = 9$, $W = 6$

13. Evaluate $\dfrac{A}{2A - 4}$ if: $A = 6$; $A = 8$; $A = 20$

14. Evaluate $3B^5$ if: $B = 2$; $B = 6$; $B = 5$; $B = 7$

15. Find the quotient of any two non-zero numbers. **RUN** the program for the following pairs of numbers: 12 and 6; 45 and 15; 48 and 96; 22 and 88.
   ```

THE IF...THEN STATEMENT

OBJECTIVE
To write programs using the **IF . . . THEN** statement

Consider the following department store advertisement.

To find the cost of 6 tapes you must make a decision.
Do you multiply by $7.00 or $4.00?
Since 6 is more than 5, you must use $4.00. The cost is $4.00 × 6 or $24.00.

A computer program for finding the total cost of the tapes must tell the computer how to "decide" whether to multiply by 7 or 4.

Example   Write a program to find the cost of the tapes described above. Allow for 3 calculations. Type and **RUN** the program to find the cost of 2 tapes, 8 tapes, 5 tapes.

The computer will have to "decide" which of two formulas to use.

Total = 7 times Number bought   OR   Total = 4 times Number bought

$T = 7 * N$ $T = 4 * N$

Note that in the program below the symbol for *is greater than* is >.

```
                                    10   FOR K = 1 TO 3
                                    20   INPUT "TYPE IN NUMBER OF TAPES ";N
    The computer checks if N > 5.   30   IF N > 5 THEN 60
    If so, it skips down to line 60. 40   LET T = 7 * N
    If N ≤ 5 it goes to line 40.    50   GOTO 70
                                    60   LET T = 4 * N
                                    70   PRINT "TOTAL COST IS "T
                                    80   PRINT : PRINT
                                    90   NEXT K
                                    100  END
```

The **RUN** for this program is shown on page 430.

THE IF . . . THEN STATEMENT

You type in 2 since 2 < 5. Line 40 is used: 7 * 2 = 14.	]RUN TYPE IN NUMBER OF TAPES 2 TOTAL COST IS 14
Line 80 **PRINT:PRINT** produces 2 lines of separation between results. This corresponds to K = 2 and 8 > 5. Line 60 is used: 4 * 8 = 32.	TYPE IN NUMBER OF TAPES 8 TOTAL COST IS 32
Third and last calculation 5 is not greater than 5. So, line 40 is used: 7 * 5 = 35.	TYPE IN NUMBER OF TAPES 5 TOTAL COST IS 35

In general, the **IF . . . THEN** statement has the following form.

the word **IF** . . . the word **THEN**

30 IF N > 5 THEN 60

line number   any line number other than that of the statement

Two arithmetic expressions joined by a decision symbol such as > comparing N with 5.

BASIC Decision Symbols

Symbol	Meaning
=	is equal to
>	is greater than
> =	is greater than or equal to
<	is less than
< =	is less than or equal to
< >	is not equal to

Exercises

Identify each of the following IF . . . THEN statements as correct or incorrect. If it is incorrect, copy and correct it.

1. 10 IF Y > 0 THEN 90
2. 20 IF A + M THEN 100
3. 40 IF G < 6 THEN 40

4. Write a program to find the cost of pens as advertised: "Special sale on pens: $1.50 each; only $1.25 each if you buy more than 6." Allow for three calculations. Then type and **RUN** the program to find the cost of 2 pens; 8 pens; 6 pens.

Use the following program to answer Exercises 5–10.

```
10   FOR J = 1 TO 4
20   INPUT "WHAT IS THE VALUE OF X? ";X
30   IF X < 4 THEN 60
40   LET Y = 3 * X
50   GOTO 70
60   LET Y = 12 / X
70   PRINT "THE VALUE OF Y IS "Y
80   PRINT : PRINT : PRINT
90   NEXT J
100  END
```

5. What is the significance of line 10?

6. Predict without **RUN**ning the program the value of Y, if X = 6.

7. Predict without **RUN**ning the program the value of Y, if X = 3.

8. At what point in the program will the computer carry out line 100?

9. Type and **RUN** the program for X = 2, 4, 8, and 12.

10. You can delete a line from a program by typing only the number of the line. Omit or delete line 50 by typing 50. Now **RUN** the program for the same values of X: 2, 4, 8, and 12. What happens? Why?

11. Bill is a salesman. If the total of his sales for each of two weeks is greater than $750, his commission is $50. Otherwise he is paid no commission. Complete the following program for determining whether or not Bill is paid a commission.

    ```
    10 FOR K = 1 TO 3
    20 INPUT "TYPE IN SALES FOR THE FIRST WEEK "; F
    30 INPUT _____
    40 LET T = F + _____
    50 IF _____ THEN _____
    60 PRINT "NO COMMISSION"
    70 GO TO _____
    80 PRINT _____
    90 NEXT _____
    100 END
    ```

    Type and **RUN** the program for sales of $475 and $319; sales of $295 and $125; sales of $335 and $435.

12. A teacher wants to write a program to **PRINT** the average of 4 grades. If the average is less than 70, the computer should also **PRINT "SEND A WARNING NOTICE"**. Write a program allowing for two sets of grades. Type and **RUN** the program to average: 72, 60, 80, 60; 80, 90, 90, 80.

THE IF . . . THEN STATEMENT

SEQUENTIAL INPUT

OBJECTIVE
To write programs involving input of numbers that are in sequence

Example 1   Write a program to find the square of any 6 numbers.

First write a formula in **BASIC**.

LET S = Number squared

LET S = N^2

LET S = N $\wedge$ 2

The computer will loop back 6 times.	10 FOR I = 1 TO 6
Note that I and N are not the same.	20 INPUT "TYPE NUMBER BEING SQUARED ";N
	30 LET S = N ^ 2
This identifies the output.	40 PRINT "SQUARE IS "S
	50 NEXT I
	60 END

Now suppose you want to write a program to square all numbers in *sequence* from 1 to 100; that is, $1^2, 2^2, 3^2, 4^2, 5^2, 6^2, 7^2, \ldots, 99^2, 100^2$.

When the program is **RUN** you would have to enter 100 values for **N**.
There is a simpler way to write the program when the *input* is *sequential*.

Example 2   Write and **RUN** a program to square the numbers from 1 to 5 using the *sequential input* method.

This provides a heading for the output.	10 PRINT "NUMBER","SQUARE OF NUMBER"
INPUT statement not needed.	20 FOR N = 1 TO 5
It is N itself being squared.	30 LET S = N ^ 2
Thus, when N = 1, 1 is squared, and so on.	40 PRINT N,S
	50 NEXT N
	60 END

The output of **RUN** is shown on the next page.

```
            ]RUN
            NUMBER          SQUARE OF NUMBER
            1               1
            2               4
            3               9
            4               16
            5               25
```

Note the use of the comma in line 40 of the program: **PRINT N, S**. This resulted in the printing of two columns; the number in one column and the square of the number in the second column.

The numbers 3, 7, 11, 15, 19, 23, . . ., and so on are in sequence. Successive numbers of the sequence differ by 4. The number after 11 is **STEP**ped up by 4, or increased by 4. The numbers—30, 25, 20, 15, 10—are also in sequence. Successive numbers are decreased by 5. The number following 25 is **STEP**ped down (decreased) by 5. The use of **STEP** in sequential looping is illustrated in the next Example.

Example 3 Write a program to evaluate $2A + 5$ for the following values of A: 3, 5, 7, 9, 11, 13, 15.

The values of A are in sequence; successive numbers are **STEP**ped up (increased) by 2.

FOR A = 3 to 15 STEP 2
- First number
- Last number
- Increases numbers by 2

```
10  PRINT "2A + 5","VALUE OF 2A + 5"
20  FOR A = 3 TO 15 STEP 2
30  LET V = 2 * A + 5
40  PRINT A,V
50  NEXT A
60  END
```

RUN the program on your own.

Exercises

Write and RUN a program to find each of the following.

1. The square of the numbers 4, 7, 10, 13, 16.
2. The value of X^3, if $X = 1, 2, 3, 4, 5, 6, 7, 8, 9$.
3. The value of $4B + 3$, if $B = 6, 8, 10, 12, 14, 16$.
4. The value of $X^2 + 3$, if $X = 1, 7, 13, 19, 25, 31, 37$.
5. The squares of the numbers 14, 12, 10, 8, 6, 4, 2. (*Hint:* the numbers are successively **STEP**ped down by 2; **STEP** -2.)

SEQUENTIAL INPUT

INT

OBJECTIVE
To use **INT** to determine if one number is a factor of another

The **INT**eger part of the number 8.679543 is 8.
In **BASIC**, the instruction **PRINT** the **INT**eger part of 8.679543 is written as **PRINT INT(8.679543).**

Example 1 Find each: **INT(3.45649)**   **INT(3.87569)**   **INT(3)**

INT(3.45649) = 3   **INT(3.87569)** = 3   **INT(3.0)** = 3

Example 2 Use **INT** to determine whether 4 is a factor of 15.

$$\begin{array}{r}3.75\\4\overline{)15.00}\end{array}$$

Does 4 divide into 15 evenly?

Does 15/4 = **INT(15/4)**?

Does 3.75 = **INT(3.75)**?

No. 3.75 ≠ 3

So, 4 is not a factor of 15.

Example 3 Write a program to determine whether a number is a factor of 528.

```
10   INPUT "TYPE IN A POSSIBLE FACTOR OF 528 ";F
20   IF 528 / F = INT (528 / F) THEN 50
30   PRINT F" IS NOT A FACTOR OF 528."
40   GOTO 60
50   PRINT F" IS A FACTOR OF 528."
60   END
```

Exercises

1. Find each of the following: **INT(32.18976)**; **INT(10 * 0.3455654 + 1)**.

2. Modify the program above with a loop so that you can test whether any 5 different numbers are factors of 528. Then **RUN** the program to determine which of the following numbers are factors of 528: 4, 17, 24, 96, 176.

RND

Objective
To represent random numbers

When tossing a die (one of a pair of dice), the die may come up a 5 or a 6 or a 3, or any one of 6 possible numbers. A computer can be programmed to simulate this by choosing *random* numbers from 1 to 6.

Most computers use **RND**(1) to generate random numbers between 0 and 1. So, the command **10 PRINT RND(1)** would result in the printing of numbers such as .779343355 or .103117626.

Example 1 Show that **INT(6 * RND(1) + 1)** produces random numbers from 1 through 6. Use the two values of **RND**(1) above.

INT(6 * RND(1) + 1)	**INT(6 * RND(1) + 1)**
INT(6 * .779343355 + 1)	**INT(6 * .103117626 + 1)**
INT(4.67606013 + 1)	**INT(.618705756 + 1)**
INT(5.6760606013)	**INT(1.618705756)**
5	1

So, **INT(6 * RND(1) + 1)** generates numbers from 1 through 6 such as 1 and 5.

Example 2 Write an expression to produce random numbers from:

1 through 4	1 through 23
INT(4 * RND(1) + 1)	**INT(23 * RND(1) + 1)**

Exercises

1. Which of the following could be a result of printing **RND**(1)?

    a. 4.678567857
    b. 5.678678905
    c. 0.9987765554

2. Write a representation for random numbers from 1 through 49.

RND

Extra Practice

PRACTICE ON SOLVING ADDITION AND SUBTRACTION EQUATIONS: CHAPTER 1

Evaluate for the given values of the variables.

1. $x + 15; x = 5$
2. $35 - a; a = 18$
3. $a + b - c; a = 15, b = 20, c = 5$
4. $a + b + c; a = 3, b = 6, c = 8$
5. $c - a + b; a = 3, b = 9, c = 1$
6. $a - b + c; a = 10, b = 6, c = 7$

Rewrite to make computation easier. Then compute.

7. $4 + 72 + 96$
8. $17 + 13 + 183$
9. $3 + 96 + 297$
10. $15 + 10 + 85$
11. $45 + 65 + 55$
12. $346 + 49 + 54$

Which, if any, of the values is a solution of the open sentence?

13. $a + 3 = 10; 5, 13$
14. $42 = x - 4; 37, 46, 47$
15. $a + 12 = 4 + 9; 0, 1, 25$
16. $y - 18 = 32 + 8; 22, 42, 58$
17. $x + 12 = 12; 0, 1, 12$
18. $b - 8 = 14 - 8; 6, 10, 12$

Solve each equation. Check the solution.

19. $x + 5 = 10$
20. $x - 6 = 15$
21. $x + 15 = 26$
22. $14 = a - 12$
23. $a + 15 = 17$
24. $b + 6 = 19$
25. $y - 10 = 15$
26. $m + 12 = 20$
27. $a + 32 = 40$
28. $x - 12 = 18$
29. $y + 16 = 36$
30. $y - 17 = 3$

PROBLEM SOLVING

31. If the cost of a pair of shoes is decreased by $8, the price is $38. Find the cost.

32. The sum of Karen's salary and a $40 commission is $430. Find her salary.

33. 18 is the same as a number decreased by 10. What is the number?

34. 15 more than Merv's age is 38. How old is Merv?

Find P.

35. $P = a + b + c$

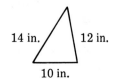

36. $P = l + w + l + w$

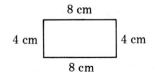

PRACTICE ON SOLVING MULTIPLICATION AND DIVISION EQUATIONS: CHAPTER 2

Evaluate for the given values of the variables.

1. $8a; a = 4$
2. $\frac{12}{x}; x = 6$
3. $\frac{6t}{4}; t = 6$
4. $\frac{144}{b}; b = 16$
5. $15a; a = 15$
6. $\frac{12x}{5}; x = 10$
7. $6x + 5; x = 3$
8. $\frac{a}{2} + 5h + 2; a = 4, b = 2$
9. $4 + 5m + 2n; m = 4, n = 5$
10. $4xy + 5; x = 8, y = 5$
11. $10ab + a - b; a = 6, b = 4$
12. $40ab - 20ab; a = 2, b = 3$
13. $25 - 4x + 5xy; x = 4, y = 12$
14. $16 + 8m + 2mn; m = 5, n = 10$

Rewrite to make computation easier. Then compute.

15. $5 \cdot 32 \cdot 20$
16. $2 \cdot 9 \cdot 50$
17. $4 \cdot 8 \cdot 5 \cdot 5$
18. $50 \cdot 70 \cdot 2$
19. $2 \cdot 14 \cdot 5 \cdot 10$
20. $8 \cdot 9 \cdot 25$

Solve each equation. Check the solution.

21. $8x = 40$
22. $72 = 9m$
23. $12 = \frac{x}{5}$
24. $36 = 4y$
25. $6a = 12$
26. $4x = 16$
27. $\frac{a}{10} = 7$
28. $\frac{x}{32} = 3$
29. $15 = \frac{y}{6}$
30. $45 = \frac{x}{2}$
31. $3n = 90$
32. $\frac{a}{10} = 12$

PROBLEM SOLVING

33. 8 times a number is 32. Find the number.
34. 12 times a number is 60. Find the number.
35. The $90 selling price of a coat is twice the cost. Find the cost.
36. Jamie's age of 6 is the same as her aunt's age divided by 5. How old is her aunt?
37. Cathy's salary of $800 per month is the same as her mother's salary divided by 3. Find her mother's salary.
38. Rico's 36 base hits this season is twice his number of hits last season. How many base hits did he have last year?
39. The total cost of a restaurant bill divided among 5 people is $4 each. Find the total cost.
40. Three times O'Neil's salary is $315. What is his salary?
41. $A = l \cdot w$. Find A if $l = 8$ cm and $w = 6$ cm.
42. $P = 2l + 2w$. Find P if $l = 12$ cm and $w = 4$ cm.

PRACTICE ON SIMPLIFYING EXPRESSIONS AND SOLVING EQUATIONS: CHAPTER 3

Simplify.

1. $4 \cdot 3m$
2. $8m \cdot 7$
3. $3(4x + 3)$
4. $12(2a - 1)$
5. $5a + 6a$
6. $6x - 3 + 4x$
7. $3a + 2a - 5$
8. $3m + 2n + 3 - 2m$
9. $5(3a - 3)$
10. $5 + 2a + 4a$
11. $8x + 5y + x + y$
12. $14 + 5x + 2y - 12 + x + y$

Simplify. Then evaluate for these values of the variables.

13. $4x + 9 + 3x; x = 5$
14. $7a + 3 + 16a; a = 10$
15. $10 + a + 6a; a = 8$
16. $2x + 3y - 10; x = 5, y = 10$
17. $15m + 10m - 3m + 2; m = 5$
18. $9b + 6a - 3a + 10b; a = 2, b = 3$
19. $5x + 3x + 8y - 2y; x = 4, y = 5$
20. $12 + 3a + 5 + 2b + 6a; a = 6, b = 7$

Solve each equation. Check the solution.

21. $3n + 6 = 15$
22. $2x - 7 = 9$
23. $12 = \frac{x}{5} - 3$
24. $21 = 7x + 7$
25. $36 = 8n + 4$
26. $\frac{a}{6} + 7 = 13$
27. $10y - 9 = 81$
28. $\frac{y}{12} + 7 = 11$
29. $11m - 4 = 73$
30. $32 = 9a - 22$
31. $16 = 5a - 19$
32. $46 = 5y + 11$

Find the value.

33. 3^2
24. 2^4
35. 1^7
36. 6^2
37. 5^3
38. 4^3
39. 3^5
40. 7^3
41. 2^6
42. 3^3
43. 5^4
44. 4^5

Write using exponents.

45. $5 \cdot 5 \cdot 5$
46. $a \cdot a \cdot a \cdot a$
47. $3 \cdot 3 \cdot 4 \cdot 4 \cdot 4 \cdot 4$
48. $m \cdot m \cdot n \cdot n \cdot n$
49. $6 \cdot 6 \cdot x \cdot x \cdot x \cdot x \cdot x$
50. $4 \cdot 4 \cdot 4 \cdot m \cdot m$
51. $x \cdot x \cdot x \cdot y \cdot y \cdot y \cdot y$
52. $2 \cdot 2 \cdot 2 \cdot b \cdot b \cdot b \cdot b$

PROBLEM SOLVING

53. 3 less than 5 times a number is 32. Find the number.
54. 18 is the same as twice a number decreased by 6. Find the number.
55. 45 is the same as 3 times a number increased by 6. Find the number.
56. 12 more than a number divided by 6 is 23. Find the number.

PRACTICE ON MULTIPLYING AND DIVIDING FRACTIONS: CHAPTER 4

Multiply.

1. $5 \cdot \frac{1}{3}$
2. $6 \cdot \frac{1}{5}$
3. $4 \cdot \frac{2}{7}$
4. $\frac{2}{3} \cdot 7$
5. $3 \cdot 2\frac{4}{5}$
6. $2\frac{2}{3} \cdot 2$
7. $10 \cdot \frac{3}{7}$
8. $4 \cdot 1\frac{2}{5}$
9. $\frac{2}{3} \cdot \frac{4}{5}$
10. $\frac{3}{2} \cdot \frac{5}{4}$
11. $\frac{1}{5} \cdot 3\frac{1}{2}$
12. $6\frac{1}{5} \cdot \frac{2}{3}$

Simplify.

13. $\frac{4}{10}$
14. $\frac{10}{15}$
15. $\frac{12}{28}$
16. $\frac{36}{15}$
17. $\frac{5}{12} \cdot \frac{4}{15}$
18. $\frac{8}{3} \cdot 1\frac{7}{8}$
19. $8 \cdot \frac{5}{24}$
20. $3\frac{3}{7} \cdot \frac{14}{8}$
21. $\frac{2}{3}$ of 15
22. $\frac{3}{5}$ of 25
23. $2\frac{4}{5} \cdot 5\frac{5}{7}$
24. $1\frac{3}{4} \cdot 3\frac{3}{7}$

Divide.

25. $\frac{3}{4} \div \frac{12}{5}$
26. $5\frac{1}{3} \div 4$
27. $20 \div 3\frac{1}{5}$
28. $\frac{7}{4} \div 5\frac{1}{4}$
29. $16 \div 1\frac{1}{3}$
30. $\frac{4}{9} \div \frac{14}{15}$
31. $3\frac{1}{5} \div 8$
32. $\frac{3}{10} \div \frac{15}{24}$
33. $1\frac{1}{2} \div 2\frac{2}{3}$
34. $\frac{4}{5} \div 1\frac{1}{4}$
35. $1\frac{7}{8} \div \frac{5}{9}$
36. $3\frac{1}{3} \div \frac{2}{5}$
37. $2\frac{2}{9} \div 3\frac{1}{5}$
38. $2\frac{4}{7} \div 4\frac{2}{7}$
39. $1\frac{1}{6} \div 2\frac{5}{8}$
40. $4\frac{2}{3} \div 6\frac{2}{3}$

Solve.

41. $\frac{2}{3}x = 30$
42. $\frac{4}{5}x = 24$
43. $\frac{7}{6}x = \frac{21}{15}$
44. $\frac{1}{2}x = \frac{8}{7}$
45. $\frac{5}{3}x = \frac{25}{12}$
46. $\frac{4}{9}x = \frac{12}{27}$
47. $\frac{3}{4}x = \frac{1}{2}$
48. $\frac{7}{8}x = \frac{3}{4}$
49. $\frac{3}{7}x = \frac{2}{3}$
50. $8 = \frac{3}{2}x$
51. $\frac{4}{7} = \frac{5}{14}x$
52. $36 = \frac{6}{11}x$

PROBLEM SOLVING

Find A or V in terms of π, where possible.

53. $A = \frac{1}{2}bh; b = \frac{10}{3}, h = 12$
54. $A = \frac{1}{2}bh; b = \frac{9}{5}, h = 5$
55. $A = \pi r^2; r = 1\frac{2}{3}$
56. $A = \pi r^2, r = \frac{3}{5}$
57. $V = \frac{1}{3}\pi r^2 h; r = 1\frac{1}{5}, h = 4$
58. $V = \frac{1}{3}\pi r^2 h; r = 2\frac{2}{5}, h = \frac{1}{3}$
59. $V = \frac{4}{3}\pi r^3; r = 3\frac{1}{4}$
60. $V = \frac{4}{3}\pi r^3; r = 2\frac{1}{3}$

EXTRA PRACTICE

PRACTICE ON ADDING AND SUBTRACTING FRACTIONS: CHAPTER 5

Compute. Simplify if possible.

1. $\frac{4}{5} + \frac{3}{5}$
2. $\frac{3}{7} + \frac{2}{7}$
3. $\frac{5}{9} - \frac{4}{9}$
4. $\frac{7}{11} - \frac{2}{11}$
5. $\frac{4}{5} + \frac{2}{15}$
6. $\frac{4}{3} - \frac{4}{9}$
7. $\frac{2}{3} - \frac{1}{2}$
8. $\frac{5}{6} + \frac{7}{9}$

9. $\frac{5}{6} - \frac{2}{3}$
10. $\frac{5}{6} + \frac{2}{5}$
11. $\frac{15}{4} - \frac{3}{2}$
12. $\frac{1}{7} + \frac{3}{14}$

13. $\frac{3}{7} + \frac{5}{3}$
14. $\frac{7}{10} - \frac{3}{5}$
15. $\frac{7}{8} + \frac{2}{3}$
16. $\frac{5}{9} - \frac{2}{5}$

17. $\frac{2}{9} + \frac{2}{3} + \frac{3}{4}$
18. $\frac{4}{5} + \frac{4}{3} + \frac{1}{2}$
19. $\frac{6}{5} + \frac{4}{3} + \frac{3}{4}$

20. $5\frac{3}{5} + 2\frac{1}{10}$
21. $7\frac{3}{4} - 2\frac{1}{6}$
22. $5\frac{3}{8} + 7$
23. $17 - 6\frac{2}{5}$

24. $4\frac{2}{3} + 5\frac{1}{2}$
25. $5\frac{3}{4} + 4\frac{2}{5} + 3\frac{1}{2}$
26. $7\frac{1}{2} + 6\frac{1}{5} + 4\frac{3}{5}$
27. $3\frac{2}{5} - 1\frac{2}{3}$
28. $7\frac{3}{4} - 4\frac{2}{3}$
29. $16\frac{5}{8} - 8\frac{3}{16}$

Solve. Simplify if possible.

30. $x + \frac{2}{3} = \frac{4}{5}$
31. $x - \frac{1}{2} = \frac{5}{6}$
32. $x + \frac{2}{5} = \frac{1}{2}$
33. $x - 4\frac{1}{2} = 5\frac{1}{3}$
34. $x + 3\frac{1}{4} = 7\frac{2}{3}$
35. $x - 6\frac{1}{2} = 6\frac{3}{5}$

PROBLEM SOLVING

36. Harry cut a $6\frac{3}{4}$ ft board from a sheet of plywood 10 ft long. How long was the remaining piece?

37. Amy weighs $50\frac{1}{2}$ kg. How many kilograms must she gain in order to weigh $55\frac{3}{5}$ kg?

38. Susan needs $5\frac{7}{8}$ yd of wool for a skirt, $1\frac{1}{4}$ yd for a vest, and $3\frac{1}{2}$ yd for a jacket. How many yards of wool does she need for the outfit?

39. Gene jogs $5\frac{1}{2}$ mi on Monday, $6\frac{1}{4}$ mi on Wednesday, and $8\frac{1}{3}$ mi on Saturday. How many miles does he jog?

PRACTICE ON READING TABLES AND READING AND MAKING GRAPHS: CHAPTER 6

Answer questions 1–4 using the table.

BALANCE REMAINING OF $1,000 LOAN OVER 5 YEARS IF YOU ARE MAKING THE REQUIRED ANNUAL PAYMENTS.

Rate of Interest	1 Year	2 Years	3 Years	4 Years	5 Years
11%	$841	$664	$446	$246	Paid off
$11\frac{1}{2}\%$	$843	$667	$470	$248	Paid off
12%	$845	$670	$473	$250	Paid off
$12\frac{1}{2}\%$	$846	$673	$476	$253	Paid off
13%	$848	$675	$479	$255	Paid off

1. You borrowed $1,000 at 11%. How much money will you still owe after 3 yr?

2. You borrowed $1,000 at 13%. How much money will you still owe after 2 yr?

3. You borrowed $1,000 at $12\frac{1}{2}\%$. How much money will you still owe after 4 yr?

4. You borrowed $1,000 at 12%. When would this loan be paid off?

5. Construct a pictograph using the information below.
   35,000 Fans, Game 1; 40,000 Fans, Game 2;
   38,000 Fans, Game 3; 45,000 Fans, Game 4

6. Make a vertical bar graph to show these baseball scores: Game 1, 3 runs; Game 2, 12 runs; Game 3, 7 runs; Game 4, 1 run.

7. Make a horizontal bar graph to show the average monthly temperature for January in Chicago, 49°; Denver, 30°; Miami, 67°; Honolulu, 72°.

8. Make a line graph to show these test scores. First test, 95; Second test, 100; Third test, 88; Fourth test, 92; Fifth test, 98.

PROBLEM SOLVING
What is the number of births for each of the months?

9. September 1984

10. October 1982

11. November 1982

12. December 1984

13. Which of the years had the most births for Sept. through Dec.?

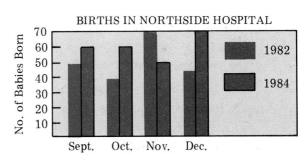

EXTRA PRACTICE

PRACTICE ON DECIMAL COMPUTATION: CHAPTER 7

Round to the nearest tenth and to the nearest hundredth.

1. 0.736
2. 7.659
3. 19.576
4. 16.435
5. 9.016
6. 97.455
7. 0.150
8. 173.497
9. 4.055
10. 10.986

Change to decimals. Round to the nearest tenth and to the nearest hundredth.

11. $\frac{658}{100}$
12. $\frac{6}{10}$
13. $\frac{86}{1000}$
14. $\frac{3}{5}$
15. $\frac{2}{3}$
16. $\frac{5}{11}$
17. $\frac{6}{7}$
18. $\frac{3}{4}$
19. $\frac{5}{12}$
20. $\frac{7}{9}$

Add.

21. 34.65 + 4.32 + 9.56
22. 8.07 + 93.654 + 9.1
23. 5.1 + 6.83 + 7.921
24. 0.35 + 0.256 + 7.3
25. 43.1 + 72.01 + 63.001
26. 41.1 + 38.23 + 0.6
27. 100.41 + 73.256 + 0.3
28. 9.101 + 12.35 + 0.721
29. 0.75 + 6.25 + 1.378

Subtract.

30. 73.47 − 41.68
31. 96.563 − 7.358
32. 127.065 − 72.198

33. Subtract 4.23 from 67.1.
34. Subtract 4.37 from 73.

Multiply.

35. (81.3)(7.45)
36. (43.75)(6.15)
37. (132.1)(0.73)
38. 81.3 × 7.5
39. 596 × 0.35
40. 68.5 × 7.39

Divide. Round to the nearest hundredth.

41. $0.03\overline{)4.987}$
42. $3.5\overline{)9.721}$
43. $0.65\overline{)72.956}$
44. $5.65\overline{)93.498}$
45. $0.71\overline{)4.1987}$
46. $2.42\overline{)91.865}$

Solve. Round to the nearest tenth.

47. $0.2x = 0.48$
48. $0.06x = 9.6$
49. $3.2x = 4.87$
50. $5.01x = 6.843$
51. $1.2x = 48.3$
52. $0.05x = 9.56$

PROBLEM SOLVING

53. Carlos ran 28.5 km in 2.5 h. Find his average speed.
54. Kathy saves $16.35 each week. How much does she save in 36 weeks?

PRACTICE ON SOLVING PROBLEMS INVOLVING METRIC UNITS: CHAPTER 8

Change as indicated.

1. 48 mm to cm
2. 9.38 km to m
3. 8 cm to m
4. 12 hm to m
5. 6 dm to m
6. 0.008 m to mm
7. 8 dm to m
8. 75 m to hm
9. 42 km to m
10. 15 cm to m
11. 40 mm to cm
12. 1,500 m to km

Make true sentences.

13. 1 m = _?_ cm
14. 8.7 m = _?_ cm
15. 3.7 km = _?_ m
16. 8 cm = _?_ mm
17. 12 km = _?_ m
18. 4 cm = _?_ m
19. 4.6 m² = _?_ cm²
20. 5 m² = _?_ cm²
21. 40,000 m² = _?_ ha
22. 8 L = _?_ mL
23. 15 mL = _?_ L
24. 7.3 L = _?_ mL
25. 30 mL = _?_ L
26. 12 L = _?_ mL
27. 25 mL = _?_ L
28. 6.2 g = _?_ gm
29. 4,500 g = _?_ kg
30. 15 kg = _?_ t

Change Celsius to Fahrenheit. Compute to the nearest degree.

31. C = 25°
32. C = 41°
33. C = 9°
34. C = 87°
35. C = 32°
36. C = 15°
37. C = 61°
38. C = 99°
39. C = 55°
40. C = 72°
41. C = 48°
42. C = 21°

Change Fahrenheit to Celsius. Compute to the nearest degree.

43. F = 100°
44. F = 40°
45. F = 52°
46. F = 73°
47. F = 77°
48. F = 51°
49. F = 85°
50. F = 33°
51. F = 64°
52. F = 49°
53. F = 96°
54. F = 200°

PROBLEM SOLVING

55. A square measures 6.8 cm on each side. Find the area in mm².

56. A rectangular field measures 210 m by 90 m. Find the area in hm².

57. A family drank $3\frac{1}{2}$ L of milk. How many milliliters is this?

58. A recipe calls for $\frac{3}{4}$ L of lemonade. How many milliliters is this?

59. A car weighs 950 kg. How many metric tons is this?

60. Lorin must take 6 mg of medicine each day. How many grams will he take in a year? (*Hint:* There are 365 days in a year.)

61. John drove his car for 75 km. How many meters is this?

62. A pole is 15 m long. How many centimeters is this?

EXTRA PRACTICE

MIXED PRACTICE: CHAPTERS 1–8

Evaluate for the given values of the variables.
1. $y + 7; y = 6$
2. $9n; n = 10$
3. $16a + 3 + 7a; a = 4$
4. $15ab + a + b + 20; a = 5, b = 4$
5. $17 + 4xy + 8y; x = 5, y = 3$

Which, if any, of the values is a solution of the open sentence?
6. $n + 5 = 12; 7, 17$
7. $a - 9 = 16 - 2; 5, 14, 24$

Solve each equation. Check the solution.
8. $y - 3 = 19$
9. $42 + x = 50$
10. $9y = 72$
11. $47 = \frac{x}{3}$
12. $4m + 9 = 53$
13. $41 = 5n - 4$
14. $\frac{2}{3}x = 20$
15. $\frac{3}{9}x = \frac{27}{2}$
16. $\frac{3}{5}x = 15$
17. $x - \frac{1}{2} = \frac{2}{3}$
18. $x + \frac{3}{5} = \frac{1}{2}$
19. $x + 2\frac{1}{2} = 4\frac{3}{5}$

Find the value.
20. 4^2
21. 5^3
22. 1^8
23. 3^5

Simplify.
24. $\frac{3}{5} \cdot \frac{15}{21}$
25. $3\frac{1}{2} \cdot \frac{16}{9}$
26. $\frac{3}{5} + \frac{8}{3}$
27. $7\frac{3}{5} - 2\frac{3}{4}$
28. $\frac{2}{5}$ of 25
29. $\frac{2}{3} \div \frac{7}{12}$
30. $3\frac{1}{2} \div 1\frac{7}{8}$
31. $16 \div 1\frac{1}{3}$

Round to the nearest tenth and to the nearest hundredth.
32. 0.437
33. 9.758
34. 12.019
35. 63.455
36. Add $36.45 + 0.316 + 147.4$
37. Subtract 3.87 from 43.96
38. Divide $48.6\overline{)493.45}$
39. Multiply $(19.6)(123.45)$

Change as indicated.
40. 63 mm to cm
41. 4.63 km to m
42. 10 cm to m
43. 9 dm to m
44. 43 m to hm
45. 13 hm to m

Change to Celsius or Fahrenheit. Find to the nearest degree.
46. $C = 30°$
47. $F = 90°$
48. $C = 80°$
49. $F = 10°$

50. 22 is the same as a number increased by 8. What is the number?
51. 9 times a number decreased by 8 is 37. Find the number.
52. $A = \frac{1}{2}bh$. Find A if $b = \frac{6}{7}$, and $h = 14$.
53. Juanita saves $21.50 each month. How much is saved after 8 months?
54. Make a vertical bar graph to show the attendance for the following games: Game 1, 1500; Game 2, 1800; Game 3, 1650; Game 4, 1200

PRACTICE ON PERCENT: CHAPTER 9

Change to decimals.

1. 50%
2. 7%
3. 400%
4. 32.4%
5. 0.7%
6. 19.7%
7. 425%
8. 3.6%
9. 0.0007%
10. $54\frac{1}{3}\%$

Change to percents.

11. 0.07
12. 0.654
13. 18
14. $\frac{3}{5}$
15. $0.12\frac{1}{2}$
16. $\frac{5}{8}$
17. $0.45\frac{3}{5}$
18. 0.02
19. 45.6
20. 9.32

Compute.

21. 15% of 743 is what number?
22. 8.7% of 48.64 is what number?
23. $6\frac{1}{4}\%$ of 156 is what number?
24. 70% of 496.50 is what number?
25. 36.4% of 945.34 is what number?
26. $5\frac{3}{4}\%$ of 90 is what number?

Compute.

27. What % of 90 is 54?
28. What % of 120 is 48?
29. What % of 155 is 73?
30. What % of 750 is 150?
31. What % of 745 is 42.6?
32. What % of 12.8 is 5.7?

Compute.

33. 24 is 60% of what number?
34. 90 is 50% of what number?
35. 18.6 is 20% of what number?
36. 43.2 is 37.5% of what number?
37. 120 is 70% of what number?
38. 16.5 is 92.3% of what number?

PROBLEM SOLVING

39. The Tigers played 30 games. They won 70% of the games. How many games did they win?

40. The Wildcats played 20 games. They lost 4 games. What % of the games played did they lose?

41. Jane got a $1,200 raise. The rate was 8% of her salary. What was her salary?

42. Darlene earns 12% commission on all sales. This week her sales were $950. Find her commission.

43. Find the cost of the following purchases at a 30% discount: $32, sweater; $48.50, shoes; $125.40, coat.

44. Find the cost of the following purchases at a "$\frac{1}{4}$ off" sale: $34.50, toaster; $125.30, fan; $64.25 radio.

PRACTICE ON WORKING WITH STATISTICS AND PROBABILITY: CHAPTER 10

Find the range. Find the mean.

1. 18, 12, 20, 16, 14, 10
2. 98, 100, 96, 98, 94, 96
3. 253, 248, 295, 309, 290
4. 1000, 1010, 995, 990, 1200
5. 50, 54, 48, 46, 43, 56, 60
6. 10, 9, 7, 8, 8, 6, 10, 9, 7, 8, 9, 7

Find the median.

7. 100, 98, 98, 97, 96, 95, 94
8. 18, 20, 19, 22, 26, 24, 21
9. 12, 10, 19, 18, 16, 14
10. 100, 98, 96, 96, 95, 98, 94, 96
11. 50, 48, 46, 48, 52, 54, 50
12. 12, 10, 8, 10, 9, 8, 8, 10, 11, 12, 10

Find the mode(s).

13. 98, 96, 100, 98, 94, 93, 92
14. 12, 10, 9, 10, 12, 8, 6, 10, 11
15. 4, 5, 9, 8, 6, 4, 5, 3
16. 100, 98, 98, 100, 96, 94, 93, 92
17. 56, 54, 50, 53, 52, 51
18. 250, 248, 246, 248, 245, 244

Use these scores for Exercises 19 and 20.

98	96	98	100	100	97	90	94
96	98	93	88	86	88	94	100

19. Make a frequency table.
20. Find the mean.

Answer the questions about the spinner.

21. How many possible outcomes?
22. How many even outcomes?
23. How many odd outcomes?
24. P(2)?   25. P(3)?   26. P(4)?
27. Spin the arrow 100 times. About how many times should it stop on an even number?

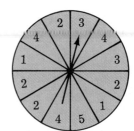

PROBLEM SOLVING

28. A deck of cards contains 52 cards. There are 4 jacks in the deck. Find P(jack).

29. A bag contains 80 red marbles and 100 white marbles. Find P(red). Find P(white).

30. There are 5 candidates for treasurer and 4 for secretary. In how many ways can these offices be filled?

31. How many three-digit numbers can be formed from the digits 2, 4, and 5, if each digit can be repeated?

PRACTICE ON SOLVING AND APPLYING RATIOS AND PROPORTIONS: CHAPTER 11

A bag contains 14 pennies, 10 dimes, 7 quarters, 4 half-dollars, and 6 silver dollars. Find each ratio. Simplify.

1. dimes to quarters
2. half-dollars to pennies
3. pennies to dimes
4. silver dollars to pennies
5. quarters to half-dollars
6. dimes to silver dollars

Solve each proportion.

7. $\frac{4}{5} = \frac{n}{10}$
8. $\frac{4}{3} = \frac{x}{6}$
9. $\frac{a}{2} = \frac{5}{10}$
10. $\frac{7}{x} = \frac{3}{6}$
11. $\frac{n}{9} = \frac{3}{27}$
12. $\frac{12}{3} = \frac{2}{y}$
13. $x:5 = 3:15$
14. $12:y = 24:36$
15. $3:5 = 18:n$
16. $\frac{n}{4} = \frac{18}{9}$
17. $\frac{8}{7} = \frac{24}{a}$
18. $\frac{8}{1} = \frac{n}{3}$

Which are true proportions?

19. $\frac{4}{9} = \frac{24}{54}$
20. $\frac{9}{2} = \frac{45}{10}$
21. $\frac{3}{15} = \frac{15}{60}$
22. $\frac{3}{10} = \frac{18}{60}$
23. $\frac{28}{35} = \frac{4}{5}$
24. $\frac{10}{30} = \frac{1}{3}$

PROBLEM SOLVING

Solve each word problem. Use a proportion.

25. William drove 180 km in 2 hours. How long will it take him to drive 300 km?
26. Tammy made $7.50 for 2 hours of cutting grass. How much did she earn for 3 hours of work?
27. Hubie bought 3 m of cloth for $18.75. How many meters of cloth can he buy for $31.25?
28. It takes 8 cups of flour to bake 3 cakes. How many cups of flour are needed to bake 5 cakes?
29. Helen bought 4 pairs of socks for $9.00. How many pairs of socks can she buy for $13.50?
30. It takes 10 eggs to make 4 quiches. How many eggs does it take to make 6 quiches?
31. Tanya earned $2,600 by working part-time. Make a circle graph to show how she spent her earnings.

Savings	$1,200	Charity	$ 100
School supplies	$ 400	Entertainment	$ 400
Clothing	$ 500		

EXTRA PRACTICE

PRACTICE ON ADDING AND SUBTRACTING INTEGERS: CHAPTER 12

Compare. Use > or <.

1. $7 \equiv 12$
2. $15 \equiv 14$
3. $-3 \equiv 1$
4. $0 \equiv -7$
5. $-9 \equiv -16$
6. $-10 \equiv 10$
7. $19 \equiv -16$
8. $9 \equiv -9$
9. $0 \equiv 10$
10. $6 \equiv -19$
11. $-21 \equiv 35$
12. $-41 \equiv -50$

Add.

13. $9 + 2$
14. $12 + (-2)$
15. $-4 + (-6)$
16. $-5 + 8$
17. $15 + (-1)$
18. $-16 + 9$
19. $7 + (-15)$
20. $16 + (-3)$
21. $-10 + (-16)$
22. $-21 + (-3)$
23. $-15 + 9$
24. $-3 + (-7)$

Copy and complete.

25. $16 + (-16) = ?$
26. $12 + ? = 0$
27. $-9 + ? = 0$
28. $19 + ? = 19$
29. $-15 + 15 = ?$
30. $20 + ? = 0$

Subtract.

31. $15 - 3$
32. $10 - 12$
33. $-6 - 5$
34. $-3 - (-5)$
35. $-10 - 9$
36. $-20 - (-10)$
37. $16 - 4$
38. $12 - (-3)$
39. $-30 - (-15)$
40. $-17 - 3$
41. $-40 - (-15)$
42. $-12 - 12$

Solve and check.

43. $x + (-6) = -3$
44. $5 + x = -4$
45. $x - 3 = 7$
46. $x - (-2) = 8$
47. $12 + x = -2$
48. $x - (-7) = -3$
49. $9 + x = -3$
50. $x + (-3) = 12$
51. $x - (-10) = 10$
52. $x - 3 = -7$
53. $x - (-2) = -8$
54. $x - 9 = 1$

PROBLEM SOLVING

55. At 7:00 A.M., the temperature was 45°F. It rose 10°F by noon and rose 23°F between noon and 5:00 P.M. What was the temperature at 5:00 P.M.?

56. At 6:00 A.M., the temperature was 8°C. It dropped 2°C by 10:00 A.M. and rose 4°C by 2:00 P.M. What was the temperature at 2:00 P.M.?

57. Fred won 200 points, lost 75 points, lost 15 points, and won 32 points. How many points did he have?

58. A football team gained 12 yards, lost 3 yards, and lost 5 yards. What was the net result?

59. In a game, Jo lost 40 points, lost 10 points, won 60 points, and won 150 points. How many points did she have?

60. Sam had $95.50 in his checking account. He wrote a check for $8.75, deposited $15.35, wrote a check for $9.50 and another check for $10.95. What was his final balance?

PRACTICE ON MULTIPLYING AND DIVIDING INTEGERS: CHAPTER 13

Multiply.

1. $15 \cdot 18$
2. $20 \cdot (-3)$
3. $-10 \cdot 35$
4. $-43 \cdot 0$
5. $-32 \cdot (-5)$
6. $-73 \cdot 42$
7. $-34 \cdot (-45)$
8. $18 \cdot (-75)$
9. $0 \cdot 98$
10. $-36 \cdot (-68)$
11. $9 \cdot (-48)$
12. $-53 \cdot (-38)$

Multiply. Use the distributive property.

13. $-5 \cdot (-3 + 6)$
14. $-6 \cdot (-2 - 4)$
15. $-13 \cdot (10 - 6)$
16. $8 \cdot (6 - 9)$
17. $-10 \cdot (15 - 3)$
18. $12 \cdot (-3 + 9)$
19. $-11 \cdot (-5 - 10)$
20. $20 \cdot (-2 + 9)$
21. $-14 \cdot (-2 - 10)$

Divide, if possible.

22. $-45 \div 1$
23. $65 \div (-1)$
24. $-44 \div 11$
25. $-33 \div (-3)$
26. $0 \div 43$
27. $-400 \div 40$
28. $-200 \div (-10)$
29. $96 \div (-12)$
30. $-500 \div (-25)$
31. $-36 \div (-9)$
32. $90 \div (-9)$
33. $-72 \div 0$

Divide.

34. $\frac{2}{3} \div \frac{7}{5}$
35. $-\frac{3}{4} \div \left(-\frac{2}{5}\right)$
36. $\frac{5}{6} \div \left(-\frac{3}{7}\right)$
37. $-\frac{9}{5} \div \frac{7}{4}$
38. $1\frac{1}{2} \div \left(-\frac{3}{8}\right)$
39. $-\frac{3}{4} \div 2\frac{1}{4}$
40. $-\frac{3}{8} \div \left(-5\frac{1}{7}\right)$
41. $2\frac{2}{5} \div 1\frac{5}{9}$
42. $-2\frac{1}{5} \div 4\frac{1}{2}$
43. $-1\frac{1}{3} \div \left(-2\frac{1}{5}\right)$
44. $\frac{8}{3} \div \left(-5\frac{1}{2}\right)$
45. $-6\frac{1}{3} \div 5\frac{1}{2}$

Simplify.

46. $3a + (-2a)$
47. $6x - (-3x)$
48. $-5 \cdot 4x$
49. $-6 \cdot (-5m)$
50. $4(2a - 5)$
51. $-(3 + 2y)$
52. $-(6 - 5)$
53. $-2a - (-3a)$
54. $-7(4x + 3)$
55. $-2(5a - 3)$
56. $-(3 - 2y)$
57. $-4(6a - 3)$
58. $-6n + (-5n)$
59. $-(-3 + 2y)$
60. $-4a - 7a$

Solve and check.

61. $-6x = 36$
62. $-5y = -45$
63. $4a = -48$
64. $-11x = -33$
65. $7n = -21$
66. $-10y = 120$
67. $-9x = -9$
68. $5a = -65$
69. $\frac{a}{-3} = 9$
70. $\frac{m}{4} = -10$
71. $\frac{x}{-9} = -3$
72. $\frac{y}{-2} = 12$
73. $\frac{n}{-5} = 10$
74. $\frac{x}{-9} = -20$
75. $\frac{y}{10} = -10$
76. $\frac{a}{-1} = -6$ 6
77. $-2x + 5 = 9$
78. $4y - 3 = -15$
79. $-3n - 1 = 14$
80. $-6a - 7 = -1$
8.1 $\frac{c}{4} - 5 = -3$
82. $\frac{a}{-3} + 2 = -1$
83. $\frac{x}{-2} - 4 = -2$
84. $\frac{y}{5} + 6 = 3$

EXTRA PRACTICE

PRACTICE ON SOLVING EQUATIONS AND INEQUALITIES: CHAPTER 14

Solve.

1. $-3y - 15 = 5y + 25$
2. $6a - 3 = 13 - 2a$
3. $-7x - 3 = -4x + 3$
4. $2(x - 5) = 12$
5. $-3(a - 2) = -6$
6. $-5(2 - 2y) = 2(4y - 6)$
7. $n - 3 > 7$
8. $a - 4 \leq -6$
9. $3h \geq -15$
10. $-2y < -14$
11. $\frac{1}{2}x \leq -2$
12. $-\frac{y}{5} > -3$

Evaluate.

13. a^5 if $a = -2$
14. $-4x^6$ if $x = -1$
15. $3a^2b$ if $a = 4$ and $b = -1$
16. $8a^2b^3$ if $a = -3$ and $b = 8$

Simplify.

17. $a^5 \cdot a^3$
18. $x \cdot x^6$
19. $(y^2)^3$
20. $(x^5)^2$
21. $(xy)^3$
22. $(a^2b)^3$
23. $(-3a^2)^2$
24. $(-4ab^2)^3$
25. $(2x^2y^3)^3$
26. $3x^2 - 4x - 1 - 2x^2 + 5x - 4$
27. $-4c + 6c^2 + 5 - 8c - 7c^2 + 7$
28. $5y^2 + 7y + 14 - 6y + 8y^2 + 3$
29. $6a^6 + 5 - 3a^4 + a^6 + 5a^4 + 9$
30. $-3d + 6d^2 + 7d^4 - 3 + 5d + 8d^2 - d^4$

Multiply.

31. $a(-3a^2 + a)$
32. $-3x(x^2 - 2x + 5)$
33. $-2y(-y^2 - 3y - 2)$
34. $-6x(5 - 3x - 2x^2)$
35. $2m(m - 3m^2 + 1)$
36. $5(-3 + 6a - 3a^2)$

Factor.

37. $3x - 9$
38. $3a^3 - 12a^2 + 18a$
39. $6a^3 + a$
40. $4x^2 - 8x + 24$

PROBLEM SOLVING

41. Twice a number, decreased by 8, is equal to 14. What is the number?
42. Six times a number, increased by 10, is equal to 19. What is the number?
43. Four times a number, decreased by 7, is equal to 21. Find the number.
44. Three times a number, increased by 2, is equal to 4. Find the number.
45. Ten times a number, increased by 5, is equal to 35. Find the number.
46. Five times a number, decreased by 15, is equal to 15. Find the number.
47. Twice a number, increased by 16, is equal to 48. Find the number.
48. Nine times a number, decreased by 10, is 71. Find the number.

PRACTICE ON CLASSIFYING GEOMETRIC FIGURES AND ON FINDING SQUARE ROOTS: CHAPTER 15

Complementary or Supplementary?

1. 50°, 40°
2. 130°, 50°
3. 90°, 90°
4. 60°, 30°

In the figure at the right, line $m \parallel$ line n. Name two pairs of congruent angles for each of the following angles.

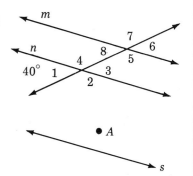

5. vertical
6. alternate interior
7. corresponding
8. congruent
9. Determine the measures of angles labeled 2 through 8.
10. Construct a line perpendicular to line s that passes through point A.

How many sides does each of the following have?

11. triangle
12. pentagon
13. hexagon
14. quadrilateral

Draw each of the following.

15. square
16. triangle
17. pentagon
18. hexagon

Classify the triangles according to the measures of their angles.

19. 43°, 52°, 85°
20. 80°, 80°, 20°
21. 60°, 30°, 90°

Find the measure of the third angle, given the measures of two angles of a triangle.

22. 38°, 72°
23. 40°, 120°
24. 10°, 165°

Give two square roots of each number.

25. 36
26. 81
27. 100
28. 25
29. 144

Approximate to the nearest tenth. Use the table on page 473.

30. $\sqrt{5}$
31. $\sqrt{32}$
32. $\sqrt{18}$
33. $\sqrt{3}$
34. $\sqrt{98}$

For each right triangle, find the missing length to the nearest tenth.

35. $a = 3, b = 4$
36. $a = 5, c = 13$
37. $b = 5, c = 7$

PROBLEM SOLVING

38. One of two supplementary angles is 3 times the size of the other. What are the measures of the two angles?

39. The lengths of two sides of a right triangle are 5 in. and 8 in. Find the length of the hypotenuse.

EXTRA PRACTICE

PRACTICE ON GRAPHING INTEGERS, EQUATIONS, AND INEQUALITIES: CHAPTER 16

Graph.

1. the integer between -1 and 4
2. the integers greater than 2
3. the integers less than 1
4. the integers greater than -1
5. $2x - 5 = -3$
6. $3b + 6 = 15$
7. $-5n + 2 = -3$
8. $6y + 2 = 12 - y$
9. $-3a - 4 = 2a + 11$
10. $4x - 5 = 3x + 5$
11. $x > -2$
12. $x \leq 3$
13. $x \geq 4$
14. $y \leq 0$
15. $y > 1$
16. $y < 2$
17. $2x - 2 > 6$
18. $3y + 5 \geq -4$
19. $5x - 1 < 4$
20. $4a - 9 \geq 16 - a$
21. $7x - 3 > 9 - 5x$
22. $y - 6 \leq 5y - 2$

Give the ordered pair for each point.

23. A
24. B
25. C
26. D
27. E
28. F
29. G
30. H

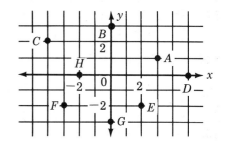

Graph each point.

31. $A(3, 2)$
32. $M(-2, 4)$
33. $F(2, -1)$
34. $Y(0, 3)$
35. $B(-3, -4)$
36. $C(-4, 0)$

Graph.

37. $y = x - 1$
38. $y = -x + 2$
39. $y = x + 3$
40. $y = 2x$
41. $y = 2x + 1$
42. $y = 3x - 2$

Find the slope of $\overleftrightarrow{AB}$ for the given pairs of points.

43. $A(2, 3); B(5, 7)$
44. $A(-4, 9); B(2, -3)$
45. $A(0, 5); B(4, 3)$
46. $A(8, 3); B(6, 1)$

Find the slope and the y-intercept of the line with the given equation.

47. $y = 4x - 2$
48. $y = 2x + 9$
49. $y = \frac{3}{5}x - 1$
50. $y = \frac{5}{3}x + 1$
51. $y = \frac{1}{4}x + \frac{3}{4}$
52. $y = -2x$

Solve each system of equations by graphing.

53. $y = x - 3$
    $y = 3x + 1$
54. $y = 2x + 12$
    $y = -4x$
55. $y = -5x$
    $y = -x - 4$

MIXED PRACTICE: CHAPTERS 9–16

Change to decimals.

1. 60%
2. 8%
3. 500%
4. 89.6%
5. 0.0065%

Change to percents.

6. 0.09
7. 0.537
8. $\frac{2}{5}$
9. $0.15\frac{1}{2}$
10. 48.35

Compute.

11. 12% of 648 is what number?
12. 6.5% of 54 is what number?
13. What % of 80 is 56?
14. What % of 890 is 356?
15. 43.2 is 80% of what number?
16. 270 is 60% of what number?
17. Find the range, mean, median, and mode.
    12, 10, 8, 12, 6, 8, 14

Solve and check.

18. $\frac{3}{5} = \frac{n}{10}$
19. $\frac{8}{x} = \frac{3}{6}$
20. $\frac{4}{9} = \frac{8}{y}$
21. $\frac{c}{8} - 4 = 5$
22. $x - 5 = 7$
23. $8 + x = 16$
24. $x + (-3) = 6$
25. $-3a = 15$
26. $-12 = -2n + 6$
27. $-2n + 9 = 7$
28. $-3(a - 1) = 6$
29. $-2y + 12 = 3y - 3$
30. $5a - 2 = 13 + 2a$
31. $-6(3 - 2y) = 18$
32. $a + 3 \geq 7$
33. $-2x < -18$
34. $3h \leq -21$

Solve for x.

35. $x + a = b$
36. $x - q = p$
37. $-x + c = -d$

Simplify.

38. $16 + (-2)$
39. $-12 + 6$
40. $-20 + (-3)$
41. $18 - 3$
42. $-20 - (-6)$
43. $-8 - 8$
44. $-15 \cdot (-3)$
45. $60 \cdot (-2)$
46. $-18 \div 2$
47. $-45 \div -9$
48. $\frac{2}{3} \div \frac{4}{9}$
49. $-\frac{3}{5} \div \frac{21}{10}$
50. $a^6 \cdot a^3$
51. $(x^2y)^2$
52. $(-2xy^3)^4$
53. $(-2a^2)(9a^3)$
54. $n(-3n^2 + n)$
55. $-2x(x^2 - 3x + 5)$
56. $3y(-y + 2y^2 - 3)$
57. $-(m^2 - 2m + 6)$
58. $-4a^2 + 3a - 2 - (2a - 3a^2 + 1)$

Evaluate. $a = 3, b = 4, x = -2,$ and $y = -1$

59. $3x^2y$
60. $-4x^3y^2$
61. $-(3b)^2$
62. $(-2a)^2$

Graph.

63. $A(-4, 2)$
64. $B(3, -4)$
65. $C(-1, -3)$
66. $y = 3x - 1$
67. $y = x - 4$
68. $y = 2x$

Factor.
69. $4a^2 - 12a + 36$
70. $10x^3 - 15x^2 + 25x$

Find the measure of the third angle, given the measures of two angles of a triangle.
71. 42°, 68°
72. 50°, 100°
73. 170°, 5°

74. A bag contains 120 white marbles and 60 red marbles. Find P(white). Find P(red).

A baseball team won 10 games and lost 4. Find each ratio.
75. Wins to total
76. Wins to losses
77. Losses to total

Find each ratio. Leave answers in fractional form.
78. sin A   79. cos B
80. tan A   81. tan B

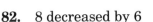

Write in mathematical terms.
82. 8 decreased by 6
83. 3 more than 6 times a number

For each right triangle, find the missing length to the nearest tenth. a and b are legs, c is the hypotenuse.
84. $a = 12$, $b = 7$
85. $b = 8$, $c = 15$

86. Classify each of the triangles in the picture according to its sides. Tell why you classified each as you did.

a.   b.    c.

Approximate to the nearest tenth. Use the table on page 473.
87. $\sqrt{53}$
88. $\sqrt{14}$
89. $\sqrt{3}$
90. $\sqrt{75}$

Find the slope of $\overleftrightarrow{AB}$ for the given pairs of points.
91. $A(3, 4)$; $B(6, 8)$
92. $A(-2, 3)$; $B(7, 8)$

Find the slope and the y-intercept of the line with the given equation.
93. $y = \frac{2}{3}x + 3$
94. $y = -\frac{4}{5}x - 5$

Solve each system by graphing.
95. $y = 3x - 1$
    $y = -2x + 4$
96. $y = -x + 4$
    $y = 2x + 1$

97. Measure each angle and classify as acute, right, or obtuse.

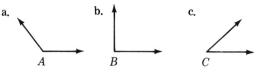

98. One of two complementary angles is four times the other. Find the measures of the two angles.
99. One of two supplementary angles is five times the other. Find the measures of the two angles.

Answers to Practice Exercises

Pages 1–2
1. 7 **2.** 10 **3.** 1 **4.** 11 **5.** 14
6. 95 **7.** 17 **8.** 52 **9.** 46
10. 80 **11.** 38 **12.** terms: 11, p; variable: p; constant: 11 **13.** terms: t, 4, u; variables: t, u; constant: 4
14. terms: x, 8, y; variables: x, y; constant: 8 **15.** terms: x, 9, y, 4; variables: x, y; constants: 9, 4

Pages 3–4
1. associative **2.** commutative
3. identity **4.** $49 + 1 + 28 = 78$
5. $93 + 7 + 62 = 162$ **6.** $198 + 2 + 19 = 219$

Pages 5–7
1. not open; false **2.** open **3.** not open; true
4. 9 **5.** 21 **6.** 0, 1
7. 0, 1, 2

Pages 8–10
1. 9 **2.** 8 **3.** x **4.** 8 **5.** 21
6. 10 **7.** 6 **8.** 20 **9.** 15
10. 19 **11.** 10 **12.** 19

Pages 11–12
1. $6 + 7$ **2.** $5 - 3$ **3.** $x + 11$
4. $x + 4$ **5.** $6 + y$ **6.** $b - 5$
7. $t + 8$ **8.** $x - 13$ **9.** $x - 5$

Pages 13–15
1. 152 lb **2.** $75 **3.** 15 yr
4. 14 yr

Pages 18–19
1. $P = 36$ in. **2.** $P = 68$ m
3. $S = \$60$ **4.** $C = \$54$
5. $W = T - L$ **6.** $I = d + p$

Pages 26–27
1. 35 **2.** 108 **3.** 416 **4.** 322
5. 2 **6.** 4 **7.** 34 **8.** 4 **9.** 6
10. 12 **11.** 10 **12.** 3

Pages 28–29
1. associative **2.** commutative
3. identity **4.** $25 \cdot 4 \cdot 17 = 1{,}700$
5. $2 \cdot 50 \cdot 19 = 1{,}900$
6. $5 \cdot 20 \cdot 33 = 3{,}300$

Pages 30–32
1. 30 **2.** 2 **3.** 47 **4.** 4 **5.** 35
6. 25 **7.** 56 **8.** 29 **9.** 18
10. 26

Pages 33–34
1. 24 **2.** 10 **3.** 35 **4.** 69
5. 36 **6.** 26 **7.** 13 **8.** 56
9. 37

Pages 36–38
1. 8 **2.** k **3.** 25 **4.** x **5.** 5
6. 3 **7.** 2 **8.** 8 **9.** 56 **10.** 16
11. 45 **12.** 18 **13.** 5 **14.** 40
15. 5 **16.** 63

Pages 39–41
1. $2 \cdot 9$ 2. $\frac{p}{q}$ 3. $13w$ 4. 3
5. 20 yr 6. 40 7. 28

Pages 46–47
1. 322 in.^2 2. 91 ft^2 3. 128 cm^2
4. 24 cm 5. 64 ft

Pages 53–55
1. $9 \cdot 7 + 9 \cdot 5 = 108$ 2. $5 \cdot 4 - 3 \cdot 4 = 8$ 3. $6 \cdot 7 + 6 \cdot 9 = 96$
4. $8(6 + 11)$ 5. $(13 - 10)3$
6. $14(12 + 9)$ 7. $16x + 24$
8. $14b - 35$ 9. $20x - 30$

Pages 56–57
1. $9t$ 2. not possible
3. $13m + 4n + 7$ 4. $7a + 4$
5. $8p + 8$ 6. $4a + 9$

Pages 58–60
1. 19 2. 29 3. 45 4. 22
5. 77 6. 42 7. 49 8. 66

Pages 61–62
1. 3 2. 5 3. 3 4. 15 5. 48
6. 30

Pages 63–64
1. $4x + 3$ 2. $2x - 9$ 3. $\frac{n}{6} - 7$
4. $\frac{x}{7} + 6$

Pages 65–68
1. 10 2. 32 3. $30 4. 24

Pages 69–71
1. 16 2. 9 3. 125 4. 64
5. 8 6. 1 7. 216 8. 5^3 9. 8^4
10. 9^2 11. 7^3 12. 6^4 13. 5^4
14. $7^3 \cdot 8^2$ 15. $a^5 \cdot b^2$ 16. 8
17. 64 18. 243 19. 128 20. 7
21. 405

Pages 72–74
1. 36 2. 64 3. 169 4. 361
5. 16π 6. 9π 7. 100π 8. 121π
9. 200π 10. 90π 11. 400π
12. 147π 13. 54 14. 216
15. 5,400 16. 600

Pages 79–81
1. $\frac{1}{4}$ 2. $\frac{1}{9}$ 3. $\frac{3}{8}$ 4. $\frac{4}{7}$ 5. no
6. no 7. yes 8. $\frac{3}{5}$ 9. $\frac{4}{9}$ 10. $\frac{5}{6}$
11. 1 12. 1

Pages 82–84
1. 4 2. 4 3. $4\frac{1}{6}$ 4. $5\frac{1}{3}$
5. $1\frac{5}{7}$ 6. $\frac{19}{5}$ 7. $\frac{17}{3}$ 8. $\frac{22}{5}$
9. $\frac{19}{3}$ 10. $\frac{29}{6}$ 11. $\frac{39}{4}$ 12. $6\frac{2}{3}$
13. $5\frac{1}{3}$ 14. $2\frac{2}{5}$ 15. $7\frac{1}{2}$ 16. $3\frac{3}{4}$

Pages 85–87
1. $1\frac{1}{9}$ 2. $2\frac{2}{9}$ 3. $\frac{15}{28}$ 4. $1\frac{1}{6}$
5. $1\frac{7}{8}$ 6. $\frac{5}{6}$ 7. $\frac{10}{27}$ 8. $\frac{9}{28}$
9. $\frac{12}{25}$ 10. $\frac{8}{9}$

Pages 88–89
1. $3 \cdot 2$ 2. $2 \cdot 5$ 3. $3 \cdot 7$
4. $5 \cdot 7$ 5. $2^2 \cdot 5$ 6. prime
7. $2 \cdot 3^2$ 8. 2^3 9. $2^3 \cdot 3$
10. $2^3 \cdot 5$ 11. 2^5 12. $2 \cdot 3^3$

Pages 90–91
1. 4 2. 3 3. 1 4. 2 5. 10
6. 9 7. 4 8. 6

Pages 92–94

1. $\frac{3}{5}$ 2. $\frac{9}{14}$ 3. $\frac{2}{3}$ 4. $\frac{14}{25}$ 5. $\frac{2}{3}$
6. $\frac{2}{5}$ 7. $1\frac{1}{3}$ 8. $1\frac{4}{5}$ 9. $1\frac{7}{8}$
10. $1\frac{1}{2}$ 11. $1\frac{3}{4}$ 12. $1\frac{1}{5}$ 13. $\frac{3}{5}$
14. $\frac{2}{3}$ 15. $\frac{3}{4}$ 16. $\frac{4}{15}$ 17. $\frac{16}{27}$

Pages 95–97

1. $\frac{1}{3}$ 2. $\frac{2}{5}$ 3. $\frac{8}{9}$ 4. $\frac{1}{4}$ 5. 3
6. 2 7. $7\frac{1}{2}$ 8. 12 9. $10\frac{1}{2}$
10. $6\frac{2}{3}$ 11. 12 12. $13\frac{1}{3}$

Pages 99–100

1. $\frac{3}{5}$ 2. $\frac{1}{2}$ 3. $\frac{1}{3}$ 4. $\frac{1}{2}$ 5. $\frac{3}{8}$
6. 4 7. $2\frac{1}{2}$ 8. $\frac{1}{4}$

Pages 102–103

1. 20 2. 15 3. 14 4. 36
5. $3\frac{1}{3}$ 6. $8\frac{1}{3}$ 7. $1\frac{5}{7}$ 8. $1\frac{1}{2}$

Pages 104–105

1. $6\frac{2}{3}$ 2. $38\frac{1}{2}$ 3. $1\frac{7}{9}\pi$ 4. $13\frac{1}{2}\pi$

Pages 111–113

1. $\frac{6}{7}$ 2. $\frac{7}{11}$ 3. $\frac{4}{5}$ 4. $\frac{4}{9}$ 5. $\frac{2}{3}$
6. $\frac{2}{3}$ 7. $\frac{1}{2}$ 8. $\frac{1}{3}$ 9. $1\frac{1}{2}$ 10. $1\frac{1}{3}$
11. $2\frac{2}{3}$ 12. $1\frac{1}{5}$

Pages 114–115

1. 12 2. 30 3. 60 4. 30

Pages 116–118

1. $\frac{1}{2}$ 2. $\frac{5}{6}$ 3. $\frac{4}{15}$ 4. $\frac{1}{5}$ 5. $1\frac{1}{2}$
6. $1\frac{1}{3}$ 7. $1\frac{1}{5}$ 8. $1\frac{2}{3}$ 9. $\frac{8}{15}$
10. $\frac{16}{21}$ 11. $\frac{5}{14}$ 12. $\frac{1}{10}$

Pages 119–121

1. $1\frac{1}{18}$ 2. $1\frac{1}{20}$ 3. $\frac{5}{12}$ 4. $1\frac{5}{18}$
5. $\frac{7}{12}$ 6. $\frac{13}{20}$ 7. $\frac{7}{18}$ 8. $\frac{11}{15}$
9. $1\frac{2}{3}$ 10. $1\frac{5}{12}$ 11. $1\frac{1}{5}$

Pages 122–124

1. $8\frac{1}{5}$ 2. $13\frac{1}{2}$ 3. $5\frac{1}{2}$ 4. $7\frac{3}{4}$
5. $5\frac{5}{12}$ 6. $9\frac{4}{15}$ 7. $4\frac{1}{12}$ 8. $2\frac{5}{8}$
9. $11\frac{3}{4}$ 40. $8\frac{7}{8}$ 11. $3\frac{3}{5}$ 12. $7\frac{9}{10}$
13. $11\frac{2}{5}$ 14. $9\frac{1}{5}$ 15. $13\frac{3}{8}$

Pages 126–128

1. 4 2. 9 3. 7 4. $3\frac{1}{2}$ 5. $1\frac{1}{2}$
6. $2\frac{3}{4}$ 7. $2\frac{9}{10}$ 8. $3\frac{4}{5}$ 9. $4\frac{1}{4}$
10. $\frac{2}{3}$ 11. $4\frac{1}{2}$

Pages 129–130

1. $\frac{10}{11}$ 2. $\frac{4}{5}$ 3. $\frac{2}{5}$ 4. $\frac{2}{3}$ 5. $\frac{7}{12}$

Pages 132–134

1. $4\frac{1}{2}$ 2. $\frac{2}{5}$ 3. $2\frac{2}{3}$ 4. $3\frac{1}{3}$
5. $2\frac{1}{2}$ 6. $4\frac{1}{2}$ 7. $2\frac{1}{2}$ 8. $9\frac{1}{3}$
9. 20 10. $4\frac{1}{2}$ 11. 12 12. $11\frac{2}{3}$

Pages 140–142

1. $4,973
2. $6,270

3.

Investment Growth					
Rate	Year				
	0	1	2	3	4
5%	$1,000	$1,051	$1,105	$1,161	$1,221
6%	$1,000	$1,061	$1,127	$1,197	$1,271

Pages 143–144

ATTENDANCE AT SOCCER GAMES

Game	Number of Fans
1	🧍🧍🧍🧍🧍
2	🧍🧍🧍🧍🧍🧍🧍
3	🧍🧍🧍🧍🧍🧍
4	🧍🧍🧍🧍

Key: Each 🧍 represents 1,000 fans.

Pages 147–149

1. Lee: 95; Connie: 85; Roger: 70
2. grade 7: 350; grade 8: 300; grade 9: 500

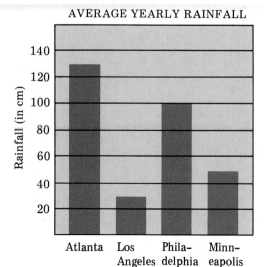

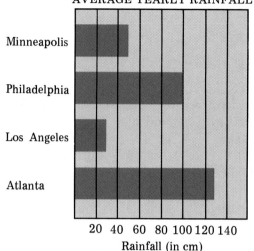

Pages 150–152

1. increased
2. stayed the same
3. decreased
4.

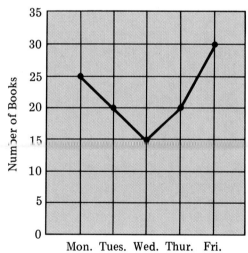

Pages 154–156

1. Marita
2. Sam
3. hogs: 1940; lambs: 1930
4. $27 for 50 kg

Pages 157–159

1. 5 cm 2. 21 3. 6 cm 4. 12 cm

Pages 165–167

1. 6.8 2. 0.000403 3. seventeen and four tenths 4. six hundred fifty-nine ten-thousandths 5. 0.7 6. 0.65 7. 0.046 8. 0.09 9. 0.6; 0.63 10. 0.7; 0.66 11. 0.7; 0.68 12. 24.2; 24.18

Pages 168–170

1. 0.6 2. 0.25 4. 0.625 4. 0.15 5. 0.7 6. 0.8 7. 0.8 8. 0.6 9. 0.57 10. 0.56 11. 0.92 12. 0.17

Pages 171–172

1. 241.40 2. 27.192 3. 19.22 4. 7.766 5. 21.85 6. 8.84

Pages 173–174

1. 149.52 2. 0.06324 3. 0.018468 4. 4.512 5. 0.63

Pages 177–179

1. 92.1 2. 3,423.1 3. 715 4. 3.8; 3.83 5. 0.6; 0.61 6. 0.8; 0.83

Pages 180–181

1. 28.7 2. 438.5 3. 9,876 4. 5,325 5. 96,257 6. 1,041.56 7. 2 8. 3 9. 2.34×10^2 10. 4.9×10^4 11. 3×10^7

Pages 182–183

1. 2.9 2. 8.61 3. 3.4 4. 2.8 5. 22.475 6. 0.5 7. 3.8 8. 0.3 9. 1.3

Pages 190–191

1. 3 cm 2. 6 cm 3. 37 mm 4. 42 mm

Pages 192–194

1. 10 2. 0.01 3. 10 4. 0.001

Pages 195–197

1. 1,250 cm 2. 7,000 m 3. 0.432 m 4. 1.5 m 5. 0.529 km 6. 0.082 km

Pages 200–201

1. 2.88 2. 3,600 3. 59,000 4. 1.589 5. 0.046 6. 2.19

Pages 202–204

1. mL 2. L 3. 8,200 4. 30,000 5. 0.68 6. 3.8 7. 600 8. 9,000

Pages 205–207

1. g 2. mg 3. 5,600 4. 3,000 5. 0.08 6. 0.4 7. 0.0045 8. 63,000

Pages 209–210

1. F = 50° 2. F = 77° 3. F = 64.4° 4. C = $26.\overline{6}$° 5. C = $15.\overline{5}$° 6. C = 10°

Pages 216–218

1. 0.73 2. 0.62 3. 0.35 4. 0.56 5. 0.85 6. 0.426 7. $0.67\frac{1}{3}$ 8. 2.25 9. 2.00 10. 0.713 11. 0.04 12. 0.004 13. 0.043 14. 0.05 15. 0.007 16. 7% 17. 46% 18. 66.7% 19. $35\frac{1}{2}$% 20. 800% 21. 40% 22. 50% 23. 75% 24. $33\frac{1}{3}$% 25. $28\frac{4}{7}$% 26. $22\frac{2}{9}$%

Pages 219–221
1. 51.1 **2.** 17.28 **3.** 1.64016
4. 46.655 **5.** 6.9375 **6.** 1.0835
7. 10 **8.** 17

Pages 222–223
1. 50% **2.** $83\frac{1}{3}$% **3.** 8%

Pages 224–225
1. 90 **2.** 20 **3.** 57 **4.** 233
5. $280

Pages 226–228
1. $11.13 **2.** $234.25 **3.** $322.50

Pages 229–231
1. $15.26 **2.** $103.96 **3.** $43.84
4. $4.64

Pages 232–233
$43.89

Pages 240–242
1. r = 12; m = 91.5 **2.** r = 6; m = 46.6 **3.** r = 8; m = 31.6 **4.** 7
5. 8.5 **6.** 90 **7.** 9 **8.** none
9. 10 and 9

Pages 243–245
1. 8.65 **2.** 12.56

Pages 248–249
1. $\frac{1}{4}$ **2.** $\frac{1}{4}$ **3.** 40 times **4.** 30 times

Pages 250–251
1. $\frac{1}{12}$ **2.** $\frac{1}{12}$ **3.** $\frac{1}{12}$ **4.** $\frac{1}{2} \cdot \frac{1}{6} = \frac{1}{12}$

Pages 252–253
1. 20 **2.** 16

Pages 260–262
1. $\frac{13}{3}$ or 13:3 **2.** $\frac{3}{13}$ or 3:13 **3.** $\frac{13}{16}$ or 13:16 **4.** $\frac{1}{7}$ **5.** $\frac{3}{7}$ **6.** $\frac{4}{7}$ **7.** $\frac{7}{7}$ or 1
8. $\frac{n}{7}$ **9.** $\frac{50}{1}$ or 50:1 **10.** $\frac{1}{35}$ or 1:35

Pages 263–265
1. yes **2.** yes **3.** no **4.** yes
5. 20 **6.** 30 **7.** 90 **8.** 20
9. 4 **10.** 9 **11.** 9 **12.** 8

Pages 266–268
1. yes **2.** yes **3.** no **4.** no
5. $54 **6.** 7 days

Pages 270–272
1. Transportation: $174; Food: $366; Savings: $120; Housing: $300; Clothing: $180; Other: $60

Fractional Part	Measure of Central Angle	Percent Planted
$\frac{50}{200}$ or $\frac{1}{4}$	$\frac{1}{4}(360°) = 90°$	25%
$\frac{25}{200}$ or $\frac{1}{8}$	$\frac{1}{8}(360°) = 45°$	$12\frac{1}{2}$%
$\frac{100}{200}$ or $\frac{1}{2}$	$\frac{1}{2}(360°) = 180°$	50%
$\frac{25}{200}$ or $\frac{1}{8}$	$\frac{1}{8}(360°) = 45°$	$12\frac{1}{2}$%
1	360°	100%

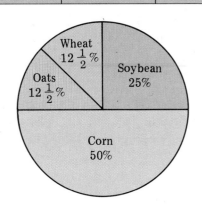

Pages 274–276

1. 60 km **2.** 37.5 km **3.** 2.25 m
4. 4.5 m **5.** 4.5 m

Pages 277–279

1. $x = 9; y = 4$ **2.** 8 m

Pages 280–282

1. 4 **2.** 3 **3.** 5 **4.** $\frac{12}{5}$ **5.** $\frac{12}{13}$
6. $\frac{5}{13}$ **7.** 1.732 **8.** 0.866
9. 0.500

Pages 288–290

1. 2 **2.** -5 **3.** 0 **4.** -1
5. -2 **6.** 6 **7.** -6 **8.** -3
9. 3 before; 5 after **10.** 0 before; 2 after **11.** -4 before; -2 after
12. -6 before; -4 after **13.** -1 before; 1 after **14.** < **15.** >
16. < **17.** < **18.** 6 **19.** 7
20. -5

Pages 291–293

1. 8 **2.** 7 **3.** -3 **4.** -6
5. -2 **6.** 3 **7.** 1 **8.** -4
9. 0 **10.** 0 **11.** 0 **12.** 0

Pages 295–297

1. 4 **2.** -5 **3.** -3 **4.** -6
5. -2 **6.** 8 **7.** 5 **8.** 7

Pages 299–300

1. -3 **2.** 8 **3.** -7 **4.** 10
5. -21 **6.** 0 **7.** 0 **8.** 25
9. -12 **10.** 25 **11.** -17 **12.** 33

Pages 301–302

1. 3 **2.** 9 **3.** -5 **4.** -6
5. -13 **6.** -13 **7.** -21
8. -27 **9.** 3 **10.** $^-9$ **11.** 2
12. 5 **13.** 10 **14.** 15 **15.** 8
16. 9

Pages 303–304

1. a gain of 5 yd **2.** no gain or loss
2. 1,700 m **4.** 2,400 m

Pages 305–306

1. -24 **2.** -5 **3.** -8 **4.** 3
5. -2 **6.** -8 **7.** 18 **8.** -6
9. -8

Pages 312–313

1. 54 **2.** -63 **3.** -48 **4.** -36
5. -504 **6.** $-1,653$ **7.** $-6,566$
8. -874 **9.** 0 **10.** 0 **11.** 0
12. 0

Pages 314–315

1. 15 **2.** 28 **3.** 27 **4.** 81
5. 3,807 **6.** 2,898 **7.** 4,140
8. 16,600

Pages 316–318

1. 36 **2.** -525 **3.** -18 **4.** -98
5. $-12 \cdot (-5) = 60$ **6.** $8 \cdot (-6) = -48$ **7.** $-7 \cdot 9 = -63$ **8.** $10 \cdot (-11) = -110$ **9.** 60 **10.** 180
11. -600 **12.** 160 **13.** 160
14. -30 **15.** -30 **16.** -8

Pages 320–322

1. 7   **2.** 7   **3.** 8   **4.** 9   **5.** −5
6. −5   **7.** −8   **8.** −8   **9.** 44
10. −36   **11.** 85   **12.** −99   **13.** 0
14. 0   **15.** not possible   **16.** not possible

Pages 323–324

1. $\frac{3}{4}$ **2.** $8\frac{1}{2}$ **3.** $-8\frac{1}{2}$ **4.** $-5\frac{1}{3}$
5. $-2\frac{2}{5}$ **6.** $1\frac{1}{2}$ **7.** $\frac{2}{3}$ **8.** $-1\frac{5}{7}$

Pages 326–327

1. $-4x$ **2.** $-2x$ **3.** $-8x$ **4.** 0
5. $4x$ **6.** $-4x$ **7.** $11x$ **8.** $-5x$
9. $21x$ **10.** $-20y$ **11.** $12y$
12. $-18x$ **13.** $6x - 3$ **14.** $-2y + 8$
15. $-3 - 4x$ **16.** $-5 + 3y$

Pages 328–329

1. −9   **2.** −7   **3.** 7   **4.** −20
5. 12   **6.** −48   **7.** −3   **8.** −1
9. −24

Pages 335–336

1. −3   **2.** −1   **3.** 5   **4.** 2

Pages 337–338

1. 4   **2.** 10   **3.** −13   **4.** 3

Pages 339–341

1. $n + 7 = 15$ **2.** $5x - 3 = 7$
3. $n - 6 = 11$; $17   **4.** $3n + 5 = 23$; 6

Pages 344–346

1. $-8 \leq -1$ **2.** $2 \geq -1$ **3.** $-4 < 20$
4. $4 > -20$ **5.** $-2 \geq -3$
6. $2 \leq 3$

Pages 347–349

1. $x > -7$ **2.** $y \leq -8$ **3.** $n \geq -4$
4. $x < -4$ **5.** $y < -5$ **6.** $y \geq 8$

Pages 351–352

1. 9   **2.** 24   **3.** 32   **4.** 64
5. −64   **6.** 1,296

Pages 353–355

1. b^4 **2.** $-12y^7$ **3.** $35r^9$ **4.** z^{10}
5. x^{12} **6.** c^8 **7.** $a^4 b^4$ **8.** $16y^2$
9. $-8s^3$

Pages 356–357

1. trinomial   **2.** monomial
3. binomial   **4.** $4y^2 + 5y - 2$
5. $-3a^4 + 6a^2 - 2a - 1$

Pages 358–359

1. $a^4 - 5a^2$ **2.** $-12c^4 - 18c^3 + 6c^2$
3. $-5x^2 - 4$ **4.** $-a^2 - 7a + 4$
5. $2c^2 - 3c - 3$

Pages 360–362

1. $5(c - 2)$ **2.** $7(3 + z)$ **3.** $6(2 - a)$
4. $6(x^2 - 3)$ **5.** $2(y^2 - 3y + 5)$
6. $2(2n^2 + 3n - 6)$ **7.** $y(y + 4)$
8. $c^2(c - 2)$ **9.** $n^2(n^2 + 3)$
10. $3y(y - 4)$ **11.** $5y(2y^2 + 3y - 1)$
12. $3y(y^3 - 2y + 3)$

Pages 368–369

1.   **2.**   **3.**

4. B, C   **5.** E, G, H

Pages 370–371

1. 90°; right   2. 130°; obtuse
3. 75°; acute   4. complementary
5. supplementary   6. supplementary
7. complementary

Pages 373–374

1. vertical angles: ∠1 and ∠4, ∠2 and ∠3, ∠5 and ∠8, ∠6 and ∠7; alternate interior angles: ∠3 and ∠6, ∠4 and ∠5; corresponding angles: ∠1 and ∠5, ∠2 and ∠6, ∠3 and ∠7, ∠4 and ∠8

Pages 377–378

1. right   2. acute   3. obtuse

Pages 382–384

1. 48 is not a perfect square.   2. 25 is a perfect square.   3. 81 is a perfect square.   4. 66 is not a perfect square.
5. 2.8   6. 8.2   7. 5.7   8. 8.7
9. $s \doteq 7.3$ m

Pages 385–387

1. 8.2 cm   2. 9.2 m   3. 7.1 cm   4. 9.4 m

Pages 393–395

1. The integers between −3 and 2.

2. The integers between −1 and 0.

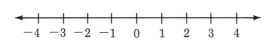

3. −3   4. no solution   5. all reals > 2
6. all reals < −1   7. all reals ≤ 4
8. all reals ≥ −2

Pages 396–398

1. all reals > −3   2. all reals < 1
3. all reals < −2   4. all reals > −3
5. all reals ≥ −3   6. all reals ≤ −6
7. all reals ≥ −5   8. all reals ≤ 9

Pages 401–403

1. (−2, 3)   2. (−4, −2)   3. (2, 0)
4. (0, 0)   5. (0, 2)   6. (3, −2)
7. –10.

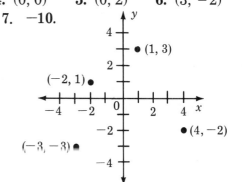

11. –14.

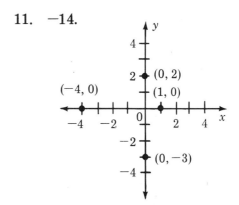

Pages 404–405

1. horizontal   2. vertical

Pages 406–408

Answers may vary. Five ordered pairs are given for each.

1. $y = x - 1$

x	$x - 1$	y	(x,y)
0	0 − 1	−1	(0,−1)
1	1 − 1	0	(1,0)
2	2 − 1	1	(2,1)
−1	−1 − 1	−2	(−1,−2)
−2	−2 − 1	−3	(−2,−3)

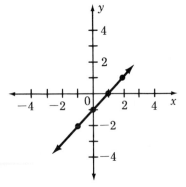

2. $y = -x + 3$

x	$-x + 3$	y	(x,y)
0	0 + 3	3	(0,3)
1	−1 + 3	2	(1,2)
2	−2 + 3	1	(2,1)
3	−3 + 3	0	(3,0)
4	−4 + 3	−1	(4,−1)

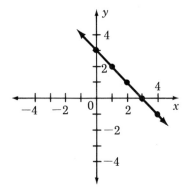

3. $y = -x - 2$

x	$-x - 2$	y	(x,y)
0	−0 − 2	−2	(0,−2)
1	−1 − 2	−3	(1,−3)
−1	−(−1) − 2	−1	(−1,−1)
−2	−(−2) − 2	0	(−2,0)
−3	−(−3) − 2	1	(−3,1)

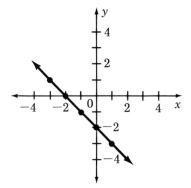

4. $y = -3x$

x	$-3x$	y	(x,y)
0	−3(0)	0	(0,0)
1	−3(1)	−3	(1,−3)
−1	−3(−1)	3	(−1,3)

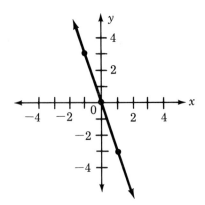

5. $y < 4x - 3$

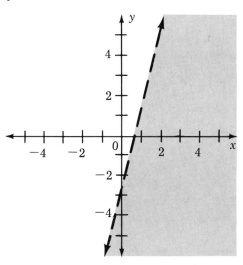

6. $y > 3x + 4$

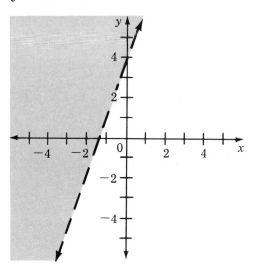

Pages 410–412

1. $\frac{4}{3}$ **2.** -1 **3.** -1 **4.** The slope is 5. The y-intercept is 3.   **5.** The slope is $\frac{1}{3}$. The y-intercept is -2.   **6.** The slope is $\frac{-5}{7}$. The y-intercept is 8.

Pages 413–414

1. yes   **2.** no

3.

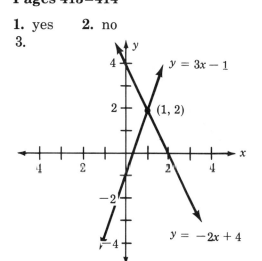

4.

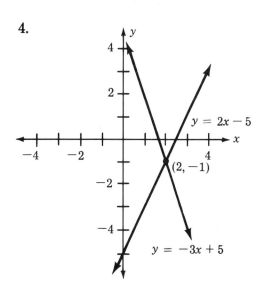

Answers to Odd-numbered Extra Practice Exercises

Chapter 1

1. 20 **3.** 30 **5.** 7 **7.** 172
9. 396 **11.** 165 **13.** none **15.** 1
17. 0 **19.** 5 **21.** 11 **23.** 2
25. 25 **27.** 8 **29.** 20 **31.** $46
33. 28 **35.** $P = 36$ in.

Chapter 2

1. 32 **3.** 9 **5.** 225 **7.** 23
9. 34 **11.** 242 **13.** 249
15. 3,200 **17.** 800 **19.** 1,400
21. 5 **23.** 60 **25.** 2 **27.** 70
29. 90 **31.** 30 **33.** 4 **35.** $45
37. $2,400 **39.** $20 **41.** $A = 48$ cm^2

Chapter 3

1. $12m$ **3.** $12x + 9$ **5.** $11a$
7. $5a - 5$ **9.** $15a - 15$ **11.** $9x + 6y$ **13.** 44 **15.** 66 **17.** 112
19. 2 **21.** 3 **23.** 75 **25.** 4
27. 9 **29.** 7 **31.** 7 **33.** 9
35. 1 **37.** 125 **39.** 243 **41.** 64
43. 625 **45.** 5^3 **47.** 4^4
49. $6^2 x^5$ **51.** $x^3 y^4$ **53.** 7 **55.** 13

Chapter 4

1. $1\frac{2}{3}$ **3.** $1\frac{1}{7}$ **5.** $8\frac{2}{5}$ **7.** $4\frac{2}{7}$
9. $\frac{8}{15}$ **11.** $\frac{7}{10}$ **13.** $\frac{2}{5}$ **15.** $\frac{3}{7}$
17. $\frac{1}{9}$ **19.** $1\frac{2}{3}$ **21.** 10 **23.** 16
25. $\frac{5}{16}$ **27.** $6\frac{1}{4}$ **29.** 12 **31.** $\frac{2}{5}$
33. $\frac{9}{16}$ **35.** $3\frac{3}{8}$ **37.** $\frac{25}{36}$ **39.** $\frac{4}{9}$
41. 45 **43.** $1\frac{1}{5}$ **45.** $1\frac{1}{4}$ **47.** $\frac{2}{3}$
49. $1\frac{5}{9}$ **51.** $1\frac{3}{5}$ **53.** $A = 20$

55. $A = 2\frac{7}{9}\pi$ **57.** $V = 1\frac{23}{25}\pi$
59. $V = 45\frac{37}{48}\pi$

Chapter 5

1. $1\frac{2}{5}$ **3.** $\frac{1}{9}$ **5.** $\frac{14}{15}$ **7.** $\frac{1}{6}$ **9.** $\frac{1}{6}$
11. $2\frac{1}{4}$ **13.** $2\frac{2}{21}$ **15.** $1\frac{13}{24}$
17. $1\frac{23}{36}$ **19.** $3\frac{17}{60}$ **21.** $5\frac{7}{12}$
23. $10\frac{3}{5}$ **25.** $13\frac{13}{20}$ **27.** $1\frac{11}{15}$
29. $8\frac{7}{16}$ **31.** $1\frac{1}{3}$ **33.** $9\frac{5}{6}$
35. $13\frac{1}{10}$ **37.** $5\frac{1}{10}$ kg **39.** $20\frac{1}{2}$ mi

Chapter 6

1. $446 **3.** $253
5. ATTENDANCE AT FOOTBALL GAMES

Game	Number of Fans
1	🚶🚶🚶🚶🚶🚶
2	🚶🚶🚶🚶🚶🚶🚶
3	🚶🚶🚶🚶🚶🚶🚶
4	🚶🚶🚶🚶🚶🚶🚶🚶

Key: Each 🚶 represents 5,000 fans.

7. AVERAGE MONTHLY TEMPERATURE

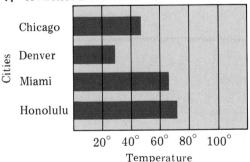

9. 60 **11.** 70 **13.** 1984

Chapter 7

1. 0.7; 0.74 **3.** 19.6; 19.58 **5.** 9.0; 9.02 **7.** 0.2; 0.15 **9.** 4.1; 4.06
11. 6.6; 6.58 **13.** 0.1; 0.09 **15.** 0.7; 0.67 **17.** 0.9; 0.86 **19.** 0.4; 0.42
21. 48.53 **23.** 19.851 **25.** 178.11
27. 173.966 **29.** 8.378 **31.** 89.205
33. 62.87 **35.** 605.685 **37.** 96.433
39. 208.6 **41.** 166.23 **43.** 112.24
45. 5.91 **47.** 2.4 **49.** 1.5
51. 40.3 **53.** 11.4 km/h

Chapter 8

1. 4.8 **3.** 0.08 **5.** 0.6 **7.** 0.8
9. 42,000 **11.** 4 **13.** 100
15. 3,700 **17.** 12,000 **19.** 46,000
21. 4 **23.** 0.015 **25.** 0.03
27. 0.025 **29.** 4.5 **31.** 77°F
33. 48°F **35.** 90°F **37.** 142°F
39. 131°F **41.** 118°F **43.** 38°C
45. 11°C **47.** 25°C **49.** 29°C
51. 18°C **53.** 36°C **55.** 4,624 mm^2
57. 3,500 mL **59.** 0.95 t **61.** 75,000 m

Mixed Practice for Chapters 1–8

1. 13 **3.** 95 **5.** 101 **7.** none
9. 8 **11.** 141 **13.** 9 **15.** 40.5
17. $1\frac{1}{6}$ **19.** $2\frac{1}{10}$ **21.** 125
23. 243 **25.** $6\frac{2}{9}$ **27.** $4\frac{17}{20}$ **29.** $1\frac{1}{7}$
31. 12 **33.** 9.8; 9.76 **35.** 63.5; 63.46 **37.** 40.09 **39.** 2,419.62
41. 4,630 **43.** 0.9 **45.** 1,300 **47.** 32°C
49. −12°C **51.** 5 **53.** $172

Chapter 9

1. 0.5 **3.** 4.00 **5.** 0.007 **7.** 4.25
9. 0.000007 **11.** 7% **13.** 1,800%
15. $12\frac{1}{2}$% **17.** $45\frac{3}{5}$% **19.** 4,560%
21. 111.45 **23.** 9.75 **25.** 344.10
27. 60% **29.** about 47% **31.** about 5.7% **33.** 40 **35.** 93 **37.** 171.43
39. 21 **41.** $15,000 **43.** sweater: $22.40; shoes: $33.95; coat: $87.78

Chapter 10

1. $r = 10$; $m = 15$ **3.** $r = 61$; $m = 279$
5. $r = 17$; $m = 51$ **7.** 97 **9.** 15
11. 50 **13.** 98 **15.** 4 **17.** none
19.

Score(s)	Tally	Frequency(f)	Sum(f · s)
100					3	300
98					3	294
97			1	97
96				2	192
94				2	188
93			1	93
90			1	90
88				2	176
86			1	86
Total		16	1,516

21. 12 **23.** 5 **25.** $\frac{1}{6}$ **27.** 58
29. $P(\text{red}) = \frac{4}{9}$; $P(\text{white}) = \frac{5}{9}$ **31.** 27

Chapter 11

1. $\frac{10}{7} = 1\frac{3}{7}$ **3.** $\frac{14}{10} = 1\frac{2}{5}$ **5.** $\frac{7}{4} = 1\frac{3}{4}$
7. 8 **9.** 1 **11.** 1 **13.** 1 **15.** 30
17. 21 **19.** yes **21.** no **23.** yes
25. 3 h 20 min **27.** 5 m **29.** 6

31.

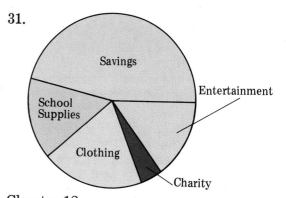

Chapter 12

1. < **3.** < **5.** > **7.** > **9.** <
11. < **13.** 11 **15.** −10 **17.** 14
19. −8 **21.** −16 **23.** −6 **25.** 0
27. 9 **29.** 0 **31.** 12 **33.** −11
35. −19 **37.** 12 **39.** −15
41. −25 **43.** 3 **45.** 10 **47.** −14
49. −12 **51.** 0 **53.** −10
55. 78°F **57.** 142 **59.** 160

Chapter 13

1. 270 **3.** −350 **5.** 160
7. 1,530 **9.** 0 **11.** −432
13. −15 **15.** −52 **17.** −120
19. 165 **21.** 168 **23.** −65
25. 11 **27.** −10 **29.** −8 **31.** 4
33. not possible **35.** $1\frac{7}{8}$ **37.** $-1\frac{1}{35}$
39. $-\frac{1}{3}$ **41.** $1\frac{19}{35}$ **43.** $\frac{20}{33}$
45. $-1\frac{5}{33}$ **47.** $9x$ **49.** $30m$
51. $-3 - 2y$ **53.** a **55.** $-10a + 6$
57. $-24a + 12$ **59.** $3 - 2y$
61. −6 **63.** −12 **65.** −3 **67.** 1
69. −27 **71.** 27 **73.** −50
75. −100 **77.** −2 **79.** −5
81. 8 **83.** −4

Chapter 14

1. −5 **3.** −2 **5.** 4 **7.** $n > 10$
9. $h \geq -5$ **11.** $x \leq -4$ **13.** −32
15. −48 **17.** a^8 **19.** y^6 **21.** x^3y^3
23. $9a^4$ **25.** $8x^6y^9$ **27.** $-c^2 - 12c + 12$ **29.** $7a^6 + 2a^4 + 14$
31. $-3a^3 + a^2$ **33.** $2y^3 + 6y^2 + 4y$
35. $-6m^3 + 2m^2 + 2m$ **37.** $3(x - 3)$
39. $a(6a^2 + 1)$ **41.** 11 **43.** 7
45. 3 **47.** 16

Chapter 15

1. complementary **3.** supplementary
For examples **5** and **7**, answers may vary. Examples are given. **5.** $\angle 1$ and $\angle 3$
7. $\angle 6$ and $\angle 3$ **9.** $\angle 2$: 140°; $\angle 3$: 40°; $\angle 4$: 140°; $\angle 5$: 140°; $\angle 6$: 40°; $\angle 7$: 140°; $\angle 8$: 40° **11.** 3 **13.** 6
15. □ **17.** ⬠
19. acute **21.** right **23.** 20°
25. ±6 **27.** ±10 **29.** ±12
31. 5.7 **33.** 1.7 **35.** $c = 5$
37. $a = 4.9$ **39.** 9.4 in.

Chapter 16

1. the integers between −1 and 4

3. the integers less than 1

5. $x = 1$

7. $n = 1$

9. $a = -3$

11. $x > 2$

13. $x \geq 4$

15. $y > 1$

17. $x > 4$

ODD-NUMBERED ANSWERS

19.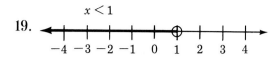

21.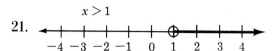

23. (3, 1)   25. (−4, 2)   27. (2, −2)
29. (0, −3)

31., 33., 35.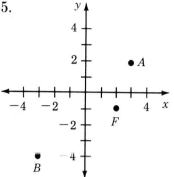

39.

x	$x + 3$	y	(x, y)
−1	−1 + 3	2	(−1, 2)
0	0 + 3	3	(0, 3)
1	1 + 3	4	(1, 4)

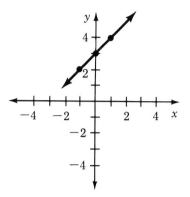

37.

x	$x - 1$	y	(x, y)
0	0 − 1	−1	(0, −1)
1	1 − 1	0	(1, 0)
2	2 − 1	1	(2, 1)

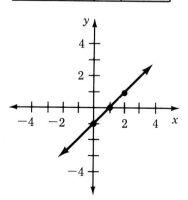

40.

x	$2x + 1$	y	(x, y)
−1	2(−1) + 1	−1	(−1, −1)
0	2(0) + 1	1	(0, 1)
1	2(1) + 1	3	(1, 3)

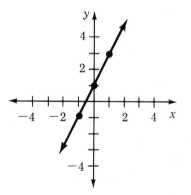

43. $\frac{4}{3}$ **45.** $-\frac{1}{2}$ **47.** 4; -2 **49.** $\frac{3}{5}$; -1 **51.** $\frac{1}{4}$; $\frac{3}{4}$ **53.** $(-2, -5)$
55. $(1, -5)$

67.

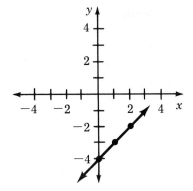

x	$x - 4$	y	(x, y)
2	$2 - 4$	2	$(2, -2)$
1	$1 - 4$	3	$(1, -3)$
0	$0 - 4$	4	$(0, -4)$

Mixed Practice for Chapter 9–16

1. 0.60   **3.** 5.0   **5.** 0.000065
7. 53.7%   **9.** 15.5%   **11.** 77.76
13. 70%   **15.** 54   **17.** 8, 10, 10, 8 and 12   **19.** 16   **21.** 72   **23.** 8
25. -5 **27.** 1   **29.** 3   **31.** 3
33. $x > 9$ **35.** $x = b - a$ **37.** $x = c + d$ **39.** -6 **41.** 15   **43.** -16
45. -120 **47.** 5   **49.** $-\frac{2}{7}$
51. $x^4 y^2$ **53.** $-18 a^5$ **55.** $-2x^3 + 6x^2 - 10x$ **57.** $-m^2 + 2m - 6$
59. -12 **61.** -144

63., 65.

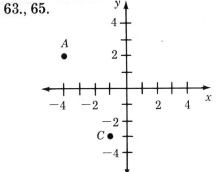

69. $4(a^2 - 3a + 9)$ **71.** 70°   **73.** 5°
75. $\frac{10}{14}$ or $\frac{5}{7}$ **77.** $\frac{4}{14}$ or $\frac{2}{7}$ **79.** $\frac{3}{5}$
81. $\frac{4}{3}$ **83.** $6n + 3$ **85.** 12.7
87. 7.3   **89.** 1.7   **91.** $\frac{4}{3}$ **93.** $\frac{2}{3}$, 3
95. $(1, 2)$ **97. a.** obtuse   **b.** right
c. acute   **99.** 30°, 150°

Table of Measures

METRIC | CUSTOMARY

Length

1 kilometer (km) = 1,000 meters
1 hectometer (hm) = 100 meters
1 dekameter (dam) = 10 meters
1 meter (m)
1 decimeter (dm) = 0.1 meter
1 centimeter (cm) = 0.01 meter
1 millimeter (mm) = 0.001 meter

1 foot (ft) = 12 inches (in.)
1 yard (yd) = $\begin{cases} 3 \text{ feet} \\ 36 \text{ inches} \end{cases}$
1 mile (mi) = $\begin{cases} 5{,}280 \text{ feet} \\ 1{,}760 \text{ yards} \end{cases}$

Mass/Weight

1 metric ton (t) = 1,000 kilograms
1 kilogram (kg) = 1,000 grams
1 hectogram (hg) = 100 grams
1 dekagram (dag) = 10 grams
1 gram (g)
1 decigram (dg) = 0.1 gram
1 centigram (cg) = 0.01 gram
1 milligram (mg) = 0.001 gram

1 pound (lb) = 16 ounces (oz)
1 ton (T) = 2,000 pounds

Capacity

1 kiloliter (kL) = 1,000 liters
1 hectoliter (hL) = 100 liters
1 dekaliter (daL) = 10 liters
1 liter (L)
1 deciliter (dL) = 0.1 liter
1 centiliter (cL) = 0.01 liter
1 milliliter (mL) = 0.001 liter

1 cup (c) = 8 fluid ounces (fl oz)
1 pint (pt) = 2 cups
1 quart (qt) = 2 pints
1 gallon (gal) = 4 quarts

1 liter = 1,000 cubic centimeters (cm^3)
1 milliliter = 1 cubic centimeter

TIME

1 minute (min) = 60 seconds (s)
1 hour (h) = 60 minutes
1 day (d) = 24 hours
1 week = 7 days

1 year (y) = $\begin{cases} 12 \text{ months} \\ 365 \text{ days} \end{cases}$
1 decade = 10 years
1 century = 100 years

Symbol List

$<$	is less than
$>$	is greater than
3^4	the fourth power of 3
x^3	the third power of x, or x cubed
$\doteq$	is approximately equal to
$\neq$	is not equal to
$0.2\overline{2}$	repeating decimal
$\overline{AB}$	line segment AB
$\cong$	is congruent to
$\overrightarrow{AB}$	ray AB
$\overleftrightarrow{TR}$	line containing points T and R
$\angle XYZ$	angle XYZ
$\triangle ABC$	triangle ABC
$m\angle A$	measure of angle A
π	pi (about 3.14)
$\overleftrightarrow{KL} \parallel \overleftrightarrow{MN}$	line KL is parallel to line MN
$\overleftrightarrow{AB} \perp \overleftrightarrow{CD}$	line AB is perpendicular to line CD
$\approx$	is similar to
$\sqrt{25}$	the square root of 25
20%	20 percent
$+6$	positive 6
-4	negative 4
$|-5|$	absolute value of negative 5
BASIC	Beginners All-Purpose Symbolic Instruction Code

Formulas

Perimeter
- rectangle $P = 2l + 2w$
- triangle $P = a + b + c$
- square $P = 4s$

Circumference
- circle $C = \pi d = 2\pi r$

Area
- rectangle $A = lw$
- square $A = s^2$
- parallelogram $A = bh$
- triangle $A = \frac{1}{2}bh$
- circle $A = \pi r^2$

Surface area
- rectangular prism $A = 2lw + 2lh + 2wh$
- cube $A = 6s^2$

Volume
- rectangular prism $V = lwh$
- cube $V = s^3$
- cylinder $V = \pi r^2 h$
- rectangular pyramid $V = \frac{1}{3}Bh$
- cone $V = \frac{1}{3}\pi r^2 h$

Table of Squares and Square Roots

No.	Square	Square Root	No.	Square	Square Root	No.	Square	Square Root
1	1	1.000	35	1,225	5.916	68	4,624	8.246
2	4	1.414	36	1,296	6.000	69	4,761	8.307
3	9	1.732	37	1,369	6.083	70	4,900	8.357
4	16	2.000	38	1,444	6.164	71	5,041	8.426
5	25	2.236	39	1,521	6.245	72	5,184	8.485
6	36	2.449	40	1,600	6.325	73	5,329	8.544
7	49	2.646	41	1,681	6.403	74	5,476	8.602
8	64	2.828	42	1,764	6.481	75	5,625	8.660
9	81	3.000	43	1,849	6.557	76	5,776	8.718
10	100	3.162	44	1,936	6.633	77	5,929	8.775
11	121	3.317	45	2,025	6.708	78	6,084	8.832
12	144	3.464	46	2,116	6.782	79	6,241	8.888
13	169	3.606	47	2,209	6.856	80	6,400	8.944
14	196	3.742	48	2,304	6.928	81	6,561	9.000
15	225	3.875	49	2,401	7.000	82	6,724	9.055
16	256	4.000	50	2,500	7.071	83	6,889	9.110
17	289	4.123	51	2,601	7.141	84	7,056	9.165
18	324	4.243	52	2,704	7.211	85	7,225	9.220
19	361	4.359	53	2,809	7.280	86	7,396	9.274
20	400	4.472	54	2,916	7.348	87	7,569	9.327
21	441	4.583	55	3,025	7.416	88	7,744	9.381
22	484	4.690	56	3,136	7.483	89	7,921	9.434
23	529	4.796	57	3,249	7.550	90	8,100	9.487
24	576	4.899	58	3,364	7.616	91	8,281	9.539
25	625	5.000	59	3,481	7.681	92	8,464	9.592
26	676	5.099	60	3,600	7.746	93	8,649	9.644
27	729	5.196	61	3,721	7.810	94	8,836	9.695
28	784	5.292	62	3,844	7.874	95	9,025	9.747
29	841	5.385	63	3,969	7.937	96	9,216	9.798
30	900	5.477	64	4,096	8.000	97	9,409	9.849
31	961	5.568	65	4,225	8.062	98	9,604	9.899
32	1,024	5.657	66	4,356	8.124	99	9,801	9.950
33	1,089	5.745	67	4,489	8.185	100	10,000	10.000
34	1,156	5.831						

Glossary

The explanations given in this glossary are intended to be brief descriptions of the terms listed. They are not necessarily definitions.

Acute angle An angle whose measure is less than 90°.

Acute triangle A triangle in which each angle measures less than 90°.

Addition property for equations If $a = b$ is true, then $a + c = b + c$ is also true for all numbers a, b, and c.

Addition property for inequalities If $a < b$, then $a + c < b + c$ for all numbers a, b, and c.

Adjacent angles Two angles with a common vertex, a common side, and no common interior points.

Alternate interior angles Two inside angles on the same side of the transversal that cuts two lines.

Altitude (of a triangle) A segment that originates at a vertex of the triangle and is perpendicular to the line containing the opposite side.

Associative property of addition For all numbers a, b, and c, $(a + b) + c = a + (b + c)$.

Associative property of multiplication For all numbers a, b, and c, $(a \cdot b) \cdot c = a \cdot (b \cdot c)$.

Average The number found from adding several numbers and then dividing the sum by the number of numbers added.

Bar graph A method of comparing quantities by the use of solid bars.

Binomial A polynomial with two terms.

Capacity To measure the capacity of an object, we measure how much it holds.

Central angle An angle whose vertex is at the center of a circle.

Circle Closed curve in a plane (flat surface) such that every point on the circle is the same distance from the center.

Circumference The measure of the distance around a circle.

Coefficient The multiplier of a variable. *Example* In $4x$, 4 is the coefficient.

Collinear points Points that are contained in one line.

Commission Amount of extra money salespeople are given for selling something. The more they sell, the more they make if they are paid a "commission."

Commutative property of addition $a + b = b + a$

Commutative property of multiplication $a \cdot b = b \cdot a$

Complementary angles Two angles the sum of whose measure is 90°.

Complete factorization A number shown as a product of prime numbers only: $36 = 3 \cdot 3 \cdot 2 \cdot 2$.

Congruent segments Segments that are of the same length.

Coordinate A number assigned to a point on the number line.

Coordinate plane Two perpendicular number lines in a plane make up a coordinate plane, or a coordinate system. Each point in a coordinate plane corresponds to an ordered pair of numbers, and vice versa.

Corresponding angles Two angles on the same side of a transversal, one on the inside, the other on the outside.

Cosine (cos) The cosine of an angle in a right triangle is the ratio of the length of the side adjacent to that angle to the length of the hypotenuse.

Counting principle One event can happen in x ways. Another event can happen in y ways. The total number of ways that both events can happen is $x \cdot y$ ways.

Cube A three-dimensional solid with all faces squares.

Decagon A polygon with ten sides.

Decimal Numbers with decimal points used to indicate place value.

Degree Celsius (°C) A metric unit of temperature.

Denominator In the fraction $\frac{3}{7}$, the number below the fraction bar is the denominator.

Dependent events Two events are dependent if the outcome of one event affects the outcome of the other.

Diameter A line segment whose endpoints are on the circle and which contains the center of the circle.

Discount A reduction applied to the price of an article.

Distributive property of multiplication over addition For all numbers a, b, and c, $a \cdot (b + c) = a \cdot b + a \cdot c$.

Divisible A number is divisible by another number if the remainder is zero, when the first number is divided by the second number.

Divisor In $7\overline{)28}$, 7 is the divisor, the number by which you are dividing.

Equation A mathematical sentence in which the $=$ symbol is used. $3x = 15$ and $x + 6 = 14$ are examples of equations.

Equation of a line $y = mx + b$ is an equation of a line. m is the slope of the line, and b is the y-intercept.

Even number Any number divisible by 2 is an even number.

Exponent In 2^4, 4 is the exponent. It tells how many times 2 is used as a factor in the product $2 \cdot 2 \cdot 2 \cdot 2$.

Extremes (in a proportion) The extremes in a proportion are the first and the fourth terms.
Example 2:3 = 4:6

or $\frac{2}{3} = \frac{4}{6}$ 2 and 6 are the extremes.

Factor A number is a factor of a given number if the given number is divisible by the number.
$5 \cdot 2 = 10$
5 and 2 are factors of 10.

Formula An equation that relates some quantities.
$A = \pi r^2$ is a formula.

Frequency The number of times that a number appears in a set of data.

Gram (g) A metric unit of weight, 0.001 of a kilogram.

Greatest common factor (GCF) The largest number by which each of a given set of two or more numbers is divisible. For example, 4 is the GCF of 8, 16, 20.

Hexagon A polygon with six sides.

Hypotenuse The side opposite the right angle in a right triangle.

Independent events Two events are independent if the outcome of one does not affect the outcome of the other.

Inequality A mathematics sentence that has the symbol > or < in it.

Integer A directed whole number, positive or negative or zero.

Irrational number A number named by a nonterminating, nonrepeating decimal. For example: 1.424224222 . . ., $\sqrt{2}$, π.

Least common denominator The same as the least common multiple.

Least common multiple (LCM) The smallest number of which each of two or more given numbers is a factor. The LCM of 3, 5, 2, is 30. 30 is the smallest number that has 3, 5, and 2 as factors.

Line segment Segment $\overline{AB}$ is the set of points on a line including A and B and all the points in between.

Liter (L) A metric unit of capacity, 1,000 mL.

Mean The mean is the average. To find the mean, add the numbers and divide the sum by the number of numbers.

Means (in a proportion) The means in a proportion are the second and the third terms.
Example   2:3 = 4:6
        or $\frac{2}{3} = \frac{4}{6}$ 3 and 4 are the means.

Median In a set of numbers arranged from least to greatest, the median is the middle number. If there is an even number of items, the median is the mean of the two middle numbers.

Meter A metric unit of length that is equal to 100 centimeters.

Metric system A system of measurement based on the number 10.

Mixed number A number that consists of the sum of an integer and a fraction. For example: $4\frac{2}{3}$.

Mode The number occurring more frequently than any other number. A set may have more than one mode.

Monomial A polynomial with one term.

Multiple Any product of the given number and a whole number.

Non-collinear points Three or more points that are not contained in the same line.

Number line A line in which the numbers and the points have been matched one-to-one.

Numerator In a fraction, the number above the fraction line is the numerator. For example, in the fraction $\frac{5}{7}$, 5 is the numerator.

Obtuse angle An angle whose measure is between 90° and 180°.

Obtuse triangle A triangle that has one obtuse angle.

Octagon A polygon with eight sides.

Odd number Any number not divisible by 2 is an odd number.

Ordered pair (2, −3) is an ordered pair of numbers. Each ordered pair of numbers corresponds to exactly one point in a coordinate plane, and vice versa.

Origin The point for zero on a number line. The point for (0, 0) in a coordinate plane.

Parallel lines Lines that lie in the same plane and do not intersect.

Parallelogram A quadrilateral with two pairs of opposite sides parallel.

Percent Indicates how many out of a hundred.
Symbol for percent: %
Example 42% means 42 out of a hundred.

Perfect square A number whose principal square root is a whole number.

Perimeter The distance around a polygon.

Perpendicular lines Lines intersecting at right angles.

Pi Pi or π is the number, approximately equal to 3.14, used in finding the circumference or area of a circle.
π is the ratio between the circumference and twice the radius.

Pictographs A graph that uses pictures to show data.

Plane A two-dimensional flat surface.

Polygon A geometric figure with three or more sides.

Polynomial A polynomial is an expression containing one or more terms.
Examples $3x, 5a^2 - 2$
$4y^2 - 3y + 1$

Prime number A whole number greater than 1 having only two factors, 1 and itself. For example, 7 is a prime. The only factors of 7 are 7 and 1.

Principal square root The positive square root of a number. For example, 5 is the principal square root of 25. $\sqrt{25} = 5$

Probability The probability of an event is the number of favorable outcomes divided by the total number of all possible outcomes.

Property of 1 for multiplication The product of any number and 1 is that number. $x \cdot 1 = x$

Proportion A proportion is an equation that states that two ratios are equal.
$\frac{a}{b} = \frac{c}{d}$ or $a:b = c:d$

Protractor A device for measuring angles.

Pythagorean theorem In any right triangle, the square of the length of the hypotenuse equals the sum of the squares of the lengths of the other two sides. If $\triangle ABC$ is a right triangle with c the hypotenuse, then $a^2 + b^2 = c^2$.

Pythagorean triple Any three numbers a, b, c such that $a^2 + b^2 = c^2$.

Quadrilateral A polygon with four sides.

Radius A segment from a point on a circle to the center of the circle.

Range The difference between the highest and lowest numbers in a set.

Ratio A ratio is a comparison of two numbers by division. A fraction is a ratio. For example, $\frac{a}{b}$.

Rational number A number that can be named by a fraction. For example, a fraction or a mixed number or a terminating or repeating decimal.

Ray A part of a straight line that has a beginning point and continues infinitely in one direction.

Real numbers The real numbers include all of the rational and irrational numbers.

Reciprocal Two numbers are reciprocals if their product is 1. 3 and $\frac{1}{3}$ are reciprocals since $3 \times \frac{1}{3} = 1$.

Rectangle A four-sided polygon with all right angles.

Regular polygon A polygon with all sides of the same length and all angles of the same measure.

Repeating decimal A decimal that repeats a digit or group of digits forever.
Example $0.4444\ldots = 0.\overline{44}$

Rhombus A quadrilateral with four sides of the same length.
Right angle An angle whose measure is 90°.
Right triangle A triangle that contains one right angle.

Scientific notation A number written as the product of a number from 1 to 10 and a power of 10.
Example $2.39 \times 10^3 = 2{,}390$
Similar triangles Triangles with the same shape but not necessarily the same size.
Sine (sin) The sine of an angle in a right triangle is the ratio of the length of the side opposite the angle to the length of the hypotenuse.
Slope of a line The slope of a line can be shown by this ratio:
$$\text{slope} = \frac{\text{Difference of y-coordinates}}{\text{Difference of x-coordinates}}$$
Sphere A round three-dimensional figure shaped like a baseball or basketball. All points on a sphere are the same distance from the center.
Square A quadrilateral in which all sides have the same measure and all angles are right angles.
Square root If $x \cdot x = n$, then x is a square root of n.
Straight angle An angle with a measure of 180°.
Supplementary angles Two angles the sum of whose measures is 180°.
System of equations Two equations with two variables form a system of equations.
Example $y = 2x - 5$
$y = -3x + 5$

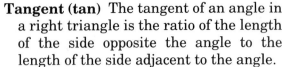

Tangent (tan) The tangent of an angle in a right triangle is the ratio of the length of the side opposite the angle to the length of the side adjacent to the angle.
Transversal A line that intersects two or more lines.
Trapezoid A quadrilateral with exactly one pair of parallel sides.
Triangle A polygon with three sides.
Trinomial A polynomial with three terms.

Variable A letter that can be replaced by numbers.
Vertical angles Angles formed by two intersecting lines that are opposite each other; they have the same measure.
Volume The measure of the amount of space occupied by a figure in three dimensions.

Weight A number that tells how heavy an object is.
Whole number The numbers 0, 1, 2, 3, 4, . . . are whole numbers.

***x*-axis** The horizontal number line in a coordinate plane.
***x*-coordinate** The first number in an ordered pair of numbers.

***y*-axis** The vertical number line in a coordinate plane.
***y*-coordinate** The second number in an ordered pair of numbers.
***y*-intercept** The *y*-intercept is the *y*-coordinate of the point of intersection of a line with the *y*-axis.

Index

Acute angle, 370
Acute triangle, 377
Addition
  checking subtraction by, 8–10
  of decimals, 171–172
  for equations, 8–10, 44, 306
  of fractions, 111–113, 116–121, 136
  of integers, 291–293
  of mixed numbers, 122–124
  of opposites, subtraction of integers by, 301–302
  properties of, 3–4
  for inequalities, 344
  of whole numbers, 3–4
Additive identity, 3–4
Algebra
  combining like terms, 56–59, 66, 335–338, 356–359
  English phrases to, 11–12, 39–41, 45, 63–64, 339–341
  graphing
    equations, 393–395, 406–408, 413–414
    inequalities, 394–398, 407–408
  order of operations, 33–34
  polynomials, 356–362
  solving
    equations, 8–10, 36–38, 44, 61–62, 102, 132–133, 182–183, 305–306, 328–329, 335–338
    inequalities, 347–349
    systems of equations, 413–414
Algebra Maintenance, 55, 142, 170, 194, 218, 253, 262, 297, 362, 369
Algebraic Expressions
  evaluating, 1–2, 26–27, 33–34, 58–59, 64, 70–71, 101, 129–130, 172–174, 331, 351–352, 355, 406
  simplifying, 54–59, 64, 326–327, 353–359
Alternate interior angles, 373

Angles
  acute, 370
  alternate interior angles, 373–374
  central, of a circle graph, 271, 379
    of a polygon, 379
  classify, 370–371, 377
  complementary, 371
  corresponding, in similar triangles, 277–279, 373–374
  measuring, 370–371
  obtuse, 370
  reflex, 370
  right, 277, 370, 375, 385
  straight, 374
  supplementary, 371
  vertical, 373–374
Approximately equal to ($\doteq$), 383, 386
Area
  of a circle, 72, 104
  in metric units, 200–201
  of a rectangle, 46
  of a right triangle, 47
  of a square, 72, 383–385
  surface, of a cube and cylinder, 73
  of a triangle, 46, 104
Arithmetic mean, 246
Arrangements, all possible, 252–253
Associative property of addition, 3–4
  multiplication, 28–29, 317

Bar graphs, 147–149
  multiple, 154, 156, 246
Bases
  exponents and, 69, 353
  of isosceles triangle, 157
BASIC, 420–435
  FOR/NEXT, 186, 211, 256, 307, 331, 389, 415, 423–425
  GOTO, 106, 186, 256, 331
  IF/THEN, 106, 186, 256, 331, 429–431

INPUT, 106, 136, 186, 211, 236, 307, 364, 389, 415, 420–422, 427–428
INT, 106, 256, 434
PRINT, 22, 49, 75, 106, 136, 161, 186, 211, 236, 256, 284, 307, 331, 364, 389, 415, 426–428
RND, 256, 435
Sequential input, 186, 432, 433
Binomial, 356

Calculator, 15, 32, 64, 91, 97, 128, 159, 172, 204, 221, 245, 265, 308, 318, 336, 355, 375, 398
Capacity, metric measure of, 202–204
Careers, 20, 68, 131, 153, 184, 208, 234, 254, 273, 298, 325, 350, 381, 409
Celsius, degrees(°C), 209–211
Centigram (cg), 205
Centiliter (cL), 203
Centimeters (cm), 190, 193
  cubic (cm^3), 202–205
  in scale drawings, 274–275
  square (cm^2), 200–201
Challenge, 7, 34, 81, 94, 124, 134, 152, 174, 179, 197, 207, 231, 242, 272, 293, 315, 329, 333, 346, 349, 384, 405
Chapter Reviews, 23, 50, 76, 107, 137, 162, 187, 212, 237, 257, 285, 308, 332, 365, 390, 416
Chapter Tests, 24, 51, 77, 108, 138, 163, 188, 213, 238, 258, 286, 309, 333, 366, 391, 417
Circle
  area of, 72, 104
  circumference of, 47
Circle graphs, 270–272
Coefficients, 56, 412
Collinear points, 368
Combinations, all possible, 252–253
Combining like terms, 56–59, 66, 335–338, 356–359
Commissions, 226–228
Commutative property of
  addition, 3–4
  multiplication, 28–29, 316
Complementary angles, 371
Composite numbers, 88–89
Computer activities, 22, 35, 49, 75, 106, 136, 161, 186, 211, 256, 307, 331, 364, 389, 415, 420–435
Cone, volume of, 105
Congruent sides, 380
Constant, 1, 412
Construction of perpendicular lines, 375
Coordinate plane, 401
Coordinates,
  difference of, to determine slope, 410–412, 415
  of horizontal and vertical lines, 404–405
  of a point, 401–403, 413–414
Corresponding angles
  of parallel lines, 373–374
  of similar triangles, 277–279
Cosine of an angle, 281–282
Counting principal, 252–253
Cubes
  of consecutive numbers, 384
  surface area and volume of, 73
Cumulative review, 109, 310, 418
Cylinder
  surface area of, 73
  volume of, 72

Data
  mean of grouped, 243–244
  organizing, 140–156, 161–163
Decagon, 380
Decigram (dg), 205
Deciliter (dL), 203
Decimals, 165–167
  addition of, 171–172
  changing fractions to, 168–170
  changing percents to, 216–218
  division of, 177–179
  in equations, 182–183, 376

multiplication of, 173–174
  by powers of ten, 180–181
as rational numbers, 323
repeating, 169
rounding of, 166–167, 169–170, 178–179
subtraction of, 171–172
Decimeter (dm), 192–193
Dekagram (dag), 205
Dekaliter (daL), 203
Dekameter (dam), 193
Dependent probabilities, 250–251
Direct variation, 266–268
Discounts, 229–231
Distributive property, 53–56, 111, 317, 326–327, 337–338, 358–361
Division
of decimals, 177–179
of fractions, 99–100
of integers, 320–322
of rational numbers, 324
using multiplication to undo, 37–39, 44
by zero, 80, 321, 331, 415
property for
  equations, 36–38, 44, 328–329
  inequalities, 345
Drawing, scale, 274–276

Endpoints of a segment, 368
Equations, 8
addition property for, 8–10, 44, 306
applying equations, 157–159
checking solutions to, 9–10, 37–38, 61–62, 102, 132–133, 182, 305, 328–329, 335, 337
decimals in, 182–183, 376
division property for, 36–38, 44, 328–329
fractions in, 102–103, 132–133
graphing, 393–395, 406–408, 413–414
integers, 305–306, 328–329
of a line, 412
multiplication property for, 36–38, 44, 328–329

slope intercept form, 412
system of, 413–414
two step, 61–62, 132–134, 183, 329, 340
with rational numbers, 328–329
using two properties for, 44
variable on both sides of, 335–338, 364
Equilateral triangle, 378
Estimation, 15
Evaluating
algebraic expressions, 1–2, 26–27, 33–34, 58–59, 64, 70–71
with decimals, 172–174
with exponents, 70–71, 351–352, 355
with fractions, 101, 129–130, 331
with integers, 406
Events, probability of
dependent, 250–251
independent, 250–251
simple, 248–249
Exponents, 69–71, 75, 88
in algebraic expressions, 70–71, 351–362
formulas with, 72–73, 104–105
product of, 354
properties of, 353–355
Extremes of a proportion, 264

Factors, 69–71, 88–89
greatest common (GCF), 90–91, 106, 360–362
polynomial, 360–362
Factorial (!), 308
Fahrenheit (°F) scale, 209–211
Favorable outcomes, 248, 250
Formulas,
business, 19, 161, 236
general, 136, 198, 209–211, 364
geometric, 18, 46–47, 72–73, 104–105, 157–158, 385, 389
Fractions, 79–81
addition of, 111–113, 116–121, 136
changing,
  to decimals, 168–170
  to mixed numbers, 82–83

to percents, 217–218
division of, 99–100
improper, 82
in equations, 102–103, 132–133
multiplication of, 82–87, 95–97
multiplied by whole numbers, 80–81
as ratios, 260–262
simplifying, 92–94, 106, 111–112, 116–117, 119–121
subtraction of, 111–113, 116–121
Frequency table, 243–245

Geometry
angles, 370–371
area, 46, 72–73, 104
circle, 72, 104
construction, 375
parallel lines, 373–374
perpendicular lines, 375–376
points, lines, and planes, 368–369
polygons, 379–380
pythagorean theorem, 385–387
scale drawing, 274–276
similar triangles, 277–279
surface area, 73
triangles, 377–378
volume, 47, 72–73, 105, 204
Gram (g), 205–207
Graphing
equations, 393–395, 406–408, 411, 413
horizontal and vertical lines, 404–405
inequalities, 394–398, 407–408
integers, 393–395
ordered pairs, 401–405
solutions, 393–398, 406–408, 413–414
systems of equations, 413–414
Graphs
bar, 147–149, 154, 156, 246
circle, 270–272
line, 150–152, 154–156
Greater than (>), 6, 289
Greatest common factor (GCF), 90–91, 106, 360–362

Hectogram (hg), 205
Hectoliter (hL), 203
Hectometer (hm), 193
Hexagon, 379
Horizontal bar graphs, 147–149
Hypotenuse of right triangle, 280, 385–387, 389

Identity property
additive, 3–4
multiplicative, 28–29, 57, 316
Improper fractions, 82
Inequalities, 289
addition property for, 344, 347, 396
checking solutions to, 347, 396
division property for, 345, 348, 396
graphing, 394–398, 407–408
multiplication property for, 345, 348
properties for, 344–346
solving, 347–349, 396–397
subtraction property for, 344, 346–347, 397
Integers, 288–290, 303–304
addition of, 291–293
division of, 320–322
equations with, 305–306
graphing on a number line, 288–290
multiplication of, 312–315
negative, 288
on a number line, 288–292
opposites of, 299–302, 306
positive, 288
as rational numbers, 323
solving problems using, 303–304
subtraction of, 295–297
by addition of opposites, 301–302
Intersecting lines, 373
Inverse operations, 8–10, 36–38
Inverse variation, 269
Irrational numbers, 383
Isosceles triangle, 157, 378

Jobs for Teenagers, 16–17, 42–43, 60, 98, 125, 145–146, 175–176, 198–199, 232–233, 247, 294, 319, 342–343, 372, 399–400

Kilogram (kg), 205–207
Kiloliter (kL), 203–204
Kilometer (km), 193–197

Least common denominator (LCD), 115–123, 127–130
Least common multiple (LCM), 114–116
Leg of a triangle, 157, 389
Less than (<), 6, 289
Lines, 368–369, 373–374
  equation of, 412
  graph of, 407, 411, 414
  graphing
    equations on, 393–395
    inequalities on, 394–398
  intersecting, 373
  horizontal and vertical, 404, 405
  parallel, 373–374
  perpendicular, 375–376, 401
Line graphs, 150–152, 154
  multiple, 155–156
Line segments, 368, 370, 379
  measuring, 190–191
Liter (L), 202–204

Magic squares, 7
Mathematics Aptitude Test, 48, 160, 255, 363
Means, 240, 242
  arithmetic, 246
  of grouped data, 243–245
  of a proportion, 264
Measurement, 190
Median, 241–242
Meter (m), 192–197

Metric prefixes, 193
Metric ton (t), 206–207
Metric units
  area in, 200–201
  capacity in, 202–204
  changing between, 195–197, 202–206
  length in, 190–197
  temperature in, 209–210
  weight in, 205–207
  volume, 204
Milligram (mg), 205–207
Milliliter (mL), 202–204
Millimeters (mm), 190–197
  cubic (mm^3), 205
  square (mm^2), 200–201
Mixed numbers, 82–84
  addition of, 122–124
  division of, 100
  in formulas, 104–105
  multiplication of, 82–87, 95–97
  subtraction of, 122–124, 126–128
Mode, 241–242
Monomial, 356
  factor from a polynomial, 360–362
  multiplication by a polynomial, 358–359
Multiples, 114
Multiplication
  of decimals, 173–174
  division undone by, 37, 38, 44
  by fractions, 82–87, 95–97
  of integers, 312–315
  of mixed numbers, 82–87, 95–97
  of monomials by polynomials, 358–359
  properties of, 28–29, 316–318
  property for
    equations, 36–38, 44, 328–329
    inequalities, 345
  by whole numbers, 39
  by zero, 313
Multiplicative Identity, 28–29, 57, 316

Negative direction, 289, 393
Noncollinear points, 369
Non-Routine Problems, 21, 74, 135, 185, 235, 283, 330, 388
Numbers,
  composite, 88–89
  irrational, 383
  percents of, 219–221
  prime, 88–89
  rational, 323–324, 394
  real, 394
Number lines, 79
  addition of integers on, 291–293
  fractions on, 79–81
  graphing on, 393–398
  integers on, 288–292
  opposites on, 299
  rational numbers on, 323–324
  real numbers on, 394–395
  subtraction of integers on, 295–297, 301
Number Problems, 339–341

Obtuse angle, 370
Obtuse triangle, 377
Octagon, 379
Open Sentence, 5–7
Opposites of integers, 299–302, 306
Order of operations, 30–34, 49, 75
Ordered pairs for points, 401–405
  graphing, 413–414
Origin
  of coordinate plane, 401
  of number line, 288

Parallel lines, 373–374, 404
Parallelogram, 380
Parentheses, 31–32, 49
  equations with, 337–338
Patterns, 32, 152, 242, 318, 384
Pentagon, 379
Percents, 216–218
  changing decimals to, 217–218
  changing fractions to, 217–218
  to decimals, changing, 216–218
  decrease, 294
  finding, 222–223
  finding the number, given the, 224–225
  increase, 294
  of a number, 219–221
  solving problems using, 218, 221, 223, 225
Perfect squares, 382–384
Perimeter, 18–19, 47, 157–159
Perpendicular lines, 375–376, 401
Pi (π), 104
Pictographs, 143–144
Plane, 368–369
  coordinate, 401
Points, 368–369
  collinear, 368
  endpoint, 368
  graphing, 401–404
  on a line, 79, 228, 323, 393–394
  noncollinear, 369
  on a plane, 401–403, 404
  ordered pairs for, 401–403
Polygons, 379–380
Polynomials, 356–362
  factor a monomial from, 360–362
  multiplication of monomials by, 358–359
  simplifying, 356–359
Positive direction, 289, 393
Possible outcomes, 248–250
Powers, 69
  of a product, 354
  product of, 353
  of a power, 354
  of ten, 180–181
Prime factorization, 88–89
  to find the GCF, using, 90–91
  to find the LCM, using, 114–115
  to simplify fractions, using, 92–94
Prime number, 88–89
Principal square root, 382
Probability, 248–249, 256

counting principal, 252–253
of dependent events, 250–251
of independent events, 250–251
of simple events, 248–249
using tree diagrams, 250–261

Problem solving
applications, 16–17, 42–43, 60, 98, 125, 145–146, 175–176, 198–199, 232–233, 247, 294, 319, 342–343, 372, 399–400
in careers, 20, 68, 131, 153, 184, 208, 234, 254, 273, 298, 325, 350, 381, 409
checking solutions to word problems, 13–14, 39–40, 43, 65–66
with equations, 13–17, 39–41, 65–67, 339–342
four basic steps to, 13–17, 39–42, 45, 65–67, 145, 198, 232, 342, 399
multiple step, 145, 175, 227
two step, 65–67, 226, 232, 399

Product
of exponents, 354
of means and extremes in proportions, 264–265
of powers, 353
power of a, 354
simplifying a, 95–97

Property
of addition, 3–4
distributive, 53–55
of equations, 8–10, 36–38, 44
of exponents, 353–355
of inequalities, 344–346
of multiplication, 28–29, 316–318

Proportions, 263–265
applications, 266–268, 274–279

Protractor, 370

Pythagorean theorem, 385–387, 389

Quadrilaterals, 379–380

Range, 240–242
Rates, 261

Rational numbers, 323–324, 394
Ratios, 260–262, 266, 277–278
trigonometric, 280–282
Ray, 370
Reading in math, 12, 121, 228, 322
Real numbers, 394
Reciprocals, 85, 99, 102–103, 133, 348
Rectangle, 380
area of, 46–47
perimeter of, 18–19, 47, 157
Rectangular solid, 47, 204
Reflex angles, 370
Regular polygon, 379
Repeating decimals, 169
Rhombus, 380
Right angle, 277, 370, 375, 385
Right triangle, 377
Rounding
in business, 175–176, 226–231
decimals, 166–167, 169, 178, 224

Sales Tax, solving problems with, 232–234, 247
Scale drawings, 274–276
Scalene triangle, 378
Scientific notation, 180–181
Segment, 368
endpoint of, 368
Similar triangles, corresponding sides, 277–280
Simple events, probability with, 248–249
Simplest form of fractions, 92–94
Simplifying algebraic expressions, 58–59, 64
with the distributive property, 54–57, 326–327
by combining like terms, 56–57, 356–357
with exponents, 353–355
Simplifying polynomials, 358–359,
by combining like terms, 356–357
Sine of an angle, 281–282
Slope of a line, 410–412

intercept form of an equation, 412
Solutions, 6, 264, 347–349
  graphing, 393–398, 406–408, 413–414
Solving
  equations, 8–10, 36–38, 44, 61–62,
    102–103, 132–134, 182–183,
    305–306, 328–329, 335–338
  inequalities, 347–349, 396–397
  systems of equations, 413–414
Squares, 380, 385
  area of, 72, 383–385
  perfect, 382–384
  units for area, 200–201
Square root, 382–384, 386
  principal, 382
Statistics, 240–246
  analyzing data, 246–247
  double bar graphs, 246–247
  frequency table, 243–245
  mean of grouped data, 243–245
  range, mean, median, mode of
    ungrouped data, 240–242
Straight angle, 374
Substitution for a variable, 1, 6, 9
Subtraction
  addition used to check, 9
  by addition of opposites, 301–302
  of decimals, 171–172
  of fractions, 111–113, 116–121
  of integers, 295–297
  of mixed numbers, 122–124
  property for, equations, 8–10, 44, 306
  inequalities, 344
  renaming in, 126–128
Supplementary angles, 371
Surface area, of a cube and cylinder, 73
Systems of equations, 413–414

Tables, 140–142, 154–156
  frequency, 243–245
Tangent of an angle, 281–282
Terms
  of an algebraic expression, 2, 56, 356

combining like, 56–57, 66, 335–339,
    356–359
of a proportion, 264
Transversal, 373
Trapezoid, 380
Tree diagram, 250–252
Triangles
  acute, 377
  area of, 46–47, 104
  classify, 377–378
  isosceles, 157–159, 378
  obtuse, 377
  perimeter of, 18–19, 157–158
  right, 280–282, 377, 385–387, 389
  similar, 277–280
  sum of the angles of, 377
Trigonometric ratios, 280–282
Trinomial, 356

Variables, 1–2, 13, 39
Variation
  direct, 266–268
  inverse, 269
Vertex of an angle, 370
Vertical angles, 373–374
Vertical bar graphs, 147–149, 154, 246
Volume,
  of a cone, 105
  of a cube, 73
  of a cylinder, 72
  of a rectangular solid, 47, 204

Weight, metric measure of, 205–207

x-axis, 401, 403–405
x-coordinate, 401, 404–405, 410–412, 415

y-axis, 401, 403–405
y-coordinate, 401, 404–405, 410–412, 415
y-intercept, 411–412

Zero, 288
  division by, 321, 331, 415
  multiplication by, 313